PRINCIPLES AND APPLICATIONS OF GAS CHROMATOGRAPHY IN FOOD ANALYSIS

PRINCIPLES AND APPLICATIONS OF GAS CHROMATOGRAPHY IN FOOD ANALYSIS

Editor

MICHAEL H. GORDON
Department of Food Science and Technology
University of Reading

ELLIS HORWOOD

NEW YORK LONDON TORONTO SYDNEY TOKYO SINGAPORE

First published in 1990 by
ELLIS HORWOOD LIMITED
Market Cross House, Cooper Street,
Chichester, West Sussex, PO19 1EB, England

A division of
Simon & Schuster International Group

Typeset in Times by Ellis Horwood Limited
Printed and bound in Great Britain
by Bookcraft (Bath) Ltd., Midsomer Norton

Exclusive distribution by Van Nostrand Reinhold/AVI London:
Australia and New Zealand:
CHAPMAN AND HALL AUSTRALIA
Box 4725, 480 La Trobe Steet, Melbourne 3001, Victoria, Australia
Canada:
NELSON CANADA
1120 Birchmount Road, Scarborough, Ontario, Canada, M1K 5G4
Europe, Middle East and Africa:
VAN NOSTRAND REINHOLD/AVI LONDON
11 New Fetter Lane, London EC4P 4EE, England
North America:
VAN NOSTRAND REINHOLD/AVI NEW YORK
115 Fifth Avenue, 4th Floor, New York, New York 10003, USA
Rest of the world:
THOMSON INTERNATIONAL PUBLISHING
10 Davis Drive, Belmont, California 94002, USA

British Library Cataloguing in Publication Data

Principles and applications of gas chromatography in food analysis.
1. Food. Gas chromatography
I. Gordon, Michael H.
644.07
ISBN 0-7476-0053-8

Library of Congress Cataloging-in-Publication Data

Principles and applications of gas chromatography in food analysis / editor, Michael H. Gordon
p. cm. — (Ellis Horwood series in food science and technology)
ISBN 0-7476-0053-8
1. Food — Analysis. 2. Gas chromatography.
I. Gordon, Michael H., 1949– . II. Series.
TX548.P75 1990
664'.07–dc20

90-31186
CIP

Table of contents

Preface

The food analyst plays an important role in modern society. Stricter control over additives in food and concern about the effects of contamination of food by industrial and agricultural chemicals are among the developments which are leading to an increasing emphasis on detailed and accurate analysis of food. However, analysis of food is required for many reasons, including detection of toxic components, monitoring legislation, detecting adulteration, formulation of controlled diets, controlling formulation during product development and detecting changes in food during storage and processing.

Foods comprise a complex mixture of components and food analysis requires efficient methods of separation with high sensitivity or specificity of detection. Although many food components are involatile or thermally labile and therefore not suitable for analysis by gas chromatography, other components are volatile and this technique is the preferred analytical method. Developments in methods of derivatization, injector design and column technology have also extended the applicability of gas chromatography to the analysis of relatively involatile compounds.

Although it is less than 40 years since gas chromatography was first described, the technique is now relatively mature. It has developed to a stage where it is flexible, efficient, widely available, reliable, easy to use and readily automated. Therefore many laboratories employ gas chromatographic methods of analysis even where alternative techniques are equally acceptable. There are many general texts available that are concerned with the principles of gas chromatography, but the aim of this book is to emphasize the features of the technique that make it valuable in food analysis with copious examples to illustrate its application in this field. I wish to thank the contributors to this book for their hard work in distilling many years of experience into the chapters.

Michael H. Gordon

List of contributors

Mr N. P. Boley, Laboratory of the Government Chemist, Queens Road, Teddington, Middlesex, TW11 OLY.

Dr A. G. W. Bradbury, General Foods Technical Center, Tarrytown, NY 10591, U.S.A.

Prof. Ing. J. Davídek, Department of Food Chemistry and Analysis, Faculty of Food Technology, Institute of Chemical Technology, 166 28, Prague, 6-Dejvice, Suchbatarova 5, Czechoslovakia.

Dr K. O. Gerhardt, Department of Biochemistry, University of Missouri, Columbia, M 65211, U.S.A.

Dr J. Gilbert, Ministry of Agriculture, Fisheries & Food, Food Safety Directorate, Food Science Laboratory, Colney Lane, Norwich NR4 7UQ.

Dr M. H. Gordon, Department of Food Science & Technology, University of Reading, Whiteknights, P.O. Box 226, Reading, RG6 2AP.

Prof. J. A. Maga, Colorado State University, Department of Food Science & Human Nutrition, Fort Collins, CO 80523, U.S.A.

Prof. Dr. H-J. Stan, Technische Universitat Berlin, Sekr. TIP 3, Institut für Lebensmittelchemie, Strasse des 17 Juni 135, D-1000, Berlin 12, West Germany.

Dr J. Velíšek, Department of Food Chemistry and Analysis, Faculty of Food Technology, Institute of Chemical Technology, 166 28, Prague, 6-Dejvice, Suchbatarova 5, Czechoslovakia.

1

Principles of gas chromatography

Michael H. Gordon

1.1 INTRODUCTION

Gas chromatography (GC) has developed rapidly since it was first introduced by James and Martin (1952). There have been many advances in column technology, detectors, injectors and data-handling techniques, and the suitability of GC for automated analyses has increased its attraction to analysts. Many food components can be analysed with great accuracy by GC and it has become one of the main techniques in analytical laboratories concerned with food analysis.

GC achieves separation of mixtures by partition of components between a mobile gas phase and a stationary phase. Although components must be stable and have significant volatility at the analytical temperature, the use of short columns and high temperatures allows the analysis of many compounds that would normally be considered relatively non-volatile. Triglycerides or steryl esters which have a vapour pressure of less than 0.05 mm at 300°C are commonly analysed by GC. Non-volatile molecules such as sugars may be converted to more volatile compounds by a simple derivatization reaction.

Since foods are complex materials, GC is usually the final stage in a series of steps in the analysis. Solvent extraction, solid-phase extraction, distillation, or other chromatographic procedures, including column chromatography, thin-layer chromatography, high-performance liquid chromatography (HPLC) or gel-permeation chromatography may be used in sample preparation preceding GC.

GC is only one of a number of instrumental procedures to be considered in the analysis of food products. In particular HPLC has developed very rapidly in recent years and has been applied in the analysis of a wide range of food components (Macrae, 1988). The development of flame ionization and mass detectors in addition to well-established detectors, such as ultraviolet and fluorescence, has extended the range of application of HPLC enormously. Analysts must consider whether HPLC or GC is the preferred technique for a particular analysis. While HPLC is clearly preferred for completely non-volatile or thermally labile molecules, and GC is better suited to the analysis of volatile compounds, there are a large number of food components which can be analysed by either technique. In selecting a method, analysts must consider availability of equipment; sensitivity and specificity of detection; the number of preliminary steps involved in sample preparation, which may affect the accuracy of the analysis; and the time taken for the full analytical procedure. The ease of automation of the procedure also becomes important when there are many samples to be analysed.

1.2 PRINCIPLES

The most common form of GC is gas–liquid chromatography (GLC), in which the stationary phase is a non-volatile liquid. The stationary phase may be coated onto small particles of an inert solid packed into a coiled glass or metal column. In the early years of GC, the vast majority of analyses were performed using packed columns of this type. However, in recent years capillary columns have become widely used. The most common form of capillary column is the wall-coated open tubular (WCOT) column, in which the stationary phase is coated or chemically bonded onto the inside wall of a long glass or fused silica capillary. An alternative procedure for GC is gas–solid chromatography (GSC), in which components are separated by virtue of differences in their adsorption onto particles of a solid adsorbent.

1.2.1 Retention of solutes

The time taken for a molecule to pass through a chromatographic column is termed the retention time, t_R. Solubility in the liquid phase (GLC) or adsorption on a solid phase (GSC) retards the analyte, and therefore t_R is greater then t_m, the elution time for a molecule which is not retained by the stationary phase. The adjusted retention time $t'_R = t_R - t_m$ is a better measure of chromatographic retention then t_R. The retention time of a solute commonly increases by about 5% for each degree centigrade reduction of temperature. The effect of temperature on chromatographic retention is evident in Fig. 1.1. A reduction of carrier gas flow or an increase in column length or stationary phase concentration also increases the retention time.

The solute capacity factor k is a useful parameter in GC. It represents the ratio of the time spent by the solute in the stationary phase to the time it spends in the mobile phase. Hence

$$k = \frac{t'_R}{t_m} \tag{1.1}$$

The retention time of any solute can be calculated from its capacity factor since

$$t_R = t_m(1 + k) = \frac{L}{\bar{u}}(1 + k) \tag{1.2}$$

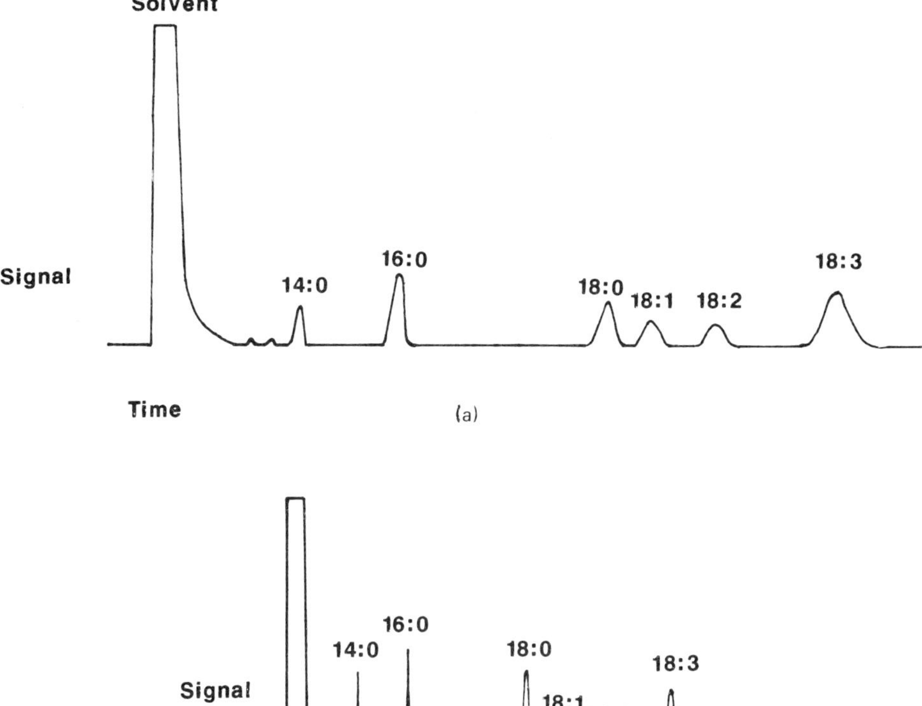

Fig. 1.1 — The effect of temperature on the analysis of a fatty acid methyl ester mixture analysed on a Carbowax 20M packed column at (a) 160°C and (b) 180°C.

where L is the column length and $\bar{u}$ is the mean linear gas velocity. The solute capacity factor, k, varies with the phase ratio, β, where

$$\beta = \frac{V_g}{V_L} \tag{1.3}$$

V_G is the volume of column occupied by the gas phase, and V_L is the volume of column occupied by the stationary phase.

The distribution constant K_D is a more fundamental parameter than k. It is defined as the ratio of the concentration of the solute in the stationary phase to its concentration in the gas phase.

$$K_D = \beta k \tag{1.4}$$

The distribution constant depends only on the type of stationary phase which is used and the temperature of the column.

Using equations (1.2)–(1.4), it is evident that

$$t_R = \frac{L}{\bar{u}}\left(1 + K_D \frac{V_L}{V_G}\right) \tag{1.5}$$

1.2.2 Band broadening

Chromatographic peaks are broadened by several kinetic processes occurring in the column. These can best be understood by considering the van Deemter equation:

$$h = A + B\bar{u}^{-1} + C\bar{u} \tag{1.6}$$

h is the height equivalent to a theoretical plate and is a measure of the column efficiency, with small values representing high efficiency. A,B,C are constant for a given column at a given temperature. The constant A is the eddy diffusion term. It describes the broadening of the peak produced by the variation in gas velocity in the porous structure of packed columns. In the case of capillary columns laminar flow occurs and hence $A = 0$. The second term, $B\bar{u}^{-1}$, represents the broadening of the peak produced by longitudinal diffusion of solute molecules in the gas phase during their passage through the column. The third term, $C\bar{u}$, is related to the resistance to mass transfer in the column, which retards the equilibration of solute molecules between the gas and the stationary phase. The theoretical van Deemter curve is compared with the experimental curves found for a capillary column using different carrier gases in Fig. 1.2. It is clear that the choice of the optimum carrier gas velocity can have an important effect on the column efficiency.

The height h is affected by the support particle size (for a packed column), the column diameter and length, the carrier gas used, the temperature, the carrier gas velocity, the sample size and the viscosity of the stationary phase (and hence the diffusion coefficient of the sample within it). A reduction in the particle diameter of the support in a packed column reduces eddy diffusion and hence reduces h, causing a reduction in peak width. Thus small particle sizes are preferred, but a reduction in particle size increases the pressure drop across the column, making it more difficult to achieve the optimum gas velocity.

The second term in the van Deemter equation becomes important at low carrier-gas flow rates, such as those used in capillary GLC. The parameter B is proportional to the diffusion coefficient of the solute in the gas phase. It varies with the solute, temperature, pressure and nature of the carrier gas. The diffusion coefficient is largest for hydrogen, with that for helium being somewhat smaller and that for

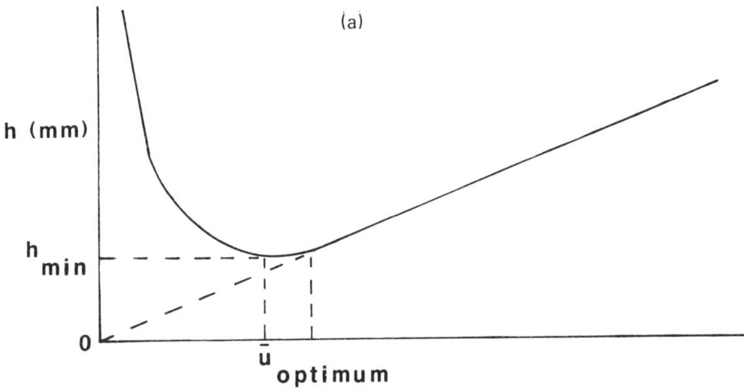

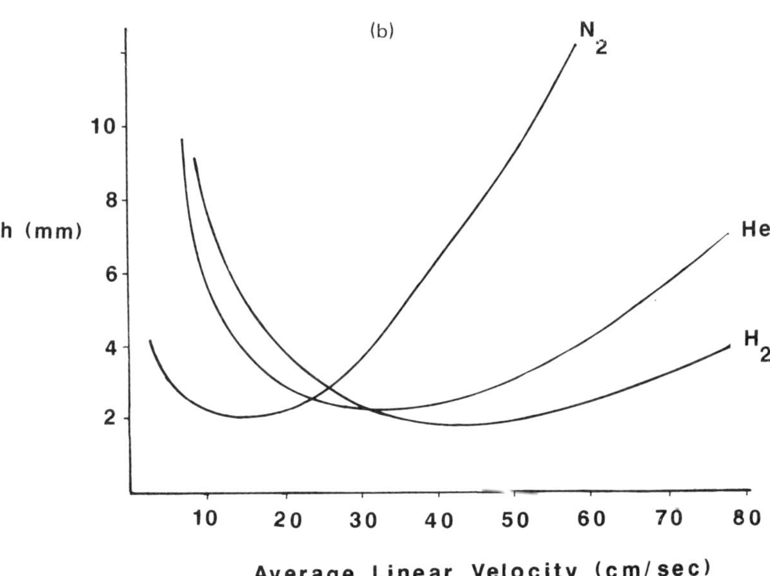

Fig. 1.2 — The effect of average linear velocity on the height equivalent to a theoretical plate of an open-tubular column. (a) theoretical van Deemter plot, (b) experimental van Deemter plots for a 50 m × 0.25 mm glass capillary with different carrier gases. Redrawn with permission from Rooney *et al.*, *American Laboratory*, volume 11, number 2, page 81–89, 1979. Copyright 1979 by International Scientific Communications Inc.

nitrogen being smaller still. Hence hydrogen is the preferred carrier gas for capillary GC, although helium is also commonly used.

The mass-transfer term, C, in the van Deemter equation is the dominant factor contributing to peak broadening at high flow rates. C corresponds to the sum of the mass transfer terms for the gaseous (C_g) and liquid phases (C_l). For open tubular columns, the following expressions have been derived:

$$C_g = \frac{1 + 6k + 11k^2}{24(1+k)^2} \frac{r^2}{D_g} \tag{1.7}$$

$$C_l = \frac{k^3}{6(1+k)^2} \frac{r^2}{K_D^2 D_l} \tag{1.8}$$

where D_g and D_l are the diffusion coefficients in the gas and liquid phases and r is the internal radius of the column.

The value of C and hence the peak width increases with the square of the column radius and inversely with the diffusion coefficients in the gas and liquid phases. An increase in the thickness of the film of the stationary phase also causes an increase in peak width because diffusion in the stationary phase is greater and k increases. The analysis of the mass-transfer term is more complex in the case of packed columns, but it can be shown that the peak width decreases with a reduction in column radius, a reduction in support particle size, and a reduction in the concentration of stationary phase.

As a consequence of these band-broadening processes, chromatographic peaks occur as symmetrical Gaussian curves (Fig. 1.3(a)). Distortion from this peak shape

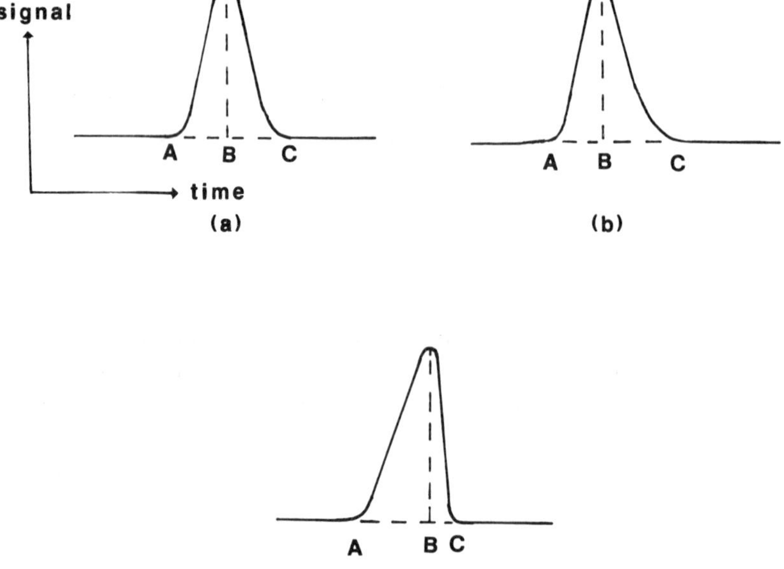

Fig. 1.3 — GC peak shapes. (a) ideal peak shape, (b) 'tailing' peak due to interaction of the solute with the support, (c) peak with leading edge due to column overload.

may occur as tailing, which arises from an interaction of the solute with the support (Fig. 1.3(b)), or as a leading edge which indicates overloading of the column (Fig. 1.3(c)).

1.2.3 Measures of column efficiency

1.2.3.1 Theoretical plate numbers

For isothermal analyses, column efficiency may be defined in terms of the number of theoretical plates, n.

$$n = \frac{(t_R)^2}{\sigma} = 16\left(\frac{t_R}{w_b}\right)^2 \tag{1.9}$$

where σ is the standard deviation of the peak, and w_b is the width of the peak at its base.

Since the peak width at half height ($w_{0.5}$) can be measured accurately, and for a Gaussian curve $w_{0.5} = 2.35\ \sigma$, the number of theoretical plates is often defined as:

$$n = 5.54\left(\frac{t_R}{w_{0.5}}\right)^2 \tag{1.10}$$

In order to describe the efficiency of a column in terms of a parameter independent of the column length, the height equivalent to a theoretical plate, h, may be used:

$$h = \frac{L}{n} \tag{1.11}$$

where L is the column length in mm.

Since the gas hold-up time, t_m, does not contribute to the retention of a solute by a stationary phase, the number of effective theoretical plates, N, is often quoted.

$$N = 5.54\left(\frac{t'_R}{w_{0.5}}\right)^2 \tag{1.12}$$

The height equivalent to an effective theoretical plate, H, is then defined as:

$$H = \frac{L}{N} \tag{1.13}$$

1.2.3.2 Separation number

The separation number or Trennzahl as it was described by the German originator, Kaiser (1962), is a useful method of describing the efficiency of a chromatographic column. The separation number, *TZ*, is based on the separation of two consecutive members of a homologous series with normal alkanes being commonly used to determine it.

$$TZ = \left[\frac{t_{R(B)} - t_{R(A)}}{w_{0.5(A)} + w_{0.5(B)}} \right] - 1 \tag{1.14}$$

A and B represent two members of a homologous series differing in molecular formula by a single methylene unit. The separation number can be considered as the number of peaks that can be placed between two successive members of a homologous series, and consequently it can be applied to both isothermal and temperature programmed conditions. Determination of the separation number with compounds of longer retention times leads to an increase in *TZ*, and *TZ* also increases as the test temperature is reduced (Jennings & Yabumoto, 1980).

Grob & Grob (1981) recommended the use of separation number in the evaluation of capillary columns but other authors prefer the number of effective theoretical plates (Jennings & Yabumoto, 1980).

1.2.4 Separation of chromatographic peaks

The relative retention of two peaks in a chromatogram is described by the separation factor, α, where $t_{R(B)} > t_{R(A)}$.

$$\alpha = \frac{t'_{R(B)}}{t'_{R(A)}} = \frac{k_B}{k_A} \tag{1.15}$$

The separation of two chromatographic peaks depends both on α and on the column efficiency. The separation can be defined as the resolution R_s where

$$R_s = 2\,\frac{(t_{R(A)} - t_{R(B)})}{w_{b(A)} + w_{b(B)}} \tag{1.16}$$

Baseline resolution corresponds to an R_s value of 1.5, but a resolution of 1.0 will achieve 94% separation and will allow satisfactory integration of the peaks. If the separation of components on a particular column is poor, the first course of action is usually to try a column with a different phase. Hundreds of stationary phases have been used in GLC and these phases vary widely in their chemical structures and also in their mode of interaction with solute molecules. Hence the separation factor is quite different for different stationary phases. If most components in a complex mixture are separated, it is often best to investigate methods of increasing the resolution in order to separate critical pairs.

The number of theoretical plates required for the separation of two components with a resolution R_s can be calculated as:

$$n_{req} = 16\,R_s^2 \left(\frac{\alpha}{\alpha - 1}\right)^2 \left(\frac{k_B + 1}{k_B}\right)^2 \tag{1.17}$$

This equation is very useful in selecting a column for a particular separation. At the optimum carrier-gas velocity the number of theoretical plates provided by a capillary column increases as the square root of the column length. Thus a column four times as long is required to achieve a doubling of n. In the case of a packed column, use of a longer column or a more densely packed column may produce the required improvement in column efficiency.

A change in temperature may be used to improve the separation of two peaks with adjusted retention times $t'_{R(B)}$ and $t'_{R(A)}$. The relative separation can be shown to be related to the enthalpies of solvation ΔH of the two components by the equation

$$\ln\left[\frac{t'_{R(B)}}{t'_{R(A)}}\right] = \frac{\Delta H_B - \Delta H_A}{RT} + C \tag{1.18}$$

where C is a constant and R is the gas constant. The component with the longer retention time generally has a higher enthalpy of solvation and therefore a reduction in temperature, T, usually leads to an increase in the relative separation of the two peaks. However, for two molecules which differ in the nature of the solute–stationary phase interactions, an increase in temperature may give improved separation. Thus Freeman & Jennings (1987) have shown that the elution sequence of *o*-chlorotoluene and *n*-propylbenzene, or of *m*-dichlorobenzene and *sec*-butylbenzene on a column coated with DB-1301 may be reversed by an increase in temperature (Fig. 1.4).

1.3 THE CHROMATOGRAPHIC SYSTEM

The basic elements required for GLC are shown in Fig. 1.5. Carrier gas from a cylinder of compressed gas is passed through a mass-flow controller, with a flow rate of 20–60 ml min^{-1} commonly being used for a packed column system. Faster flow rates may be used to accelerate the analysis if the components of the mixture are well-separated. Very low flow rates of 0.5–1.5 ml min^{-1} are commonly used in GLC with narrow-bore capillary columns, and intermediate flow rates may be used with wide-bore capillary columns. Very low flow rates are often achieved using pressure control rather than flow control. The carrier gas passes through the column, which is located in an oven, and then flows through the detector. The signal from the detector is amplified and recorded on a chart recorder or electronic integrator. The temperatures of the injector block, oven and detector block are controlled independently.

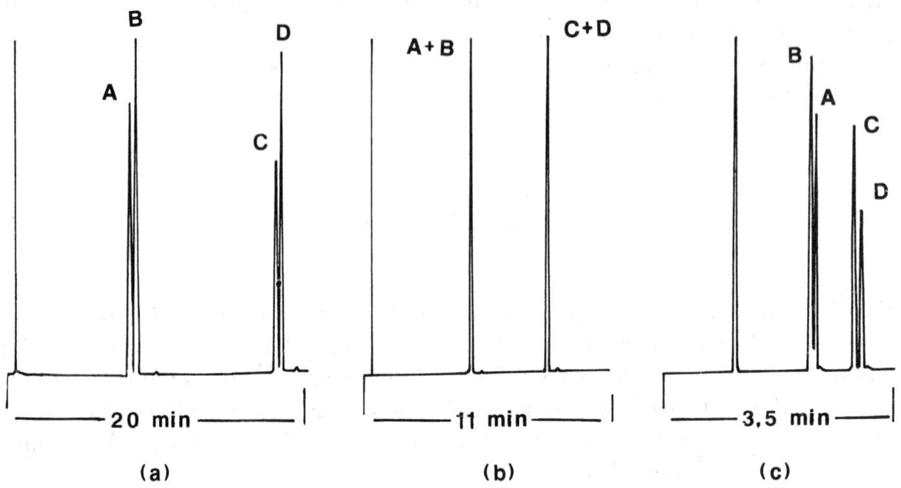

Fig. 1.4 — Effect of temperature on the selectivity of DB-1301 with dissimilar solutes. (a) Temperature programme 30°C, 3.5 min; 2°C min^{-1} to 60°C and 5°C min^{-1} to 120°C. (b) 35°C, 5 min; 6°C min^{-1} to 140°C. (c) 105°C isothermal. Solutes: A, *o*-chlorotoluene; B, *n*-propylbenzene; C, *m*-dichlorobenzene; D, *sec*-butylbenzene. Redrawn with permission from Freeman and Jennings (1987).

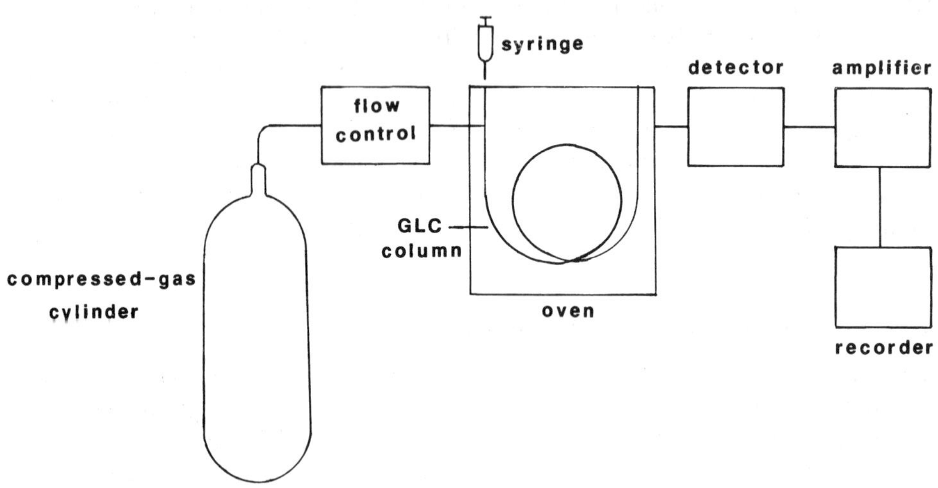

Fig. 1.5 — The chromatographic system.

Nitrogen is the most common carrier gas for packed-column analysis, although other inert gases, including argon, carbon dioxide, helium and hydrogen may also be used. Hydrogen is the preferred carrier gas for capillary GLC because it is cheap and it gives the best resolution, owing to its low viscosity and hence high diffusion coefficient. Analysts using hydrogen as a carrier gas need to be aware that it forms explosive mixtures with air. However, very sensitive gas-leak detectors are available commercially, and one of these can be used to ensure safe operation. Helium is an acceptable alternative for narrow-bore capillary columns but nitrogen should be avoided as a carrier gas for these systems, since it leads to a significant loss of resolution. This is illustrated in the analysis of fatty acid methyl esters in Fig. 1.6,

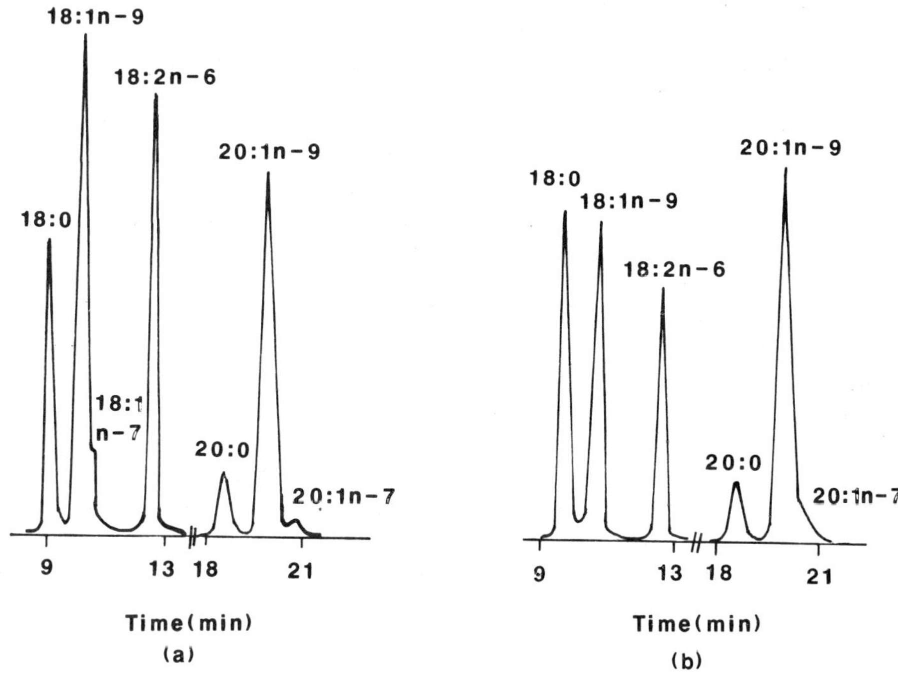

Fig. 1.6—Effect of carrier gas on the chromatogram of rapeseed oil fatty acid methyl esters: (a) hydrogen; (b) nitrogen. WCOT column (46 m×0.25 mm) with butane 1,4-diol succinate, BDS as stationary phase. Inlet pressure 3.5 kg/cm. Temperature (a) 150°C; (b) 170°C. Redrawn with permission from Mayzaud & Ackman (1976).

where the double-bond positional isomers 20:1*n*-9 and 20:1*n*-7 are separated if hydrogen is used as the carrier gas but not if nitrogen is used. The carrier gas should be dry and free from traces of oxygen, particularly if a capillary column with a bonded stationary phase is being used, since bonded phases deteriorate rapidly in the presence of traces of oxygen.

Mixtures are usually applied to GLC columns as dilute solutions in volatile solvents. The most common sample application procedure involves injection through a septum either directly onto the head of a column or into a heated space, with the carrier gas transporting the sample onto the column. Injection into a heated space generally requires the use of an inlet temperature about 50°C higher than that of the column. It is important that the sample is still stable at this elevated temperature, and this procedure is often avoided with high-boiling mixtures where decomposition may occur. Automatic samplers are widely used in industry, allowing the injection of samples when the instrument is unattended. This facility is particularly useful where a large number of routine analyses are required, and automatic injection can be performed with packed columns or capillary columns in all injection modes except direct on-column onto a narrow-bore (0.25-mm internal diameter (i.d.)) capillary column. However, the use of a 0.32-mm i.d. column or a wide-bore capillary precolumn connected to a narrow-bore capillary column allows automated on-column injection with high-resolution systems. Automatic injection may also be performed with a programmed-temperature vaporizer (PTV) injector. Injection into a cold injector is followed by rapid heating to introduce sample onto the column either in a split or splitless mode. Other methods of sample application include thermal desorption of volatiles trapped on porous polymers, which is a valuable technique in headspace analysis (see section 1.4.2.4d). Gas samples can be injected with a gas-tight syringe, but introduction of gas from a calibrated sample loop linked to a multiport rotary valve is a more precise procedure. Biological macromolecules which are too non-volatile or thermally labile to be analysed directly by GC can be pyrolysed prior to analysis (see section 1.9.3).

The column is located in an oven with a low thermal capacity to allow rapid heating or cooling with virtually no temperature overshoot. The oven may be operated isothermally or with a temperature programme. Temperature programming is required if a mixture contains components with short retention times as well as components with very long retention times at a given temperature. An increase in temperature shortens the retention time of the slow-moving components, but a low temperature is still required for the separation of the earlier components. Hence, the use of a temperature programme involving a holding period at a low temperature followed by a steady rate of temperature increase with a holding period at a higher temperature would be useful in this case.

1.4 GC COLUMNS

1.4.1 Packed columns

The vast majority of analyses in the early years of GLC were performed with packed columns, and they are still widely used. Packed columns consist of a glass or metal (usually stainless steel) column with an internal diameter of 2–5 mm and length of 0.5–8 m. Glass has the advantage that it is more inert than metal, and also the packing can be inspected visually. Metal columns have the advantage that they are not fragile, but gaps can develop in the packing during use, and these cause a loss of resolution without the cause being obvious to the analyst. In addition, metal columns can lead to decomposition or isomerization of reactive compounds. The use of glass-lined metal tubing combines the inertness of glass with the robustness of metal.

The column is packed with fine particles of an inert support coated with the stationary phase. The weight percentage of stationary phase is generally in the range

1–25%. The support must be chemically inert to avoid adsorption of eluting components which leads to tailing. It must consist of uniform particles with a large surface area and must be capable of being uniformly wetted by the stationary phase. The most commonly used supports are prepared from diatomaceous earth, although other support materials including PTFE (polytetrafluoroethylene), glass beads and porous polymers are sometimes used. The inertness of a support can be improved by acid washing and by treatment with dimethyldichlorosilane. Acid washing removes mineral impurities which can cause decomposition of the sample or stationary phase, while silylation converts polar silanol groups on the surface of the support to silyl ethers, which often improves peak shape, particularly for polar solutes. Column supports are sold with various particle-size distributions, and commonly mesh sizes 60/80 (177–250 μm diameter), 80/100 (149–176 μm) or 100/120 (125–149 μm) are used. Injection of 1 μl of a solution containing solutes at about the 1% level for each component (that is 10 μg) is suitable for packed-column analysis.

Packed columns are still commonly used in applications where high resolution is not necessary because of their relatively low cost and the fact that they can separate larger amounts of material than capillary columns.

1.4.2 Capillary columns

Capillary columns are often used instead of packed columns for the analysis of complex mixtures. They give much better resolution and retention times are also generally shorter. Two types of capillary columns are manufactured namely wall-coated open tubular (WCOT) and porous-layer open tubular (PLOT) columns. The most common are WCOT columns, which contain the stationary phase as a thin film coated onto, or chemically bonded to, the wall of a long glass or fused-silica capillary. A range of WCOT columns is available with dimensions in the range 0.05–0.75 mm i.d. in a column of length 8–100 m containing a film of thickness 0.1–6 μm. Narrow-bore WCOT columns with i.d. 0.05–0.35 mm, and film thickness 0.1–1 μm are used when the highest resolution and fastest analysis is required. Wide-bore WCOT columns with i.d. 0.5–0.75 mm, and film thickness 1–6 μm can be used with simple reducing adapters in gas chromatographs designed for packed columns. They can be considered as alternatives to packed columns, operating with short columns at similar flow rates to packed columns with similar resolution but shorter analysis times, or at lower flow rates with improved resolution. Fig. 1.7 shows the effect of flow rate on the number of theoretical plates achieved in the analysis of methyl octanoate on a 10-m wide-bore WCOT capillary column. At a flow rate of 30 ml min^{-1}, n is approximately 3000, which is similar to a 6′ (1.83-m) packed column, while in excess of 18 000 theoretical plates can be achieved at lower flow rates. Wide-bore WCOT columns give better resolution than packed columns for temperature-programmed analysis, since the temperature gradient across a packed column can lead to considerable band-broadening. The sample loading capacity of a wide-bore WCOT column is comparable to that of a packed column. This is evident from Fig. 1.8, where the mass of methyl octanoate required to increase peak width by 10% is 2.6 μg on a wide-bore WCOT column and 16 μg on a 5% coated packed column. Other advantages claimed for wide-bore WCOT columns include more reproducible injections and superior inertness. The shorter analysis time and superior inertness of a wide-bore WCOT column are shown in Fig. 1.9.

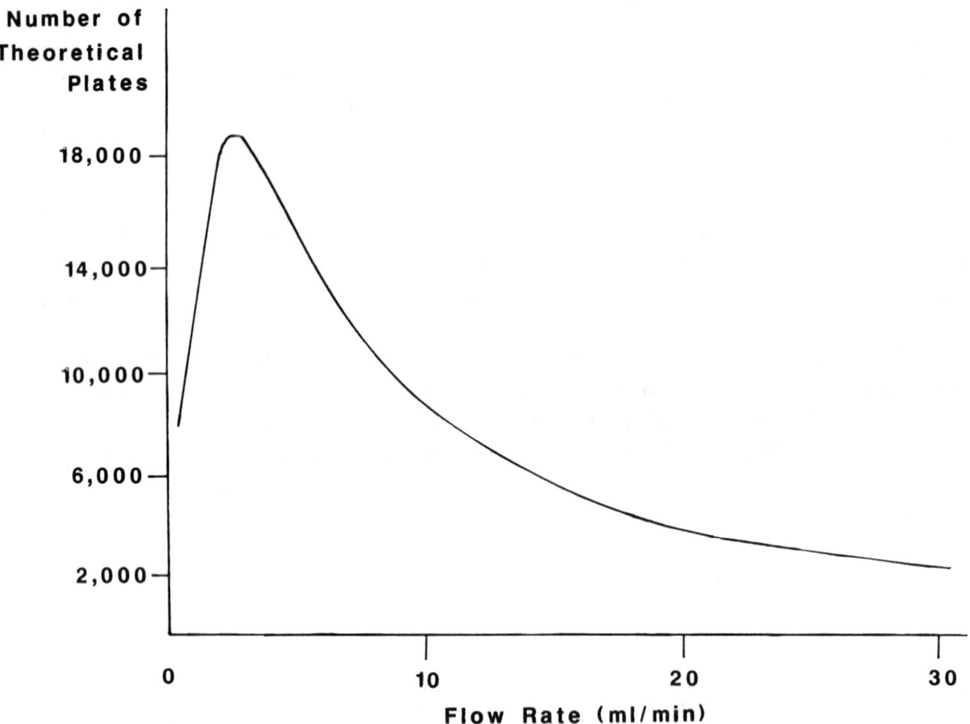

Fig. 1.7 — Effect of flow rate on the efficiency of a 530 μm methyl silicone column used for the analysis of methyl octanoate at 110 C. Redrawn with permission from Hewlett Packard.

The main advantages of capillary columns are their much improved resolution and shorter analysis times and these features are optimal in narrow-bore WCOT columns. The optimum mean linear carrier-gas velocity, $\bar{u}_{opt}$, increases with a reduction in column diameter (Fig. 1.10), and therefore the analysis time is shorter for narrower columns. The maximum theoretical efficiency of a WCOT column has been shown to be inversely proportional to column diameter (Golay, 1958), since

$$h_{min} = r\left[\frac{1+6k+11k^2}{3(1+k)^2}\right]^{\frac{1}{2}} \tag{1.19}$$

Narrow-bore columns of 0.25–0.35 mm i.d. are commonly used, since these combine very high efficiency with acceptable sample size and hence sensitivity of detection. The high efficiency of WCOT columns arises from the fact that $A=0$ in the van Deemter equation, and also from the longer columns that can be used because the pressure drop is considerably less than in packed columns. The liquid phase is also more uniformly spread as a thin film in a capillary column. Columns in which the stationary phase is coated onto the capillary have low temperature limits and short

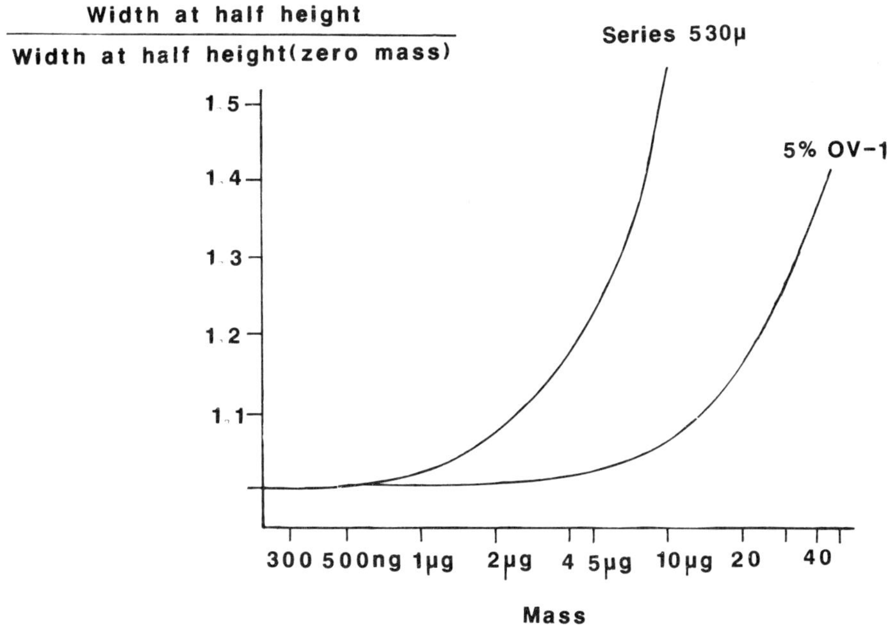

Fig. 1.8 — Effect of injection mass on the peak width of methyl octanoate analysed on a 10-m 530 μm methyl silicone capillary column. Redrawn with permission from Hewlett Packard.

lifetimes compared with packed columns because of the small amount of stationary phase present. The columns deteriorate by oxidation of the stationary phase, by loss of the stationary phase during use due to bleeding from the end of the column, or by the accumulation of high-boiling materials in the column. The first of these problems is minimized by the use of an oxygen trap in the carrier gas line to remove traces of oxygen. The development of chemically bonded WCOT columns has reduced the extent of the other two problems. The stationary phase is chemically bonded to the wall of the column, and the column can be cleaned periodically by washing with an organic solvent. Accumulation of high-boiling materials in a WCOT column containing a coated stationary phase may be minimized by cutting short lengths off the front of the column periodically.

The development of fused silica as a material for capillary columns has been a major factor in the increase in their popularity in recent years. Fused silica is made by fusing silica produced by burning SiH_4 or $SiCl_4$ in oxygen. It is significantly lower in metallic impurities than soda lime, borosilicate or fused quartz, which is prepared from the naturally occurring crystalline mineral SiO_2 (Table 1.1). Column bleed is increased by impurities in the silica (Blomberg & Wannman, 1979; Schomburg *et al.*, 1978) and therefore fused silica is the preferred column material. Borosilicate glass capillary columns are very fragile and require careful handling, while fused-silica columns are coated with a polyimide coating which makes them very flexible. However, they are inherently straight which allows them to be easily attached to the

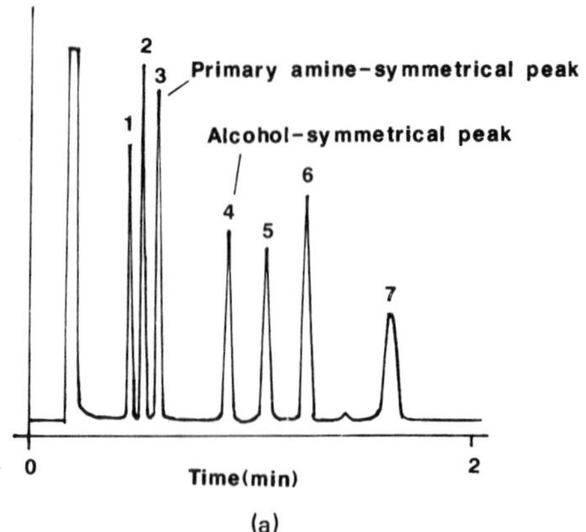

(a)

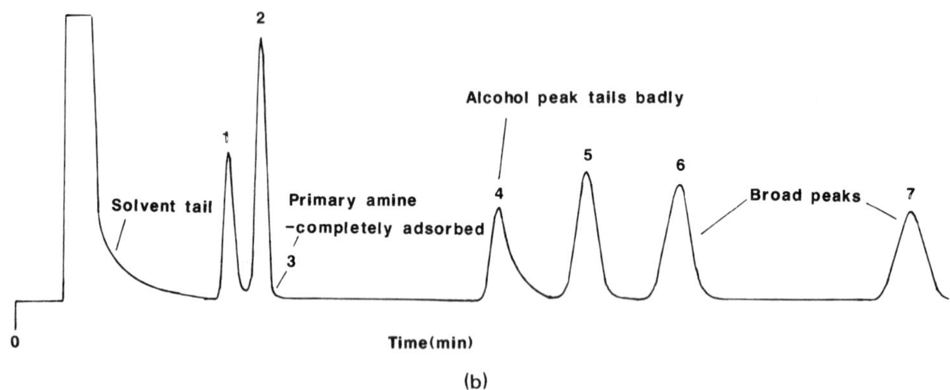

(b)

Fig. 1.9 — A comparison of (a) DB-1 530 μm wide-bore capillary column with (b) OV-101 packed column for the analysis of a mixture. 1,4-chlorophenol; 2, dodecane; 3, 1-decylamine; 4,1-undecanol; 5, tetradecane; 6, acenaphthene; 7, pentadecane. Redrawn with permission from Jones Chromatography.

injector and detector ports. Aluminium-clad fused-silica columns have been developed for use at high temperatures (above 370°C) when the conventional polyimide coating is not stable. Polyimide coatings with improved thermal stability have recently been introduced (Chrompack, 1988). PLOT columns are capillary columns in which the stationary phase has an increased surface area. This is achieved by building up a porous layer on the inside column wall either by chemical treatment or by the deposition of porous particles from a suspension. Support-coated open tubular (SCOT) columns contain a porous support layer on the inside column wall coated with a stationary phase, while open-tubular adsorption columns are available for GSC and have applications in the analysis of permanent gases (de Zeeuw *et al.*, 1987). SCOT columns have several advantages over narrow-bore WCOT columns, including higher sample capacity, higher flow rates of carrier gas can be used, fewer plates are required for the separation of early peaks, and very good efficiency can be

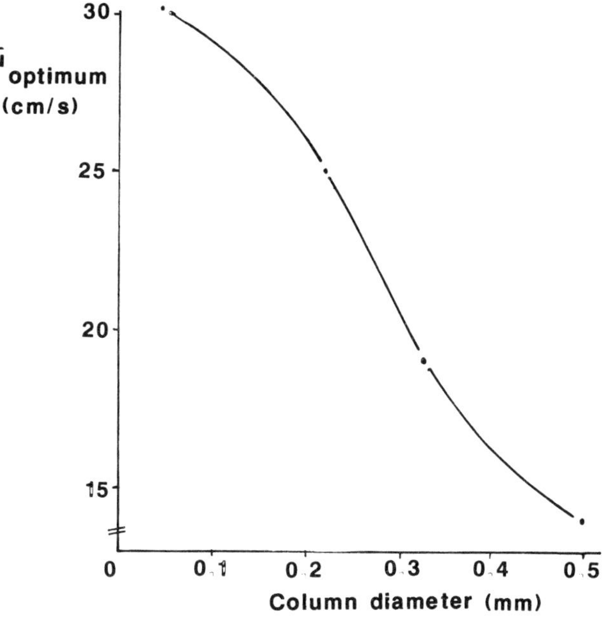

Fig. 1.10 — Optimum mean linear carrier-gas velocity, $\bar{u}_{optimum}$, as a function of the column diameter. Redrawn with permission from *Chrompack News*, 1984.

Table 1.1 — Approximate glass compositions (%) (Dandeneau *et al.*, 1979)

	SiO_2	Al_2O_3	Na_2O	CaO	MgO	B_2O_3	BaO
Soda lime	68	3	15	6	4	2	2
Borosilicate	81	2	4		—	13	—
Fused quartz	100	10–50[a]	1[a]	0.8–3[a]	0.2[a]	0.1[a]	—
Fused silica	100	0.1[a]	0.04[a]	0.1[a]	0.1[a]	0.01[a]	—

[a]Amount in p.p.m.

achieved. However, residual activity of the particles used in building up the porous layer is often observed, partly due to the very thin liquid phase film which is used. Wide-bore thicker-film WCOT columns have partly replaced SCOT columns in recent years since they have similar advantages over narrow-bore WCOT columns.

The main drawback of a capillary column is the relatively high cost. In the case of narrow-bore WCOT columns the sample size is small and several methods of introducing samples onto capillary columns have been developed to allow the sample to be introduced as a narrow band without overloading the column.

1.4.2.1 Split injection

This injection mode was developed to allow the injection of samples of normal size and concentration. A sample (approximately 1 μl) is injected by syringe through a septum into a heated space which is swept with carrier gas. The carrier gas is then split into two streams, one of which is vented while the other passes onto the column. The split ratio can be adjusted with a needle valve with ratios in the range 10:1 to 200:1 being commonly used. The procedure is very effective and excellent resolution can be achieved, since the components are focused at the head of the column by the reduction of temperature from the injector to the column. Bleed of components from the septum onto the column, which may interfere with the analysis, is minimized by a split-injection technique. This procedure should not be used for accurate quantitative analysis of mixtures containing components with a wide range of volatility, since non-linear splitting of high-boiling solutes may occur. Grob (1979) has shown split injections produce results with strong discrimination and very high standard deviations in the analysis of triglyceride mixtures.

1.4.2.2 Splitless injection

In this technique, a dilute solution is injected into a heated space followed by transport of the whole sample onto the column. Discrimination can arise from incomplete elution of high-boiling components from the syringe (Grob, 1979). Considerable tailing of the solvent may occur when splitless injection is used and this may interfere with the analysis of volatile components. In order to overcome this problem and to improve the resolution, cold trapping is commonly used. This involves injection into a heated space with the column maintained at a low temperature, commonly sub-ambient. The solvent is volatile and passes along the column while the components of the mixture are concentrated at the head of the column. After a short delay of about 30 s, the column is heated to the analytical temperature. It is generally helpful to vent residual solvent vapours at the end of the delay period.

An alternative splitless injection procedure uses the Grob solvent effect (Grob & Grob, 1978). The column temperature is maintained at a temperature about 25°C below the boiling point of the solvent during the injection. The sample condenses at the head of the column and the solvent acts as a second stationary phase concentrating the solute into a narrow band. This allows good resolution to be achieved. Residual solvent vapours are vented after about 20–40 s to avoid tailing of the solvent peak.

1.4.2.3 Cold on-column injection

Injectors which allow injection of sample directly onto the capillary column are now available. Syringes with metal or fused-silica needles are used depending on the injector design. The sample (approximately 0.3 μl) is injected rapidly as a drop directly onto the column, which is at a temperature below, but close to, the boiling point of the solvent (Grob & Neukom, 1980). This causes the solution to be applied as a narrow band giving good resolution while no discrimination of solutes occurs. This procedure is particularly useful with mixtures containing high-boiling solutes since the whole sample is applied to the column. Cold on-column injection gave much more accurate analyses of triglyceride mixtures than the alternative techniques, with a standard deviation of 1–3%, compared with around 10% for splitless injection and up to 35% for split injection (Grob, 1979). The latter two techniques also suffered from discrimination of solutes in the injector.

1.4.2.4 Thermal desorption procedures

Trapping of volatiles on solid adsorbents followed by thermal desorption into the gas chromatograph is a valuable procedure in vapour analysis because it allows the concentration of the volatiles prior to analysis. Thermal desorption from Tenax is commonly performed at 220–270°C, while desorption from other polymers, including Porapak Q, occurs at lower temperatures. Cold-trapping of desorbed volatiles in a pre-column or at the front of the analytical column is necessary to achieve good resolution.

1.4.2.5 Gas sampling valves

Gases may be introduced onto a capillary column via a gas sampling or switching valve. Switching of part of the effluent from one GC column to a second column allows heart-cutting to be performed (see section 1.7.1).

1.5 STATIONARY PHASES

A wide variety of stationary phases are used in GLC columns. Desirable characteristics include low volatility up to high temperatures, good thermal stability and low viscosity. A limited amount of 'bleed' of stationary phase from the column can be tolerated, but excessive bleed due to significant volatility causes a rapidly rising baseline during temperature-programmed analysis, and the column lifetime may be short, especially in the case of capillary columns. Several stationary phases which can be chemically bonded to the capillary wall have been developed for use in WCOT columns and these eliminate bleed from the column. If excessive bleed occurs during a temperature-programmed run and no alternative column can be used, the effect may be minimized by the use of a dual-column procedure in which two identical columns are kept in the same oven with an identical carrier-gas flow. The difference in the signal from the detectors at the end of each column is recorded. A sample is applied to one of the columns and a chromatogram with a good baseline can be obtained. Some modern gas chromatographs have an automated bleed compensation facility with a single detector system. The technique involves a temperature-programmed run in the absence of a sample, and this background signal is then subtracted from the signal produced by a sample analysed under the same conditions before the chromatogram is recorded.

Selection of a stationary phase depends on the chemical structures of the components in the mixture to be separated. In general, non-polar stationary phases separate mixtures of non-polar components mainly on the basis of boiling points, with dipole-induced dipole interactions being significant for polar solutes. Polar stationary phases have specific dipole–dipole, dipole-induced dipole and hydrogen-bonding interactions with solutes. A useful method of comparing the retention of solutes on a stationary phase is by means of the Kovats retention indices (Kovats, 1958). The Kovats retention index of a normal alkane is defined as 100 times the number of carbon atoms in the molecule. The retention index for other molecules is equal to the retention index of the hypothetical hydrocarbon with the identical corrected retention time. The retention index I_k can be expressed as:

$$I_k = 100z + 100\left[\frac{\log t'_{R(x)} - \log t'_{R(z)}}{\log t'_{R(z+1)} - \log t'_{R(z)}}\right] \quad (1.20)$$

where $t'_{R(x)}$ is the adjusted retention time of unknown x, $t'_{R(z)}$ is the adjusted retention time of a hydrocarbon with z carbon atoms.

The Kovats retention index is very useful for identifying unknown compounds when no pure standard is available. It has advantages over quoting the retention time relative to that of a reference compound, since relative retention times may vary with small changes in chromatographic run parameters such as ageing of columns. The differences in the retention indices of a set of standards between a non-polar stationary phase (squalane) and a stationary phase of interest are termed the McReynold's constants (McReynolds, 1970). These constants are published for a wide range of stationary phases and they provide a useful guide for the selection of a stationary phase for the analysis of a mixture. Comparison of the data for Carbowax 20M and Dexsil 300 illustrates the principles underlying the McReynold's constants (Table 1.2). All standards are retarded more strongly by the Carbowax 20M due to dipole–dipole and dipole-induced dipole forces. However, the elution order of 1-butanol and 2-pentanone is reversed on the Carbowax 20M phase compared with the Dexsil 300 phase because of the additional retention of the alcohol on Carbowax 20M due to hydrogen bonding.

1.6 DETECTORS

The detector converts the flow of molecules passing from a GC column into a voltage that can be monitored by a chart recorder, recording integrator or other data capture device. A wide variety of detectors are available. Some detectors, including the flame ionization detector, respond to the mass of eluent per unit time (mass flow) while other detectors, such as the thermal conductivity detector, respond to the vapour concentration within the detector cavity. Detectors also vary in their selectivity of response, with the thermal conductivity detector being truly universal, while several detectors are more or less selective in their response. Non-destructive detectors, such as the photoionization or thermal conductivity detectors, can be used in series with destructive detectors to increase the information produced from the analysis. Alternatively the effluent can be split between two destructive detectors.

1.6.1 Flame ionization detector (FID)

The FID is the most common detector in GC analysis (Fig. 1.11). Hydrogen-containing carrier gas flows out of the jet and expands. A flow of air is passed around the outside of the jet and it surrounds the hydrogen and carrier gas from the jet. The mixture of gases burns at the tip of the jet. The flow of hydrogen is commonly set at the same rate as the carrier gas in packed column GC (20–60 ml min^{-1}) while the air

Table 1.2 — McReynolds' constants and details of some common stationary phases

Phase	Structure	McReynolds' constants $(\Delta I)^a$						Applications
		$\Delta I(1)$	$\Delta I(2)$	$\Delta I(3)$	$\Delta I(4)$	$\Delta I(5)$	Sum	
Squalane	Hydrocarbon	0	0	0	0	0	0	Boiling-point separations
OV-1	Methylsilicone	16	55	44	65	42	222	Boiling-point separations
Dexsil-300	Carborane methylsilicone	47	80	103	148	96	474	Boiling-point separations
OV-17	50% phenylmethyl silicone	119	158	162	243	202	884	Semi-polar compounds, unsaturated hydrocarbons
QF1	Trifluoropropyl silicone	144	233	355	463	305	1500	Alkaloids and carbonyl compounds
XE60	Cyanopropyl phenylmethyl silicone	204	381	340	493	367	1785	Polar compounds
Carbowax 20 M	Polyethylene glycol	322	536	368	572	510	2308	Polar compounds
OV-275	Cyanosilicone	629	872	763	1106	849	4219	Fatty acid methyl esters, polarizable molecules

[a]Reference compounds: 1, benzene; 2, 1-butanol. 3, 2-pentanone; 4, 1-nitropropane; 5, pyridine.

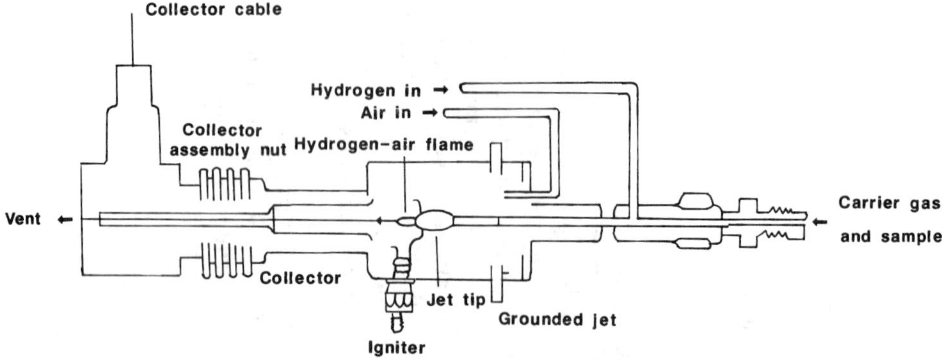

Fig. 1.11 — Flame ionization detector (Perkin Elmer). Redrawn with permission.

flow is set at a rate 5–10 times as fast. In the case of capillary GC with low carrier-gas flow rates, make-up carrier gas is usually introduced at the end of the column.

An electrostatic potential is applied between the collector, which operates at a voltage of about 400 V, and the jet, which is grounded. The collector lead is connected to an electrometer for current measurement. When carrier gas is eluting from the column there is virtually no signal and a stable baseline is achieved, but virtually all organic compounds eluting from the column burn in the flame and radicals formed suffer some ionization, leading to a current which increases with increasing mass of the compound. A wide linear dynamic range of about 10^8 is achieved. In the case of hydrocarbons, the radical reaction (1.21) occurs.

$$^{\bullet}CH + O^{\bullet} \rightarrow CHO^{+} + e^{-} \tag{1.21}$$

Only one carbon atom in 10^5 atoms in the flame leads to the formation of an ion and an electron. Most organic molecules burn in the flame and produce a response, but the detector does not respond to CO,CO_2, H_2O, NH_3 and H_2S. The response to a saturated hydrocarbon is proportional to the number of carbon atoms eluting from the column, but some substituents reduce the response and this can be allowed for by considering substituents as contributing an effective carbon number to the molecule (Table 1.3). These values for the effective carbon number allow the relative response factors of organic molecules to be predicted to within 20%. However experimental determination of relative response factors is often required for accurate quantification. Large numbers of experimental relative response factors have been reported in the literature (Watanabe *et al.*, 1982; Leibrand & Dunham, 1973).

The detection limits of the FID are approximately 5 pg s^{-1} for light hydrocarbon gases, increasing up to 10 pg s^{-1} for higher organic molecules.

1.6.2 Thermal conductivity detector (TCD)

The TCD (Fig. 1.12) consists of a cavity in a metal block with a coiled filament passing through the middle. The filament is commonly composed of platinum,

Table 1.3 — Contributions to effective carbon number (O'Brien, 1985)

Atom	Type	Effective carbon no. contribution
C	Aliphatic	1.0
C	Aromatic	1.0
C	Alkene	0.95
C	Alkyne	1.30
C	Carbonyl	0.0
C	Nitrile	0.3
O	Ether	− 1.0
O	Primary alcohol	− 0.6
O	Secondary alcohol	− 0.75
O	Tertiary alcohol, esters	− 0.25
Cl	Two or more on aliphatic C	− 0.12 each
Cl	On alkene C	+ 0.05
N	In amines	Similar to O in corresponding alcohols

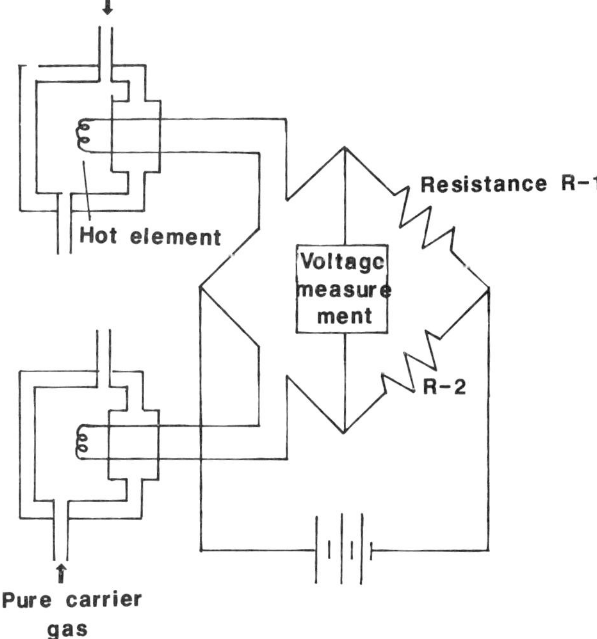

Fig. 1.12 — Thermal conductivity detector and associated circuitry.

nickel, tungsten or alloys of platinum or tungsten. It is heated by a direct current and the filament resistance is monitored. If the gas surrounding the filament contains an eluting compound the thermal conductivity changes from the background value and this causes a change in the filament temperature and hence the resistance. Hydrogen or helium is used as the carrier gas, since both have a high thermal conductivity. The TCD has the advantage that it is simple and virtually universal, responding to both organic and inorganic compounds. Compounds containing metal atoms or halogens are the main classes for which the response factor deviates from that of hydrocarbons, but for most organic compounds the response factor is similar. The main drawback to the TCD is that the sensitivity is less than that of the other common detectors, with a limit of approximately 1 ng ml^{-1}. The TCD has applications in the analysis of gases which are not detected by the FID, such as NH_3, H_2S and CO_2.

1.6.3 Thermionic detector (TD)

The thermionic detector (also termed the alkaline flame ionization detector or nitrogen-phosphorus detector) is a modification of the FID. A non-volatile alkali metal salt such as rubidium silicate is introduced into the detection system and volatilized in a flame. Alkali metal atoms are formed which ionize in the fuel-poor hydrogen flame and are subjected to an electric field. If a compound containing a heteroatom (P, N, halogens, S, As, Si) is introduced into the detector, the background current changes. The response differs in magnitude and, with certain heteroatoms, in polarity from that of the FID. The detector response to molecules that do not contain an appropriate heteroatom, or which contain a hetaroatom for which the conditions have not been optimized, is lower than that of the FID. The flow rate of hydrogen is one of the main variables, since this determines the flame temperature and hence the extent of volatilization of the metal salt and ionization in the flame. Other variables include the flame shape and size, and the composition of the salt tip. The low cost of the TD and the ease of conversion of an FID contributed to the popularity of the detector. There have been several improvements in the design of the TD in recent years aimed at reducing the high noise level and hence reducing the minimum detectable mass rate as well as improving the sensitivity and selectivity of detection. One design of TD uses a three-electrode detector to reduce the noise level by measuring the background ionization current in a separate circuit from that of ions containing a heteroatom.

Nitrogen–phosphorus detectors have been developed in which the ionization of molecules occurs in the absence of a flame. The alkali metal source is heated electrically and a low flow rate of hydrogen in the range 1–5 ml min^{-1} is used. The flow is insufficient to sustain an ordinary flame, and the detector is sometimes referred to as a flameless alkali-sensitized detector. Nitrogen–phosphorus detectors are not sensitive to halogens. The thermionic detector described by Kolb and Bishoff (1974) (Fig. 1.13) is an example of this design. An electrically heated rubidium silicate bead is located a few millimetres above the tip of the jet and below the collector electrode. A negative potential is applied to the source to prevent the loss of rubidium ions and to reduce the response of the detector in the flame ionization mode. The salt bead is electrically heated to 600–800°C, and hydrogen and air are combined to produce a plasma close to the bead. A low hydrogen flow rate, which corresponds to the flameless mode, is used for nitrogen–phosphorus detection, but

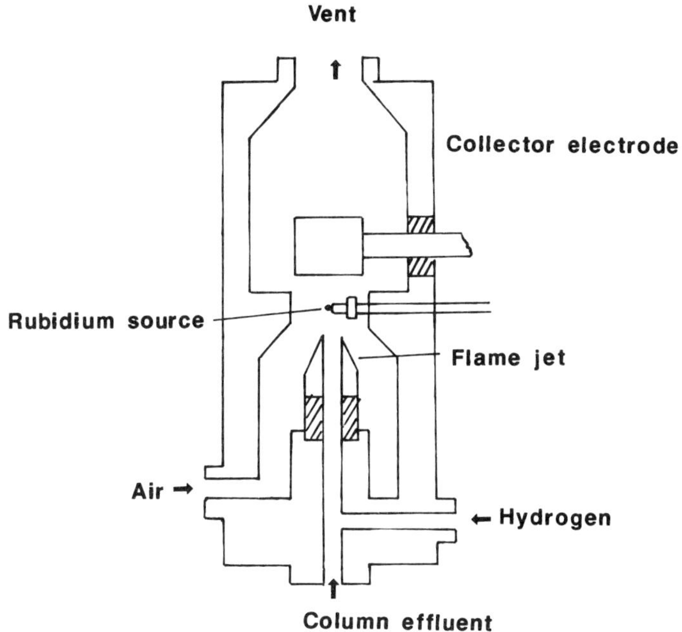

Fig. 1.13 — Thermionic ionization detector (Perkin Elmer). Redrawn with permission from Kolb & Bishoff (1974).

the more traditional alkali FID mode with a higher hydrogen flow rate (30 ml min^{-1}) is used in the phosphorus-selective variation. The flame jet is earthed to reduce the background flame ionization signal, and the negative potential of the bead deflects electrons to earth and away from the collector electrode. Sensitivity and selectivity of the detector are affected by the bead heating current, jet potential, carrier gas, air and hydrogen flow rates and the bead position. Detection limits are about 10^{-13} g s^{-1} for nitrogen and 5×10^{-14} g s^{-1} for phosphorus. The linear response range is about 10^4–10^5, and selectivity ratios 4×10^4 g C/g N and 7×10^4 g C/g P can be achieved. Contamination of the bead can reduce the response significantly.

1.6.4 Photoionization detector (PID)

When a compound AB is irradiated with UV (ultraviolet) radiation, ionization occurs if the photon energy $h\nu$ exceeds the ionization potential of the compound. Ionization can be represented as a two-step process (1.22) (Sevcik & Krysl, 1973)

$$AB + h\nu \rightarrow AB^* \rightarrow AB^+ + e^- \qquad (1.22)$$

In modern PIDs the UV lamp is separated from the ionization chamber by a window comprising magnesium fluoride or lithium fluoride (Fig. 1.14). Interchange-

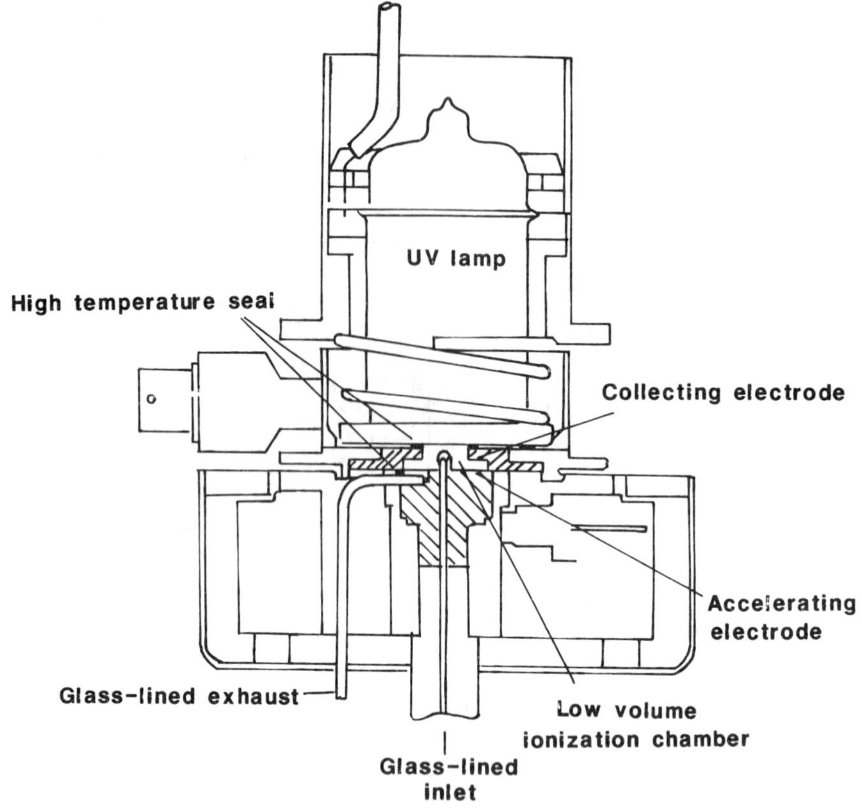

Fig. 1.14 — The photoionization detector. Redrawn with permission from HNU.

able lamps varying in energy of the estimated photons (commonly 8.3 eV, 9.5 eV, 10.2 eV and 11.7 eV) may be used as radiation sources. For maximum selectivity, a lamp producing photons with energy just sufficient to photoionize the molecules being detected should be used. A chamber adjacent to the UV source contains a pair of electrodes. A positive potential is applied to the accelerating electrode and this drives ions formed by photoionization to the collecting electrode where the current is measured. Any compound with an ionization potential less than 11.7 eV can be detected with the PID. Aliphatics (except methane), aromatics, ketones, aldehydes, esters, heterocyclics, amines and organic sulphur compounds can be detected. When used with a 10.2 eV lamp, the PID does not respond to common solvents including methanol, chloroform, dichloromethane, carbon tetrachloride, and acetonitrile.

The sensitivity of the PID with a particular lamp increases as the ionization potential of the molecules decreases. Hence, the sensitivity varies in the order (Langhorst, 1981):

aromatics$\rangle$alkenes$\rangle$alkanes

and

ketones⟩aldehydes⟩esters⟩alcohols⟩alkanes

The PID is a non-destructive detector which combines a wide linear dynamic range (10^7) with high sensitivity. The lower limit of detection is commonly 10–100 times less than that of an FID for organic molecules, e.g., 2×10^{-12} g for benzene, while the sensitivity is about 20 times greater than the FID (Driscoll, 1976). The linear dynamic range of several common detectors is compared in Fig. 1.15. Applications of

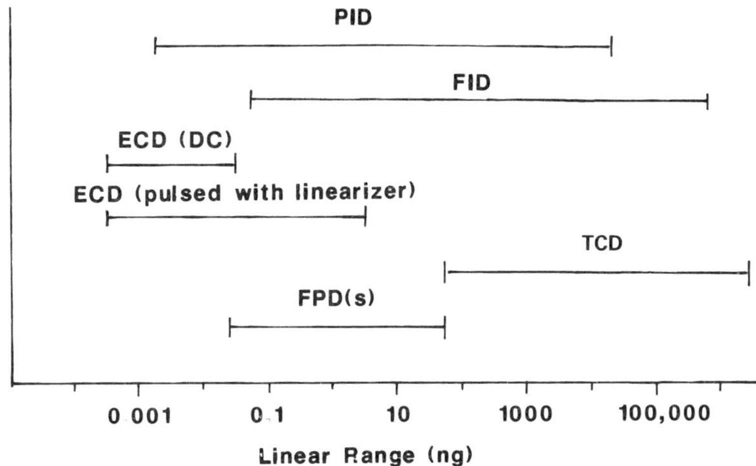

Fig. 1.15 — Comparison of linear dynamic ranges of various GC detectors. Redrawn with permission from Driscoll (1976).

the PID include analysis of pesticides, drugs, polychlorinated biphenyls, polycyclic aromatic hydrocarbons, drinking water for purgeable organic compounds, and flavour volatiles. A recent application of the PID is the analysis of the degree of unsaturation of fatty acid methyl esters by comparing the PID chromatogram with the FID chromatogram (Fig. 1.16). The PID is more sensitive to unsaturated molecules and the ratio of peak areas from the two chromatograms allows fatty acids with 0, 1 and 2 double bonds to be distinguished.

1.6.5 Flame photometric detector (FPD)

The FPD (Fig. 1.17) is a mass-flow detector operating on the principle of a flame photometer. The column effluent passes into a hydrogen-enriched low-temperature flame inside a shielded jet. Combustion of eluting compounds leads to characteristic low molecular weight species in excited states. Phosphorus-containing molecules form HPO*, while sulphur-containing molecules form S_2^*. These two species return to the ground state with the emission of radiation at 526 nm and 394 nm respectively.

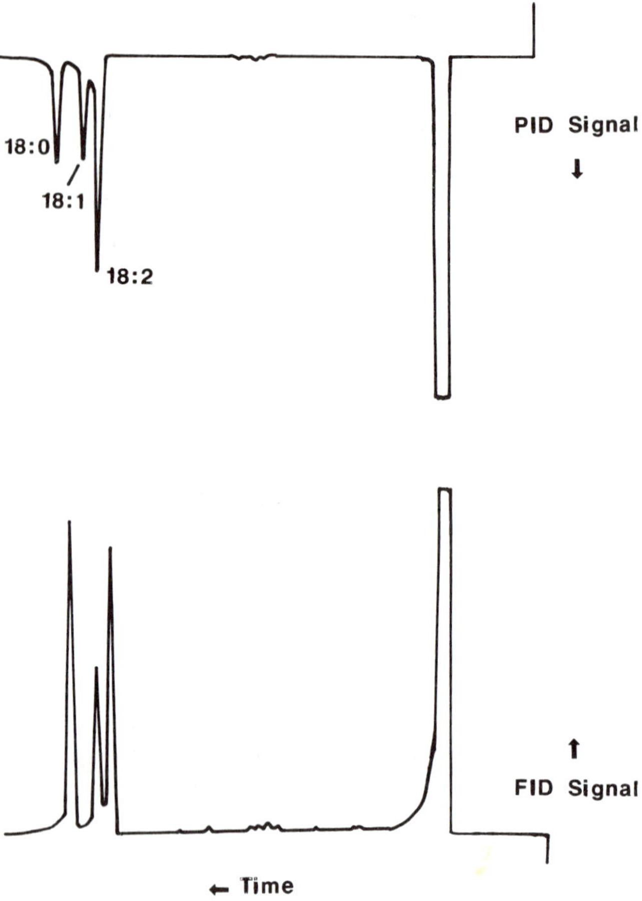

Fig. 1.16 — Analysis of fatty acid methyl esters using PID (attenuation 800) and FID (attentuation 20). Response ratio (PID:FID) corrected for attenuation is 31.1, 22.9, and 13.5 for 18:2, 18:1, and 18:0. Courtesy of HNU Systems Ltd.

Passage of the emitted radiation through a narrow bandpass filter into a photomultiplier tube allows the emission at the appropriate wavelength to be quantified. The sensitivity of the FPD to phosphorus and sulphur compounds varies with the detector design, hydrogen, air and carrier-gas flow rates, structure of the solute, concentration of the solute, detector temperature, and properties of the filter and photomultiplier. Often the carrier gas and air are mixed before entering the flame through a hole in the flame tip. The eluent burns in an atmosphere of hydrogen, with light emitted from hydrocarbons occurring in the oxygen-rich flame region close to the burner tip, while sulphur and phosphorus emissions occur in the diffuse hydrogen-rich upper flame region. The detector is designed to allow the light from the latter region to be detected. Dual flame detectors have been designed in which the sample is decomposed initially in a flame, while the emission is monitored after the combustion

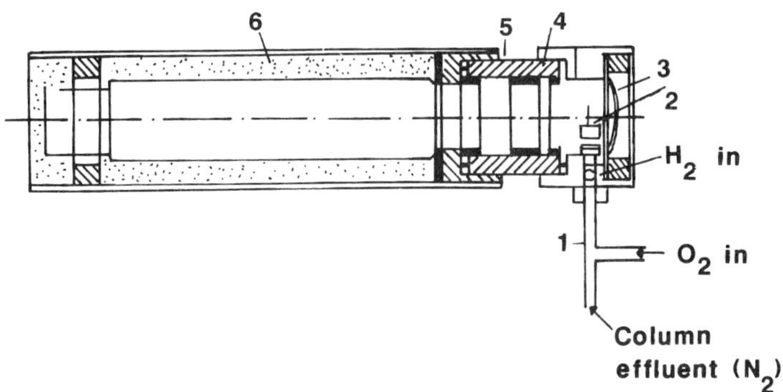

Fig. 1.17 — Schematic diagram of flame photometric detector, 1, Flame ionization burner tip; 2, burner; 3, mirror; 4, glass window; 5, optical filter; 6, photomultiplier tube. Redrawn with permission from Brody & Chaney (1966).

products have passed into a second flame (Fig. 1.18). Variation of response with structure of sulphur and phosphorus compounds, extinguishing of the flame by solvent, and hydrocarbon quenching of the emission of radiation are minimized with this design. In general the FPD response is linear for phosphorus compounds over a 10^4 concentration range, while for sulphur compounds the response R, varies with sulphur concentration [S] according to the equation $R = k[S]^n$ where n is in the range 1–2. The non-linear response with sulphur concentration arises from the fact that S_2^* is the light-emitting species. The effect of structure on the detector response to sulphur and phosphorus compounds has been studied by several workers. Sugiyama *et al.* (1973) reported that the detector response followed the order butanethiol⟩thiolane⟩thiophene⟩butylsulphide⟩thiophenol⟩benzothiopene⟩phenyl⟩sulphide⟩pentyl disulphide. The selectivity of the detector response for phosphorus relative to hydrocarbons is 10^4–10^5 g C/g P using an interference filter. The selectivity for sulphur depends on the amount of sulphur present; it varies from 10^3 g C/g S at low sulphur amounts to 10^6 g C/g S at high sulphur amounts.

1.6.6 Chemiluminescence detector

The chemiluminescence detector is mainly of significance in the analysis of *N*-nitroso compounds in foods. The detector designed for the analysis involves the catalytic decomposition of *N*-nitroso compounds to form nitric oxide and an organic radical according to (1.23)

$$\begin{matrix} R_1 \\ R_2 \end{matrix}\!\!>\! N - NO \ \overset{WO_3}{\rightleftharpoons} \ \begin{matrix} R_1 \\ R_2 \end{matrix}\!\!>\! N^\bullet + NO \qquad (1.23)$$

Tungsten trioxide adsorbed on the walls of a porous ceramic tube has been recommended as a catalyst (Gough *et al.*, 1977). The pyrolyser effluent expands into

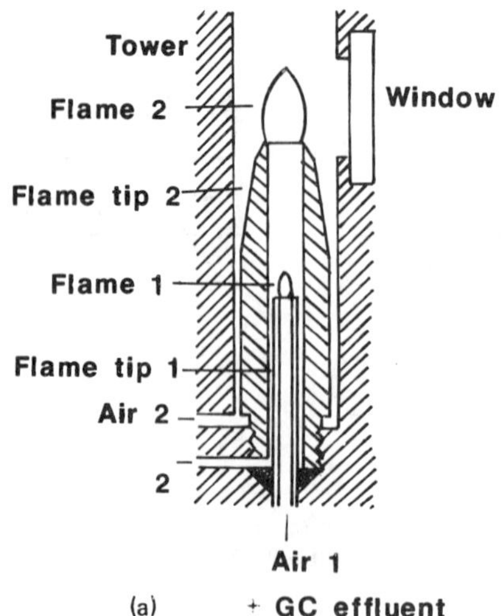

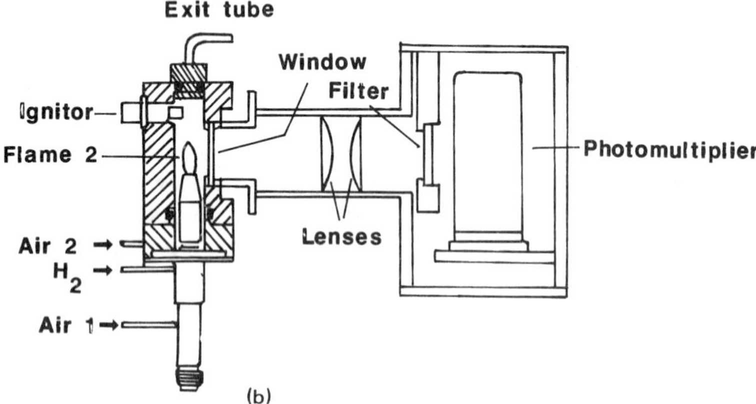

Fig. 1.18 — Dual-flame photometric detector: (a) Schematic diagram of the dual flame burner; (b) Schematic diagram showing the relationship between the burner and photometric viewing components. Reprinted with permission from Patterson *et al.*, *Analytical Chemistry*, **50**, 339. Copyright (1978) American Chemical Society.

a reaction chamber in which the nitric oxide reacts with ozone to form nitrogen dioxide in an excited state.

$$NO + O_3 \rightleftharpoons NO_2^* + O_2 \tag{1.24}$$

The excited nitrogen dioxide rapidly decays back to its ground state, emitting light in the near-infrared region.

$$NO_2^* \rightarrow NO_2 + h\nu \tag{1.25}$$

The emitted radiation is detected with a photomultiplier after passing through a red optical filter.

The detection limit for nitrosodimethylamine is about 5×10^{-11} g, and the response is linear over a concentration range of 10^5–10^6.

1.6.7 Electron capture detector (ECD)

The ECD is a detector which combines a high degree of selectivity and good sensitivity. A high response is produced mainly by halogenated organic compounds and nitroaromatics as well as some conjugated carbonyl compounds. As a consequence of its high response to polyhalogenated compounds, the ECD is often used for pesticide residue analysis.

In the ECD, β particles emitted from a radioactive source such as ^{3}H or ^{63}Ni bombard the carrier gas passing through an ionization chamber. ^{63}Ni may be used alone but ^{3}H must be adsorbed on a metal such as titanium. Collisions between the high-energy β electrons and the carrier gas produce a plasma of positive ions, radicals and a relatively large number of secondary thermal electrons, which have a lower energy in the range of 0.02–0.05 eV. A potential difference is applied to the detector electrodes either continuously or in pulses and the secondary electrons create a current. The background current which occurs in the presence of carrier gas is reduced when a molecule which can capture a secondary electron passes into the detector. Electron capture produces either a negative molecular ion (equation (1.26)) or a fragment if the molecular ion dissociates (equation (1.27)).

$$e^- + AB \rightarrow AB^- \tag{1.26}$$

$$e^- + AB \rightarrow A^\bullet + B^- \tag{1.27}$$

The ECD is designed to allow secondary electrons to be collected while the slower-moving ions are not.

Early designs of the ECD incorporated the electrodes as parallel plates but modern instruments are usually based on a coaxial cylinder or an asymmetric design in which the coaxial cylinder is displaced. The parallel-plate ECD allowed the relative position of the radioactive foil and the collector electrode to be varied, but the detector volume was large, particularly when ^{63}Ni was used as a source. In the coaxial design (Fig. 1.19), the anode is positioned inside the cylinder formed by the source. The distance between the source and the anode must be sufficient to allow the β electrons to be deactivated by collisions without colliding with the anode. The detector volume of 2–4 ml in the coaxial design may be reduced further if an asymmetric configuration is used. This involves a cylindrical cathode, which may serve as the detector body, separated by a glass, ceramic or PTFE insulator from a small anode. The applied electric field is concentrated near the anode, which is upstream from the ionized gas. Collection of energetic electrons by the anode is avoided in this configuration and small diameters are possible. Tritium has advantages as a radiation source because it provides lower energy electrons than nickel (18 keV for ^{3}H, 67 keV for ^{63}Ni), and foils with a higher specific activity that provide denser radiation can be manufactured. However, tritium foils lose activity at

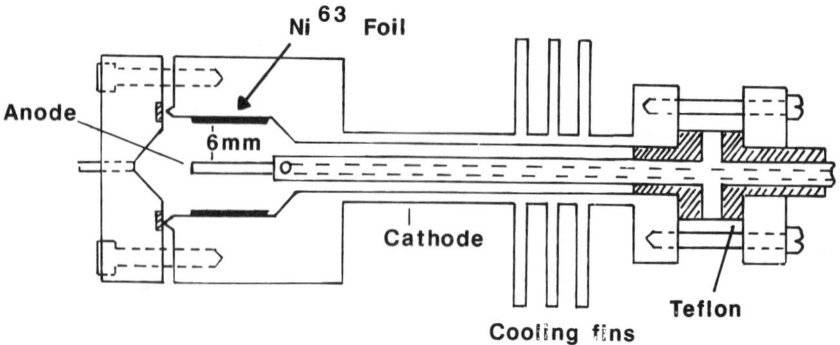

Fig. 1.19 — Coaxial high-temperature Ni electron-capture detector. Redrawn with permission from Fenimore *et al.*, *Analytical Chemistry*, **43**, (14), 1971. Copyright (1971). American Chemical Society.

elevated temperatures and therefore the temperature limit is only 200–225°C for a titanium–^{3}H foil compared with 400°C for a ^{63}Ni foil. A high detector temperature reduces contamination of the source by column bleed or involatile components from the GC column, and it may also affect the detector sensitivity. ^{63}Ni is often used as a source in commercial instruments.

Application of a constant potential to the detector electrodes can cause anomalous detector behaviour. The positive ions drift slowly to the cathode while the secondary electrons produced by the collision of the electrons with the carrier gas move more rapidly and are collected by the anode before they can react with solute molecules. This space charge effect may also be accompanied by the generation of a contact potential due to adsorption of solute molecules on the electrode surface. Negative molecular ions produced by electron capture may also be collected at the anode, although this effect is reduced by the use of an asymmetric detector. Non-electron-capturing ionization processes may also occur (Karasek & Denney, 1974b).

Commercial instruments use voltage pulses to overcome these effects. The voltage pulse is applied as a square-wave pulse with amplitude (50 V) and pulse width (1–3 μs) selected to allow all the secondary electrons to be collected at the anode after each pulse. The secondary electron concentration is restored by the ionizing β radiation, and the electron energy reaches thermal equilibrium under voltage-free conditions (typically for 100–150 μs). The electrons do not drift out of the plasma and negative ion formation occurs in the region where positive ions are also present. Detector sensitivity increases with increasing intervals between the pulses because positive and negative ions may recombine, but the interval must not be sufficient to allow recombination of positive ions and electrons. Most commercial ECDs maintain the detector current constant by varying the pulse frequency rather than by measuring the cell current at constant pulse frequency. This technique is claimed to give a wider linear response range, about 5×10^{4}. A typical calibration curve for an ECD in the constant current pulse mode is given in Fig. 1.20. The linear dynamic range is dependent on carrier-gas flow rates, solute structure and detector design. In

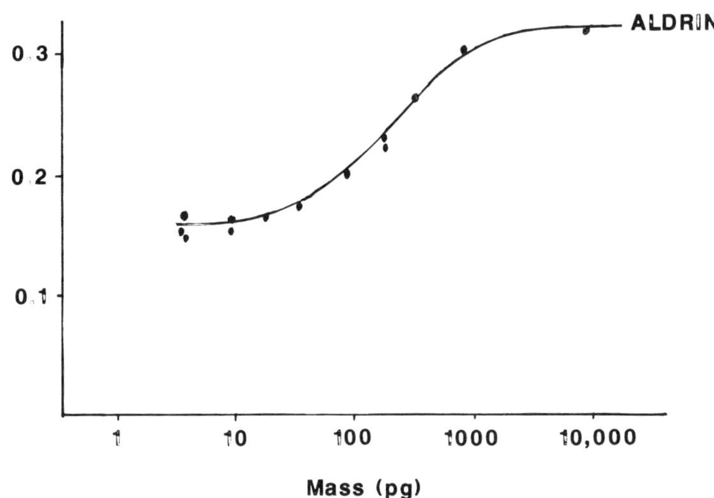

Fig. 1.20 — Constant-current ECD calibration graph for aldrin. Redrawn with permission from Sullivan & Burgett (1975).

some detectors, the carrier gas may also affect the linear dynamic range with argon containing 5–10% methane being preferred.

Nitrogen or argon containing 5–10% methane are the most common carrier gases for packed column analyses and the make-up gases used with capillary columns with the ECD. Pure argon or helium are unsuitable because they form metastable ions which can ionize solute molecules on collision. Methane deactivates the metastable ions before they can collide with solute molecules and also maintains the thermal equilibrium of the secondary electrons. Oxygen and moisture must be completely removed from gases, and hydrogen should be avoided with a tritium detector, except at low flow rates as a capillary column carrier gas, because it strips tritium from the detector at high temperatures. The response of the ECD to halogenated compounds increases strongly with the number of halogen atoms, and the carbon atom to which the halogen is attached determines the response in the order allylic$\rangle$saturated$\rangle$vinyl. Conjugated carbonyl compounds including benzophenones, quinones, phthalate esters and coumarins also produce a high detector response. The minimum detection limit for compounds with high electron-capture coefficients is less than for other GC detectors. Detection limits are 5×10^{-14} g s^{-1} for *tert*-butyliodide, 1×10^{-14} g s^{-1} for chloropyrifos, 5×10^{-13} g s^{-1} for benzophenone and 1×10^{-13} g s^{-1} for aldrin. Response data for a wide range of compounds are reported by Zlatkis and Poole (1981).

Detector temperature has a strong effect on the ECD sensitivity. Molecules that capture an electron without dissociation (equation (1.26)) show an increased response at lower detector temperatures while those that react with dissociation (equation (1.27)) show increased sensitivity at higher temperatures.

The ECD response to certain compounds can be increased by the addition of electron-capturing reagents to the carrier gas. This phenomenon, known as selective electron-capture sensitization, has been exploited in recent years by the addition of either nitrous oxide or oxygen to the carrier gas (Dressler, 1986). The response of the ECD to vinyl chloride is increased 760 times by the presence of 30 p.p.m. nitrous oxide in the nitrogen carrier gas, whereas the response to dichloromethane is reduced by factor of three (Fehsenfeld *et al.*, 1981).

1.6.8 Infrared detector

In recent years, infrared (IR) detectors have been developed for use with both capillary and packed-column analyses. The development of Fourier transform infrared spectrometers (FTIR) was an essential step in the use of IR for GC detection.

FTIR spectrometers acquire data using an interferometer, with the Michelson interferometer (Fig. 1.21) being the simplest type. Radiation from a source is

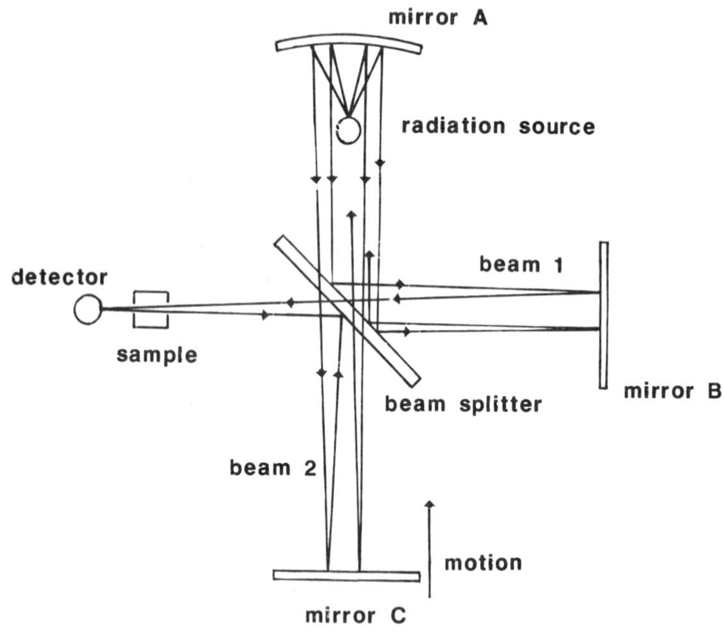

Fig. 1.21 — The Michelson interferometer.

collimated by mirror A and the beam is reflected onto a beam splitter, which reflects half the radiation, beam 1, onto mirror B, and allows the rest of the radiation, beam 2, to pass through to mirror C. Mirrors B and C reflect the radiation back to the beam splitter, which again reflects half the radiation and transmits half. Thus, half of each beam passes through the sample to the detector. For light of wavelength λ, beams 1 and 2 will be in phase at the detector if the path lengths are equal or differ by a multiple of λ. Constructive interference occurs if this condition is met, while destructive interference occurs at half-integer values of λ.

Mirror C is moved towards the beam splitter during the analysis. Since the source emits radiation over a range of wavelengths, the detector signal varies in a complex manner as mirror C is moved. Absorption of radiation at specific wavelengths by the sample modifies the variation of signal with wavelength. The resultant pattern is called an interferogram and can be converted by Fourier transformation into a spectrum. Fourier transformation is a mathematical technique involving the separation of the interferogram into a combination of a large number of sine and cosine terms varying in frequency.

FTIR spectrometers have the advantages of rapid data acquisition since all frequencies of the spectrum are recorded at all times during the experiment; high sensitivity with a good signal-to-noise ratio since there are no slits to attenuate the signal; no stray light and hence good quantitative accuracy; and no tracking errors, which also leads to good wavelength and photometric accuracy.

In a typical GC–FTIR system (Fig. 1.22), the effluent from the GC column is fed

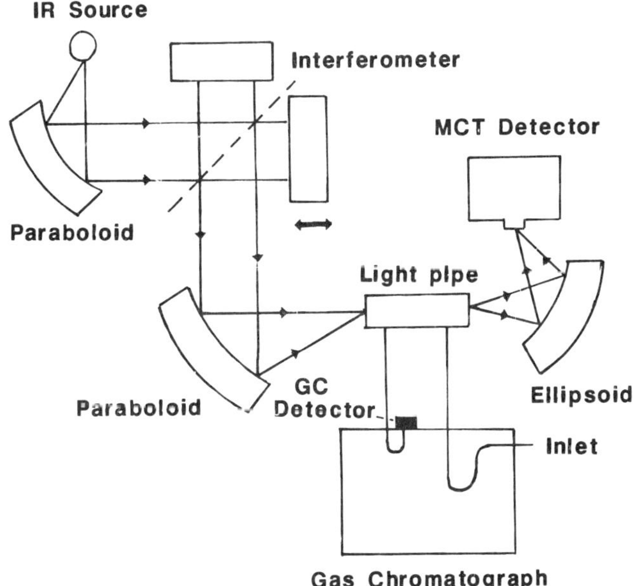

Fig. 1.22 — Schematic diagram of GC–FTIR instrument. MCT, mercury–cadmium telluride photodetector. Redrawn with permission from Griffiths *et al.*, *Analytical Chemistry*, **55**, 1361A. Copyright (1983). American Chemical Society.

into a heated IR measuring cell called the light pipe. A packed column requires a heated transfer line while a fused-silica capillary column can feed directly into the light pipe. The light pipe (Fig. 1.23) consists of a gold-coated glass tube with IR-

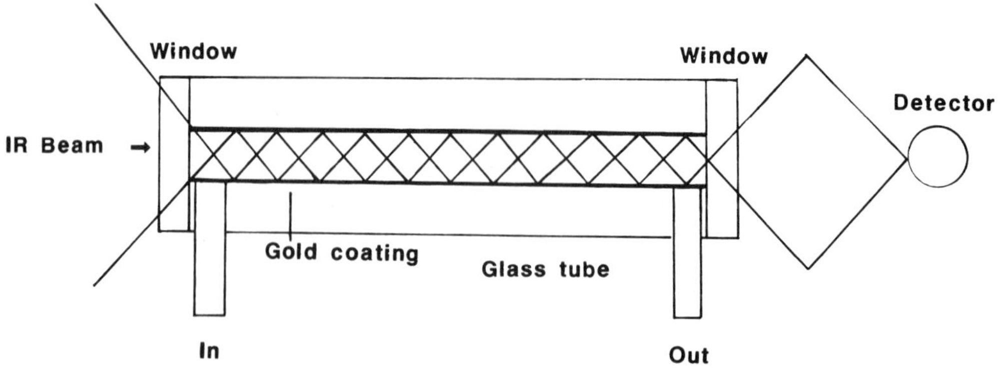

Fig. 1.23 — Schematic diagram of a light pipe for a FTIR detector for capillary gas chromatography.

transparent windows commonly of potassium bromide or zinc selenide. At the light pipe entrance, accelerating gas is added to the GC effluent to maintain the linear gas velocity through the GC–IR interface. Part of the effluent is often split to an FID by a fused silica transfer line at the end of the light pipe. The IR beam undergoes multiple reflections by the gold coating of the light pipe and it is then focused onto the element of a mercury–cadmium–tellurium detector cooled in liquid nitrogen. The volume of the light pipe for high-resolution GC-FTIR work is less than 0.1 ml, although larger volumes may be used if packed column work is also performed on the same instrument.

The mass of compound detectable by FTIR with a light pipe is about 10 ng for molecules with strong IR absorption, but the detection limit can be reduced to less than 1 ng if a matrix isolation technique is used (Fig. 1.24). In the matrix isolation technique, the effluent from the GC is directed onto a gold-plated copper disk with argon. The disk is kept at cryogenic temperatures, about 12 K, and the argon and GC components condense in a solid spiral band while the helium carrier gas is removed by the vacuum. The optical beam from the spectrometer passes through the sample matrix, is reflected by the gold disk and passes back through the sample to a second mirror and thence to the IR detector. All the IR measurements are performed after the chromatograms have been deposited. Each molecule of solute is isolated from the others in the argon matrix and therefore intermolecular interactions which give rise to band broadening in the IR spectrum are eliminated. The chromatogram can be obtained from the acquired data either by the Gram–Schmidt reconstruction of the interferogram or by the integration of the absorbance of the IR spectrum. Selective detection of components with specific absorption frequencies, analogous to selected-ion monitoring in GC–MS (mass spectrometry), has been reported (Hangac *et al.*, 1983).

1.6.9 Electrolytic conductivity detector (ELCD)

Electrochemical detectors have been developed in which the solute molecules are decomposed under either oxidizing or reducing conditions. Oxygen or hydrogen is

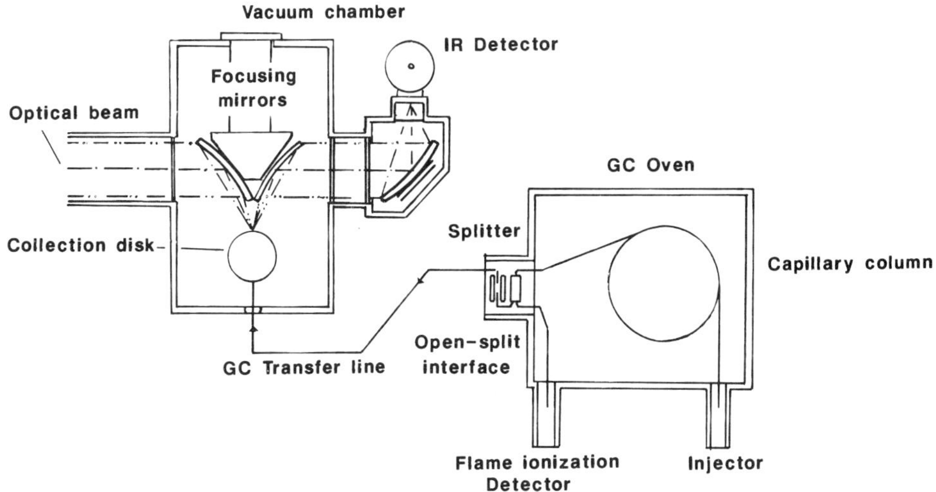

Fig. 1.24 — Schematic diagram of Cryolect GC–IR and collection chambers. Redrawn with permission from Bourne *et al.*, *American laboratory*, volume 16, number 6, page 90–101, 1984. Copyright 1984 by International Scientific Communications, Inc.

mixed with the GC column effluent, which is then passed into a pyrolysis tube or a catalytic converter comprising a nickel or quartz tube with a nickel wire inside at 500–1000°C. The reaction products are swept into a mixing chamber where they are mixed with an appropriate solvent, which is often an alcohol alone or mixed with water. The liquid phase is separated and passed through a conductivity cell with platinum electrodes. The early design of Coulson (Coulson, 1965) has been superseded by the Hall microdetector (Hall, 1974) which is smaller and more sensitive. The use of a bipolar pulse differential detector (Anderson & Hall, 1980), in which the solvent conductivity measured in one cell is subtracted from the conductivity of solvent containing decomposition products in a second cell, provides an increased sensitivity.

The ELCD can be used as a selective detector for halogen, sulphur and nitrogen-containing compounds. A nitrosamine-selective detector has been developed in which the effluent is heated to 700°C under reducing conditions in the absence of a catalyst (Anderson, 1979). The nitrosamine is reduced according to equation (1.28):

$$\begin{matrix} R_1 \\ R_1 \end{matrix} > N - NO + 3H_2 \rightarrow \begin{matrix} R_1 \\ R_1 \end{matrix} > NH + NH_3 + H_2O \qquad (1.28)$$

The selectivity over other nitrogenous compounds is in the range 200–500 when a water–*n*-propanol (1:1) electrolyte is used. The detection limit for nitrosamines is about 10 pg with this detector (Anderson, 1979).

1.6.10 Atomic emission detector (AED)

The AED recently introduced by Hewlett-Packard allows simultaneous independent detection of many elements, including C, O, N, S, P and Cl. The effluent from the

GC column is energized by a plasma, which causes the molecules to decompose and separate into excited atoms. The electrons in the atoms return to a low energy state by emitting light, which passes into a spectrophotometer. The light emitted is then separated by a diffraction grating into characteristic wavelengths and transmitted to a photodiode array detector, where it is converted into an electrical signal and plotted. The detector is capable of detecting 10^{-12}-g levels of the elements, and it appears to be highly promising for several areas of food analysis, including pesticide residue and aroma analysis, where selective detectors are required.

1.6.11 Other detectors

A mass spectrometer is a very powerful detector both in terms of its sensitivity and its power of identification. GC–MS has advanced rapidly in the last few years and there is a considerable amount of literature on the subject. Therefore GC–MS is discussed in detail in Chapter 2. Several other detectors may be used in GC analysis, including the plasma or ion mobility detector, which has been used in the analysis of aliphatic *N*-nitrosamines (Karask & Denney, 1974a) and polychlorinated biphenyls (Karasek, 1971); the helium detector, which can be used with GSC in the analysis of gases such as CO_2 and O_2; and the radioactivity detector. The reader is referred to more detailed reviews, including Adlard (1975) and Dressler (1986) for discussion of these detectors.

1.7 MULTIDIMENSIONAL CHROMATOGRAPHY

In recent years analysts have developed techniques involving a combination of two or more different chromatographic columns, identical columns under different operating conditions or two detectors which differ in their selectivity. These analyses are described as two-dimensional chromatography or, if more than two parameters are varied, multidimensional chromatography. The aim is usually to (i) improve the resolution of part of a chromatogram, (ii) reduce the analysis time or (iii) increase the sensitivity of the analysis. Analyses involving the use of two columns in series to modify separation characteristics are not usually considered as two-dimensional chromatography since the separation is similar to that of a single column with a mixed stationary phase (Lamb & Purnell, 1975).

There are several versions of multidimensional chromatography.

1.7.1 Heart-cutting

This involves the preliminary separation of a mixture on one column, with a portion of the column effluent, which elutes over a fixed time period, being removed to a second column for further separation of the components. Fig. 1.25 illustrates a typical application of heart-cutting in the analysis of ethyl carbamate in alcoholic beverages. Heart-cutting provides several benefits:

(1) Improved separation of components in a particular section of the chromatogram. Identification of components also becomes more certain because their retention time depends on retention by two stationary phases.
(2) Reduced analysis time.
(3) Good sensitivity for the analysis of components present in trace levels, since the

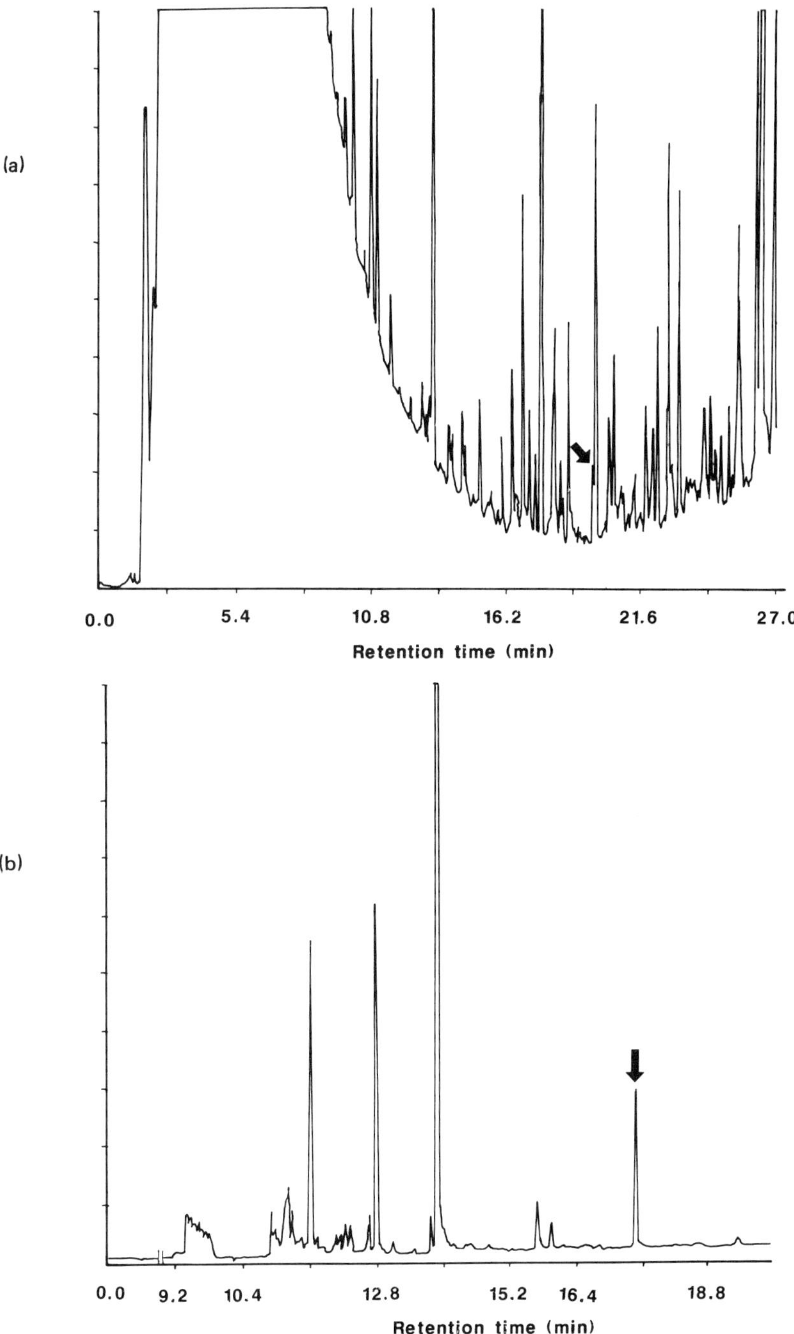

Fig. 1.25 — Advantages of heart cutting. (a) Determination of ethyl carbamate in a whisky (containing 230 p.p.b.) by capillary GC on a dB wax column. (b) Determination of ethyl carbamate in the whisky by initial separation on a CP-SIL 5 CB fused-silica column (0.53 mm i.d., 10 m), with heart cutting to a CP-WAX 52 fused-silica column (0.22 mm i.d., 25 m). Redrawn with permission from van Ingen *et al.* (1987).

first column may be overloaded while achieving adequate separation of groups of components.

(4) Contamination of the analytical column is reduced, since undesirable sample components including derivatising agents are not transferred to this column.
(5) Solvents which are incompatible with the detector are removed from the analytical sample.
(6) Improved quantitative analysis of components eluting on the tail of the solvent. The bulk of the solvent can be eliminated by the precolumn, while the final separation is achieved on the analytical column.

1.7.2 Backflushing

The analysis time of a mixture containing components with a long retention time can be reduced by backflushing. The more volatile components are eluted from the column into the detector, but the direction of the carrier-gas flow through the column is then reversed and the strongly retained components are flushed out into the detector without passing through the column. Non-volatile components may be vented if quantification of this fraction is not required.

1.7.3 Foreflushing

This technique is used for the analysis of high-boiling components in a volatile matrix. The volatile components are vented via a pre-column or a sample loop, and the gas flow is then reversed to transport the components of interest onto the analytical column.

1.7.4 Dual detector analyses

Sections of a single chromatogram may be analysed by two different detectors by switching the effluent path at the appropriate time. Alternatively the column effluent may be split continuously between two detectors, or a non-destructive detector may be fixed in series with a second detector.

1.7.5 Cryogenic trapping

This technique involves the use of cryogenic coolants to trap volatile components, primarily for the concentration of dilute samples requiring multiple injections. Cryogenic trapping may be applied either pre- or post-column. It it is applied pre-column, improved resolution is achieved, as in the related technique of cryogenic focusing (or cryofocusing), which involves the use of low-temperature coolants to focus the sample into a plug at or near the head of the column. An uncoated section of capillary column, producing a retention gap, is recommended for this technique in order to avoid the splitting of peaks (Grob & Muller, 1982). Pre-column trapping may be used in the accurate determination of retention indices.

1.7.6 Serial coupled columns at different temperatures (SECAT)

Two columns coupled in series but maintained at different temperatures may be required for the analysis of complex mixtures (Kaiser & Reider, 1979).

Most multidimensional techniques require the path of carrier gas flow to be changed during the analysis. The simplest way of achieving this is to use a mechanical switching valve. The use of a four-port switching valve to backflush high-boiling components to the detector is illustrated in Fig. 1.26. Switching valves must have minimal internal volume and be chemically inert and leakproof. They are commonly used in cryogenic focusing since the transfer time is lengthy in this technique.

An alternative method of altering the path of carrier-gas flow involves pneumatic switching (Deans, 1968). The direction of flow between columns is changed by opening and closing external valves and no moving parts are located within the gas chromatograph. Deans' switching mechanism is illustrated in Fig. 1.27.

This system employs taps a, b and c outside the oven. Taps a and c are used for setting up purposes only. Carrier gas enters the chromatograph through pressure controller 1 and tap a, which is open. It then passes through the two columns in series and into the detector. Although tap c is normally open, there is negligible inlet of carrier gas through this line because pressure controller 2 is set at the same pressure as that produced at the tee connection by the carrier gas passing from column 1. Tap b is closed for normal operation, but is opened to allow heart-cutting. With tap b open, carrier gas enters through pressure controller 1, flows through column 1 and out through tap b and the needle valve. At the same time, carrier gas enters through pressure controller 2 and tap c. Most of this carrier gas passes through column 2, although a small part flows up and exits from tap b through the needle valve.

1.8 DERIVATIZATION

Derivatization prior to gas chromatography is usually performed in order to increase the volatility or thermal stability of components or to improve their chromatographic properties, as in the elimination of tailing, which is commonly observed for polar components. Derivatization may also be applied in order to aid identification or improve the sensitivity of detection with selective detectors.

Derivatization reagents must react quantitatively with functional groups. For some compounds, selective derivatization reagents are required and rapid reaction with minimum sample work-up is preferred. In many cases, the derivatizing reagent does not interfere with the chromatogram and the product mixtures may be injected directly onto the gas chromatograph.

A wide variety of derivatization reactions may be applied. Silylation, which involves the replacement of an active hydrogen by a silyl group, is a popular method of derivatization of alcohols, amines, carboxylic acids and thiols.

The most common silyl group is trimethylsilyl, $-Si(CH_3)_3$, although higher trialkyl silyl derivatives may be prepared in order to aid separation or identification by mass spectrometry. Bis-(trimethylsilyl)trifluoroacetamide (BSTFA) is highly effective as a silylating agent, with a silylation rate of ten times that of bis (trimethylsilyl) acetamide (BSA). Trimethylsilylimidazole (TSIM) is less effective than BSA or BSTFA in general but is useful for silylating hindered hydroxyl groups and can also be used in the presence of water, unlike most silylating agents. Other derivatives, including acetates or higher esters, are also commonly prepared from hydroxy or thiol substituents. Carboxylic acids are often analysed as esters, as in the analysis of fatty acid methyl esters.

Carbonyl compounds are commonly analysed without derivatization, although they may be derivatized to hydrazones or oximes when required. Perfluoro deriva-

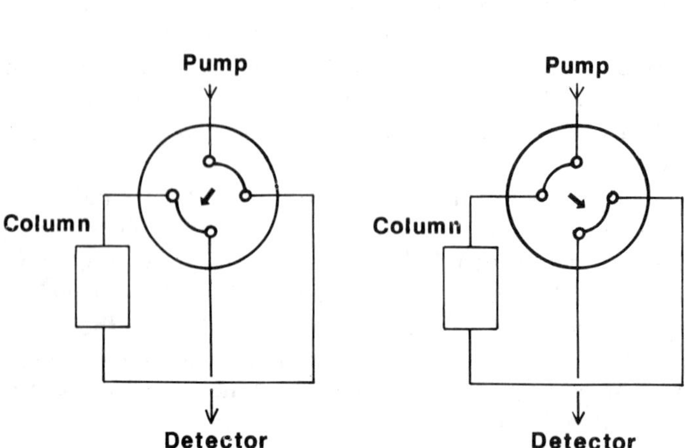

Fig. 1.26 — Use of a four-port switching valve for backflushing. Position A — sample is applied to the column. Position B — carrier gas flow is reversed for rapid elution of strongly retained components to the detector.

tives are commonly prepared in order to allow molecules to be detected by an ECD. The most common perfluoro derivatizing groups are trifluoroacetyl, pentafluoropropionyl, perfluorobenzene and heptafluorobutyryl. Derivatization reactions are discussed in detail by Knapp (1979), and many examples are included in later chapters.

1.9 REACTION GAS CHROMATOGRAPHY

Reaction GC is applied to molecules in order to aid identification, simplify the analysis of complex mixtures, or to allow analysis of molecules which are too unstable to be analysed without reaction. Molecules may be chemically modified, removed by specific reactions or thermally decomposed.

1.9.1 Chemical modification

Organic molecules may be chemically modified by various catalytic reactions prior to GC analysis. Catalytic hydrogenation may be used to determine the chemical skeleton of an unknown molecule (Beroza & Sarmiento, 1963). Hydrogen is used as the carrier gas and the sample is swept through a hot tube packed with a catalyst. Multiple bonds in the sample become saturated, and hydrogen replaces any halogen, sulphur, oxygen and nitrogen atoms. The hydrocarbon skeleton may then be identified by its retention time in the GC. Catalytic deoxygenation (Thompson *et al.*, 1960a) or catalytic desulphurization (Thompson *et al.*, 1960b) may be used to prepare hydrocarbons from molecules containing heteroatoms.

1.9.2 Subtraction loops

Irreversible chemical absorption on specific in-line absorbers has been used to selectively remove compounds containing specific functional groups as an aid to GC

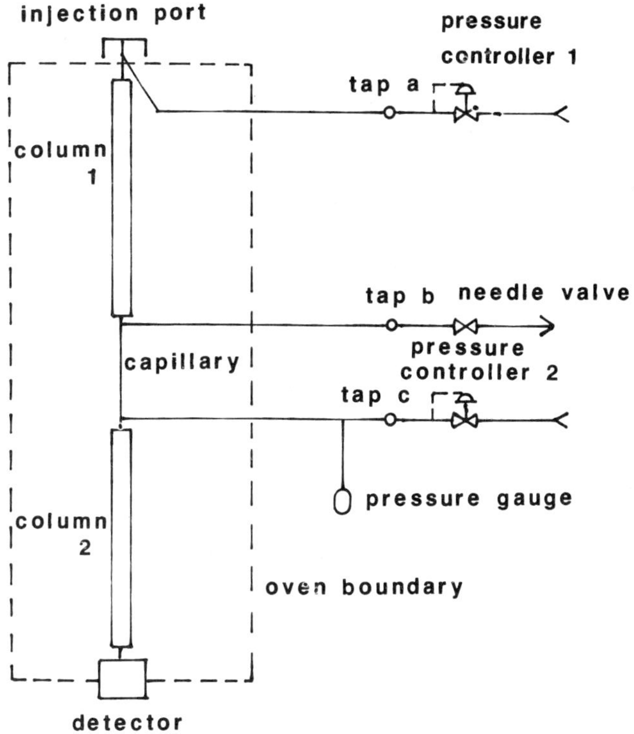

Fig. 1.27 — Deans' pressure-switching system. Redrawn from Deans (1968).

analysis (Kalo, 1981). These absorbers, which are termed subtraction loops, may be placed before or after an analytical column in order to selectively remove compounds containing specific functional groups. Subtraction loops usually consist of a short packed or PLOT column containing an appropriate reagent deposited on a diatomaceous earth support. The subtracting agent may alternatively be deposited as an ordinary liquid phase on a length of WCOT capillary column. Subtracting agents reported in the literature include *o*-dianisidine for aldehydes; benzidine, sodium hydrogen sulphite and semicarbazide for aldehydes and ketones; boric acid and 3-nitrophthalic anhydride for epoxides and nitrogen containing bases; and zinc oxide for carboxylic acids.

1.9.3 Pyrolysis–gas chromatography

The most common form of reaction GC is pyrolysis–GC. Materials that are not sufficiently volatile or thermally stable for GC without degradation may in some cases be analysed by this technique, which can provide a fingerprint corresponding to the degradation products produced under carefully controlled conditions. Pyrolysis is most valuable if the degradation products are: (1) as nearly unique to the sample as possible (2) reproducible and (3) capable of successful separation and elution in the GC.

Pyrolysers should be capable of heating almost instantaneously to temperatures in the region 500–850°C, where pyrolysis is usually performed. The degradation products often vary considerably with temperature and therefore slow rates of heating do not produce a sufficiently characteristic pyrogram. Pyrolysers operate either in a pulse mode or a continuous mode. In a pulse-mode pyrolyser, the sample is in contact with a central heating source, commonly a metal wire, which is capable of being heated rapidly to the pyrolysis temperature. When the pyrolysed molecules vaporize from the heat source, they are transported by carrier gas into cooler regions and subsequently to the GC for analysis. Pulse-mode pyrolysers include both coiled filaments of electrically resistive wire, which are heated by an electrical current, and Curie-point pyrolysers, which are heated by a high-frequency induction coil surrounding the sample. Different pyrolysis temperatures are obtained with Curie-point pyrolysers by using alloys containing differing amounts of ferromagnetic metals, namely iron, cobalt and nickel, since these alloys vary in the Curie point, which is the temperature at which the metal loses its ferromagnetic properties and becomes paramagnetic.

Continuous-mode pyrolysers, including the furnace type, are commonly used. The heat source is the wall of the pyrolyser, which is maintained at the pyrolysis temperature. The residence time of the sample in the pyrolyser determines the extent of secondary reactions after pyrolysis, and careful design is required to ensure rapid heating of the sample and removal of pyrolysis products.

1.10 QUANTIFICATION IN GAS CHROMATOGRAPHY

1.10.1 Manual integration

In the early days of GC, integration of peak areas was usually performed by recording the chromatogram on a chart recorder and determining the areas manually. For an isothermal separation, the area of a chromatographic peak may be determined by multiplying the peak height by the peak retention time measured from injection. If the separation is not isothermal, the peak area may be determined as the product of the peak height, and peak width measured at one-half peak height or half the peak base width. An alternative procedure involves cutting out the chromatographic peak (or a photocopy), with the peak area being proportional to the mass of paper under the peak. The latter technique is preferred for non-symmetrical peaks. Determination of peak areas by drawing a triangle through the sides of the peak and determining its area as half height times base width has also been used for symmetrical peaks.

1.10.2 Electronic integrators and computers

The use of electronic devices for data acquisition and integration is now widespread. These include stand-alone integrators, microcomputers with integrator software, large LIMs (laboratory information management systems) with integration as a part, and chromatographs with the integrator built in. Standard features of integrators include measurement of peak areas, heights and retention times; performance of standard calculations such as area per cent, internal standard and external standard; and self-calibration, in which analyses of standard solutions are used to calculate response factors, which are stored and subsequently used in the analysis of unknowns. The most modern systems allow the use of multiple internal standards; multilinear calibration; least mean squares averaging of results from multiple standard solutions; and validation checks, such as the measurement of column efficiency and peak shape. Whole chromatograms may be stored to allow reprocessing, including baseline correction and optimization of integration parameters. Screen display and manipulation of chromatograms, and battery-driven memory protection are also useful features of some integration systems.

1.10.3 Principles of electronic integration

The analogue signal from the chromatograph is fed into a voltage-to-frequency converter, which generates an output pulse rate proportional to the peak area. When the slope detector senses a peak, the pulses from the voltage-to-frequency converter are accumulated and counted to produce a value proportional to the peak area. The integrator is required to define the baseline, and identify the start, apex and end of the peak. Peaks must be discriminated from noise or a drifting baseline as well as from each other. The integrator only produces accurate data if variables such as peak width, sampling rate, threshold peak height or threshold peak area values have been set correctly. Incorrect settings may lead to peaks being confused with noise, incorrect start or finishing points or incorrect fixing of the baseline.

GC peaks are often not resolved from all other peaks. A peak may be not quite resolved from its neighbour, the overlap may be severe, or the peak may occur as a shoulder on the tail of another peak or the solvent. A perpendicular can be dropped to the interpolated baseline from the valley between two overlapping peaks that are similar in area without significant error. However, the error arising from this procedure increases as the ratio of peak heights differs greatly from unity. When the overlap between two peaks has increased to the extent that one peak represents a shoulder on the other peak, a tangent may be drawn to the side of the larger peak as a the new baseline for the shoulder peak. This technique is known as tangent skimming.

Accurate GC analysis is dependent on the injector, the column, the detector and the integrator. No discrimination between components should occur in the injector; the column should give adequate separation of components; losses of components during chromatography should be minimal and constant; accurate response factors should be used to allow for losses of components and the differential sensitivity of the detector; and optimal integration parameters should be used. Quantitative analysis by GC is discussed in detail by Novak (1988).

REFERENCES

Adlard, E. R. (1975). A review of detectors for GC, *CRC Crit. Rev. Anal. Chem.*, **5**, (1), 1–36

Anderson, R. J. (1979). Nitrogen selective detection in GC. *Tracor chromatography*, Application 79–3, Tracor Instruments, Austin, TX.

Anderson, R. J. and Hall, R. C. (1980). Hall bipolar pulse, differential electrolytic conductivity detector for GC: design and applications. *Am. Lab.* **12**, 108–124.

Beroza, M. and Sarmiento, R. (1963). Determination of the carbon skeleton and other structural features of organic compounds by gas chromatography. *Anal. Chem.*, **35**, 1353.

Blomberg, L. and Wannman, T. (1979). *In situ* synthesis of highly thermostable non-extractable methylsilicone gum phases for glass capillary gas chromatography. *J. Chromatogr.*, **168**, 81–88.

Bourne, S., Reedy, G., Coffey, P., Mattson, D. (1984). Matrix isolation GC/FTIR. *Am. Lab.* **16** (6), 90–101.

Brody, S. S. and Chaney, J. E. (1966). Flame photometric detector. Application of a specific detector for phosphorus and for sulphur compounds sensitive to sub-nanogram quantities. *J. Gas Chromatogr.*, **4** (2), 42–46.

Chrompack (1988). Polyimide HT: a new high temperature column coating. *Chrompack News*, **15** (2), 8.

Coulson, D. M. (1965). Electrolytic conductivity detector for GC. *J. Gas Chromatogr.*, **3**, 134–137.

Dandeneau, R., Bente, P., Rooney, T. and Hiskes, R. (1979). Flexible fused silica columns: an advance in high resolution GC. *Am. Lab.* **11** (9), 61–69.

Deans, D. R. (1968). A new technique in heart cutting in gas chromatography. *Chromatographia*, **1**, 18–22.

Dressler, M. (1986). *Selective gas chromatographic detectors*, Elsevier, Amsterdam.

Driscoll, J. N. (1976). The photoionization detector. *Am. Lab.* **9**, 71–79.

Driscoll, J. N. (1977). Evaluation of a new photoionisation detector for organic compounds. *J. Chromatogr.* **134**, 49–55.

Fehsenfeld, F. C., Goldan, P. D., Phillips, M. P. and Sievers, R. E. (1981). In Zlatkis, A. and Poole C. F. (eds), *Electron capture—theory and practice in chromatography*, Chapter 4, Elsevier, Amsterdam, pp. 69.

Fenimore, D. C., Loy, P. R. and Zlatkis, A. (1971). High temperature tritium source for electron capture detectors. *Anal. Chem.*, **43** (14), 1972–1975.

Freeman, R. R. and Jennings, W. (1987). Optimising gas chromatographic separations. *J. High Resolut. Chromatogr. Chromatogr. Commun.*, **10**, 231–234.

Golay, M. J. E. (1958). In *Gas Chromatography* Desty, D. H. (ed.), Butterworths, London, pp. 36.

Gough, T. A., Webb, K. S. and Eaton, R. F. (1977). Simple chemiluminescent detector for the screening of foodstuffs for the presence of volatile nitrosamines. *J. Chromatogr.*, **137**, 293–303.

Griffiths, P. R., de Haseth, J. A., and Azzarraga, L. V. (1983). Capillary GC/FTIR. *Anal. Chem.* **55**, 1361A–1387A.

Grob, K. Jr. (1979). Evaluation of injection techniques for triglycerides in capillary gas chromatography. *J. Chromatogr.*, **178**, 387–392.

Grob, K. Jr. (1982). Band broadening in space and the retention gap in capillary gas chromatography. *J. Chromatogr.*, **237**, 15–23.

Grob, K. and Grob, K. Jr. (1978). Splitless injection and the solvent effect. *J. High Resolut. Chromatogr. Chromatogr. Commun.*, **1** (1), 57–64.

Grob, K. Jr. and Grob, K. (1981). Evaluation of capillary columns by separation number or plate number. *J Chromatogr.*, **207**, 291–297.

Grob, K. Jr. and Muller, R. (1982). Some technical aspects of the preparation of a retention gap in capillary gas chromatography. *J. Chromatogr.*, **244**, 185–196.

Grob, K. Jr. and Neukom, H. P. (1980). Should the septum part of vaporizing injectors be kept at lower temperatures? *J Chromatogr.* **198** (1), 64–69.

Hall, R. C. (1974). Highly sensitive and selective microelectrolytic conductivity detector for gas chromatography. *J. Chromatogr. Sci.*, **12** (3), 152–160.

Hangac, G., Hohne, B. A. and Isenhour, T. L. (1983). Accurate assignment of Kovats retention indexes for GC/FTIR data. *J Chromatogr. Sci.*, **21** (6), 241–245.

van Ingen, R. H. M., Nijssen, L., van den Berg, F. and Maarse, H. (1987). Ethyl carbamate in alcoholic beverages by two dimensional chromatography. *J. High Resolut. Chromatogr. Chromatogr. Commun.* **10**, 151–152.

James, A. T. and Martin, A. J. P. (1952). Gas–liquid partition chromatography: the separation and microestimation of volatile fatty acids from formic acid to dodecanoic acid. *Biochem. J.* **50**, 679–690.

Jennings, W. and Yabumoto, K. (1980). Effect of test temperature on separation number. *J. High Resolut. Chromatogr. Chromatogr. Commun.*, **3**, 177–179.

Kaiser, R. (1962). New developments in gas chromatography from the 1961 literature. *Z. Anal. Chemie*, **189**, 1–14.

Kaiser, R. E. and Reider, R. I. (1979). Polarity change in capillary GC by serial column temperature optimisation. *J. High Resolut. Chromatogr. Chromatogr. Commun.*, **2**, 416–422.

Kalo, P. (1981). Glass capillary reaction loops for detection of alcohols, aldehydes and ketones by subtraction. *J. Chromatogr.*, **205**, 39–47.

Karasek, F. W. (1971). Plasma chromatography of the polychlorinated biphenyls. *Anal. Chem.* **43**, 1982–1986.

Karasek, F. W. and Denney, D. W. (1974a) Detection of aliphatic *N*-nitrosamine compounds by plasma chromatography. *Anal. Chem.* **46**, 1312–1314.

Karasek, F. W. and Denney, D. W. (1974b) Detection of 2,4,6-trinitrotoluene vapours in air by plasma chromatography. *J. Chromatogr.* **93**, 141–147.

Knapp, D. (1979). *Handbook of analytical derivatisation reactions*. John Wiley & Sons, New York.

Kolb, B. and Bischoff, J. (1974). New design of a thermionic nitrogen and phosphorus detector for GC. *J. Chromatogr. Sci.*, **12** (11), 625–629.

Kovats, E. (1958). Gas Chromatographische Charakterisierung organischer Verbindungen. *Helv. Chim. Acta*, **41**, 1915–1932.

Lamb, R. J. and Purnell, J. H. (1975). Criteria for the use of mixed solvents for gas liquid chromatography. *J. Chromatogr.*, **112**, 71–79.

Langhorst, M. L. (1981). Photoionization detector sensitivity of organic compounds. *J. Chromatogr. Sci.*, **19**, 98–103.

Leibrand, R. J. and Dunham, L. L. (1973). Preparing high efficiency packed GC columns. *Res. develop.*, **24** (9), 32–38.

Macrae, R. (1988). *HPLC in Food Analysis*, 2nd edition, Academic Press, London.

McReynolds, W. O. (1970) Characterization of some liquid phases. *J. Chromatogr. Sci.*, **8** (12), 685–691.

Mayzaud, P. and Ackman, R. G. (1976). Some empirical observations on the choice of carrier gas in the GC analysis of fatty acid methyl esters using wall-coated open tubular columns. *Chromatographia*, **9** (7), 321–324.

Novak, J. (1988). *Quantitative analysis by gas chromatography*, 2nd edition, Marcel Dekker, New York.

O'Brien, M. J. (1985). Detectors, Chapter 6 in *Modern Practice of Gas Chromatography*, 2nd edition, Grob, R. L. (ed.) John Wiley & Sons, New York, pp. 211–291.

Patterson, P. L., Howe, R. L. and Abu-Shurnays, A. (1978). Dual-flame photometric detector for sulfur and phosphorus compounds in gas chromatograph effluents. *Anal. Chem.* **50**, 339–344.

Rooney, T. A., Aetmayer, L. H., Freeman, R. R. and Zerenner, E. H. (1979). Van Deemter curves for an open tubular column. *Am. Lab.* **11** (2), 81–89.

Schomberg, G., Dielmann, R., Borwitzky, H. and Husmann, H. (1978). Capillary gas chromatography on compounds of low volatility. Temperature stability of stationary liquids on various glass surfaces. *J. Chromatogr.*, **167**, 337–354.

Sevcik, J. and Krysl, S. (1973). Photoionization detector. *Chromatographia*, **6**, 375–380.

Sugiyama, T., Suzuki, Y. and Takeuchi, T. (1973). Intensity characteristics of molecular sulfur emission for sulfur compounds with flame photometric detector. *J. Chromatogr. Sci.*, **11**, 639–641.

Sullivan, J. J. and Burgett, C. A. (1975). Non-linearity in constant current electron capture detection. *Chromatographia*, **8** (4), 176–179.

Thompson, C. J., Coleman, H. J., Ward, C. C., Hopkins, R. L. and Rall, H. T. (1960a). Identification of oxygen compounds in GLC fractions by catalytic deoxygenation. *Anal. Chem.* **32**, 1762–1770.

Thompson, C. J., Coleman, H. J., Ward, C. C. and Rall, H. T. (1960b). Desulfurization as a method of identifying sulfur compounds. *Anal. Chem.*, **32**, 424–430.

Watanabe, C., Tomita, H., Sato, K., Massada, Y. and Hashimoto, K. (1982). Accuracy and reproducibility in splitless, packed and open tubular cool on-column injections. *J. High Resolut. Chromatogr. Chromatogr. Commun.* **5** (10), 630–632.

de Zeeuw, J., de Nijs, R. C. M., Buitjen, J. C., Peene, J. A. and Mohnke, M. (1987). Fused silica PLOT columns with porous polymer coatings. *Int. Lab.*, December, 52–57.

Zlatkis, A. and Poole, C. F. (eds) (1981). *Electron capture—theory and practice in chromatography*, Elsevier, Amsterdam.

2

Gas chromatography–mass spectrometry

Klaus O. Gerhardt

2.1 INTRODUCTION

Combined gas chromatography–mass spectrometry (GC–MS) is probably the most comprehensive instrumental analytical technique available to the scientist in food analysis at present. The technique is well established in food science, and a predominant area of application is in food safety, where reliable information on food contaminants, e.g. pesticides, mycotoxins and veterinary drug residues, is of vital consequence. Information to be obtained can be both the unequivocal identification or confirmation of the contaminant and the quantification.

A different major area where GC–MS has been a powerful technique for the food chemist is in the separation and identification of volatile flavour components of foodstuff (food aroma).

Flavour plays a very important role in food acceptability and consequently in food marketing. It is essential to gain knowledge of what flavour is and how it is affected by food harvesting, processing and storage. Modern sophisticated analytical methods, including GC–MS, have been essential in our understanding of this difficult and complex field. In addition, knowing the pattern of flavour components may reveal, for example, falsification of the characteristics of a commercial product.

Gas chromatography (GC) provides high resolving power for the separation of complex mixtures of compounds that can be volatilized either directly or after derivatization. However, a shortcoming of GC is that identification of the constituents is based on the correlation of retention time to reference compounds; thus, without prior information, chromatographic data cannot provide reliable structural information.

In mass spectrometry (MS) a wealth of structural information can be obtained, generally including the molecular ion indicative of the molecular weight. Usually, MS of a mixture of components provides little or no useful information. However, on-line coupling of both GC and MS yields a most powerful and versatile analytical technique for the separation and identification of the constituents of complex mixtures of volatile compounds. Combined GC–MS offers a hybrid instrument of high specificity, high sensitivity, rapidity of analysis and a wide range of applicability. An advantage is the small sample requirement: unknowns may be identified from approximately 0.5 μg of material, or from 100 pg (full scan) to 25 ng of sample component present; known constituents in mixtures may be measured at lower than 50 pg levels.

The introduction of high-performance fused silica capillary columns has enormously expanded and improved the applicability of GC–MS to a wide array of analyses. In addition, sophisticated data systems interfaced with modern GC–MS instruments facilitate significantly instrument operation, as well as data acquisition, manipulation and evaluation. Exensive spectral data bases containing typically 44 200 spectra (NBS/NIH/EPA) are part of the state-of-the-art instrumentation simplifying the confirmation or identification of unknown constituents.

2.2 GC–MS INTERFACES

The coupling of a gas chromatograph with a mass spectrometer is associated with an obvious problem because the carrier gas (He) containing the analytes exits the gas chromatographic column at about atmospheric pressure while the mass spectrometer must be operated at or below 10^{-5} torr (1.3×10^{-3} Pa). The introduction of an interface was essential in removing the carrier gas (He) and reducing the pressure to approximately 10^{-5} to 10^{-6} torr (1.3×10^{-3} to 1.3×10^{-4} Pa) in the ion source housing.

A number of enrichment devices, also called molecular separators, have been developed for the selective removal of the carrier gas from the sample. The separation depends on the difference in the physico-chemical properties between the carrier gas and the sample molecules. The separation yield (efficiency) depends on operating parameters (carrier-gas flow rate, temperature, molecular weight range, etc.). However, some loss of sample is associated with all separator designs. Some of the more common types of interfaces used are the effusion (Biemann–Watson), semipermeable membrane (Llewellyn) and jet (Ryhage) separator. Fig. 2.1 illustrates the three more general types of interfaces. The direct coupling technique will be discussed in more detail later. A main application of the various separators was with packed columns. Of the separators, the Ryhage or jet (single-stage) separator is the more common interface used at present to accommodate GC–MS work with packed columns. A very detailed description of GC–MS interfaces and the associated vacuum technology has been given by McFadden (1973, 1979) and Gudzinowicz *et al.* (1977).

2.2.1 Direct coupling

With the introduction of *flexible* fused silica capillary columns the direct coupling of the GC column with the ion source of the mass spectrometer has become a much simpler construction. Importantly, the comparatively low carrier-gas flows of capillary columns (1 to 5 ml min^{-1}) can be readily handled by high-capacity diffusion pumps of modern mass spectrometers with pumping speeds of 500 l s^{-1} and greater. The capillary column can be coupled to the ion source basically in two ways: as direct or open-split connection. The direct connection routes the analytical column from

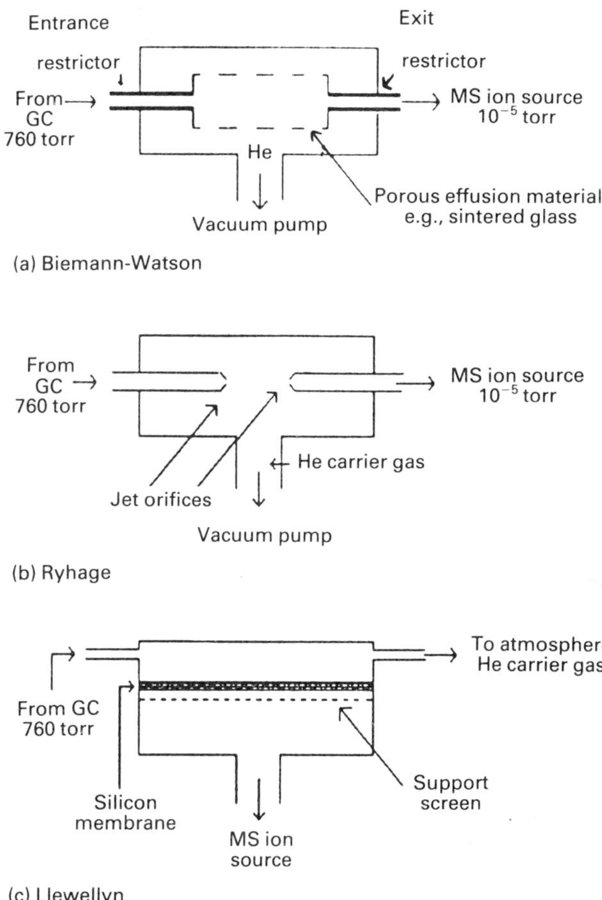

Fig. 2.1 — Schematics of the three most common GC–MS interfaces: (a) effusion separator (Biemann–Watson), (b) single-stage jet-orifice separator (Ryhage), (c) silicon-membrane separator (Llewellyn).

the GC oven through the heated 'interface region' into the ion source chamber of the MS. This way the full sample is transferred and optimal sensitivity can be achieved by eliminating interface-related active sites that could cause adsorptive losses and degradation of sensitive compounds. Alternatively, the analytical column (fused silica or glass) can be terminated inside the GC oven and the effluent can then be passed into the ion source via a deactivated uncoated capillary column, which is coupled through a very low dead volume union (located in the GC oven) to the analytical column (Friedli, 1981; Peltonen *et al.*, 1988). The latter approach allows an easier and quicker change of analytical columns because there is no need to completely vent the mass spectrometer (usually, modern mass spectrometers have source isolation valves) and for aligning of the column exit in the ion source chamber (Giang, 1984). By direct connection of the capillary GC column to the ion source, the column exit is at vacuum outlet pressure. It is often claimed that the decreased column outlet pressure has a deleterious effect on optimum column efficiency (Hatch & Parrish, 1978; Vangaever *et al.*, 1979) resulting in separation efficiency losses of up to 30%. However, a study by Cramers *et al.* (1981), calculated the column efficiency to decrease by 12.5% at most, which was in agreement with experimental observations. Moreover, the optimum carrier-gas velocity was found to be shifted to higher values. The gain in the speed of analysis is dependent on the nature of the carrier gas and increases strongly with lower (sub-atmospheric) optimum inlet pressures (Cronin & Caplan, 1987). The resulting narrower and thus higher chromatographic peaks improve the detection limits. While the direct connection is very convenient it has the disadvantage that all of the eluate enters the ion source, which in the case of on-column injections (trace analysis), can be amounts of up to 5 μl injected onto the column. In addition, bulk impurities, including derivatizing reagents and sample matrix, can contaminate the ion source and reduce the sensitivity.

A variation of the direct coupling is the so-called open-split device introduced by Henneberg *et al.* (1975). In this arrangement a narrow gap between the outlet of the GC column and the inlet of the transfer line to the ion source of the mass spectrometer is maintained at about atmospheric pressure by an adjustable flow of helium make-up gas shown schematically in Fig. 2.2. Different materials have been

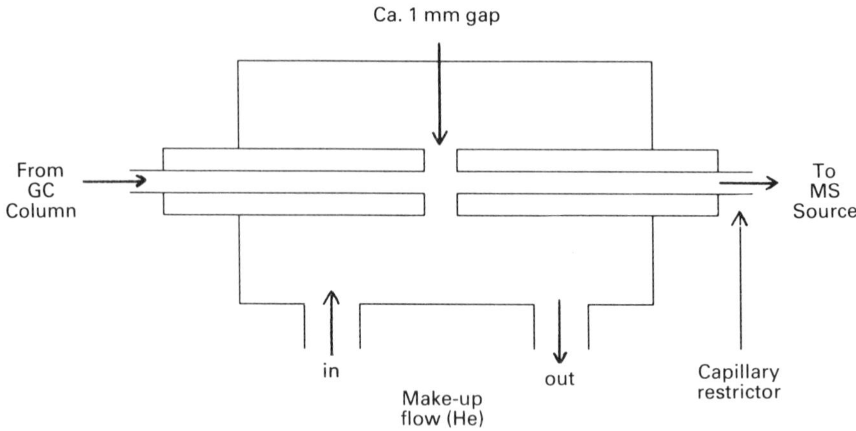

Fig. 2.2 — Open-split interface.

used for the transfer line (Stan & Abraham, 1978; Wetzel *et al.*, 1982) to accommodate the analysis of some classes of critical compounds. However, the introduction of flexible fused-silica capillaries presented very attractive features: flexibility, good mechanical properties, inactivity and cheapness (Koller & Tressl, 1980). Significant advantages with the open-split interface are that the chromatography is not affected by the mass spectrometer vacuum, thus, retention times from GC and GC–MS runs can be readily compared. Columns can be changed without venting or the use of an isolation valve system. The bulk of large solvent peaks can be vented to waste by increasing the flow of make-up gas.

2.3 MASS SPECTROMETER DESIGNS

Mass spectrometers are ion-optical instruments that function as a group of subsystems, each of which performs an operation on the sample in sequential order: vaporizing the sample (inlet system), producing a beam of gaseous ions from the sample vapors (ion source), separating the resulting mixture of ions according to their mass-to-charge ratios, m/z, (mass analyser) and providing output signals to the detector, from which the nominal or exact mass and abundance of each ion species detected may be determined (Roboz, 1986). In support of the overall operation of the mass spectrometer are two additional subsystems; the essential vacuum system and the data system (Grayson, 1986). Of the many different designs of mass spectrometers in existence there are two principally different types of instruments that are most commonly used in GC–MS work: magnetic sector field mass spectrometers and quadrupole mass filters. The distinction is based on their modes of separation of the ions. A recent introduction as a gas chromatographic detection device is the ion-trap detector (ITD). The ion trap can be conceived as a three-dimensional quadrupole device (Brodbelt & Cooks, 1988; Hübschmann & Schubert, 1986).

2.3.1 Sample inlet system

In the majority of cases analytical mass spectrometers incorporate two types of inlet systems: a batch inlet system for gases, liquids and solids of moderately high vapour pressure, and a direct inlet (solids probe) system for higher molecular weight, less-volatile solids. The solids introduction probe consists of a rod with a heated tip which is introduced through a vacuum lock system into the mass spectrometer. A detailed description of inlet systems is presented by McFadden (1973).

In the early 1980s gas chromatographs with mass-selective detection (GC–MSD) were introduced. These relatively low-priced bench-top models have become very popular. However, the GC–MSD instruments are usually not equipped with a batch or direct-probe inlet system. The gas chromatograph serves with both the GC–MSD and any other GC–MS combination as the sample introduction system. However, the primary purpose of the gas chromatography is to present purified components separated from all other components in a mixture to the ion source. In addition, the sample molecules are vaporized. Care has to be taken that the interfaces remain evenly heated to prevent condensation of the sample constituents.

2.3.2 The ion source

2.3.2.1 Electron impact (EI)

The oldest and most widely used ionization method is by electron impact (EI) (Ligon, 1979). An electron beam emitted from a conductively heated rhenium or tungsten filament (cathode) *in vacuo* interacts with the vaporized sample in the ion source chamber by dislodging electrons and forming a positive ion beam representative of the sample.

$$M+e^- \rightarrow M^{+\bullet}+2\,e^-$$

The majority of ions formed has a single positive charge. However, the rather non-specific electron beam-molecule interaction is accompanied to a lower extent by other reactions (electron capture, multiply charged ions, etc.). By adjusting source parameters (e.g. repeller), only a beam of positively charged ions will be monitored. Negatively charged ions will be discharged, unionized sample molecules (more than 99.9%) and neutral fragments will be pumped away. A schematic of a typical source is shown in Fig. 2.3. The vaporized sample molecules are introduced into a small

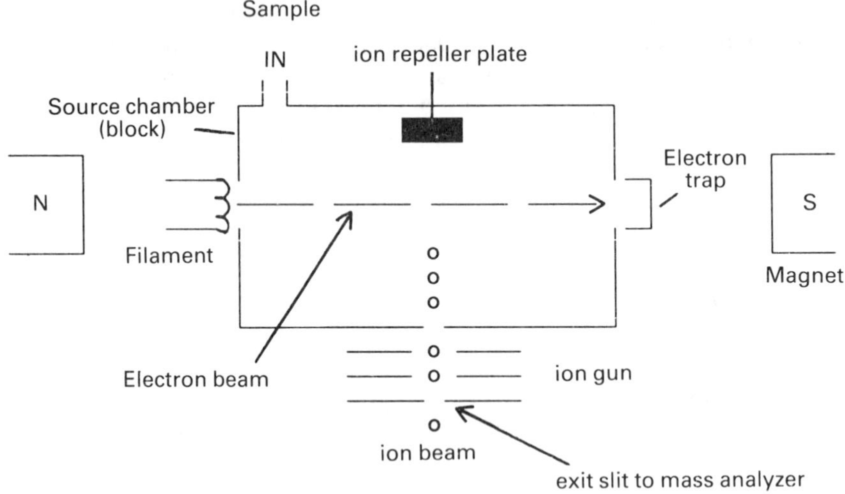

Fig. 2.3 — Schematic of a typical electron-impact (EI) source.

metal box of several millilitres' volume. The source chamber has several holes; through one of them the electron beam enters the box. By applying a potential difference variable from 5 to 100 V between the cathode and the source chamber the electrons are accelerated across the chamber. An electrode, called electron trap, located opposite the filament, registers the flow of electrons. Incorporating the filament and electron trap into an appropriate electronic circuit provides a means of controlling the emission from the filament. A constant flow of electrons is essential for a consistent and reproducible formation of ions leading to a reproducible fragmentation pattern. In addition, a set of small permanent magnets collimates the electrons into a tight spiralling 'beam' on their path to the trap. Thus, the longer path increases the probability of molecule–electron beam interactions, thus yielding a narrower, well-defined ion beam. The positive ions formed are 'pushed' (extracted) from the ion source chamber by a repeller of a slightly higher positive voltage than the chamber toward the exit hole of the box. Outside the ion source chamber the ions are accelerated by a strong electrical field (ion gun) passing through an exit slit into the mass analyser for separation according to their mass-to-charge ratio. Accelerating voltages vary and can range up to 10 keV, depending on the mass spectrometer design. However, the accelerating voltage for quadrupole instruments is low, about 10 eV.

During their travel through the ion source the electrons do not only interact with the sample but also with molecules that could originate from other sources, e.g. background gases, carrier gas, and previous samples. The bulk of the sample molecules is continuously removed by the vacuum pump of the source housing, typically maintaining a pressure of 10^{-5} torr (1.3×10^{-3} Pa). Heating of the source chamber is not only essential for keeping molecules of interest in the vapour phase but it also counteracts the contamination of the source chamber.

The electron energy used for EI-ionization of organic compounds is by convention 70 eV, which is more than sufficient energy to eject an electron(s) from higher-energy molecular orbitals. Most organic compounds have ionization potentials of 8–16 eV. The amount of ionization or ionization efficiency increases sharply with the increase in electron energy of the electron beam, reaching a plateau around 70 eV. However, mass spectra may be obtained at lower electron energy, e.g. 15 eV, reducing the extent of fragmentation but also decreasing the ion yield.

Electron impact ionization of the sample molecules in high vacuum (sample pressures of about 10^{-5} torr (1.3×10^{-3} Pa)) produces in a primary process positive molecular ions (odd electron ions) having an array of internal energies, ranging from 0 to 10 eV, relative to the ground state of the molecular ion (Harrison, 1983). If the critical internal energy required for the fragmentation of the molecular ion is extremely low (or zero) the decomposition will proceed very readily. In this event the initially formed molecular ions may not survive the short time span of about 10^{-5} s needed for the ion to reach the detector. Typically, these primary odd electron ions undergo subsequent fragmentation to an assembly of lower mass ions. Under EI conditions negative ions are also formed predominantly by the ion-pair mechanism and at much lower abundance. The majority of negative EI spectra for organic compounds are devoid of molecular ions (Hunt *et al.*, 1976) and are commonly less useful for molecular weight and structure determination due to lack of specificity and poor sensitivity.

2.3.2.2 *Chemical ionization (CI)*

Some classes of compounds do not produce a stable molecular ion under EI ionization. An alternative much gentler ionization technique, introduced by Munson and Field (1966), is chemical ionization (CI), usually minimizing fragmentation and enhancing the probability of forming ion species from which the molecular weight can be derived. Thus, GC–MS using chemical ionization is an important complementary technique to EI ionization. This approach can greatly facilitate the identification of compounds of interest by providing clear evidence of the compounds' molecular weight. In addition, CI has generally the advantage of being a more selective and sensitive technique because it produces a few but more intense ions in the high mass region of the mass spectrum, which is useful in the detection (identification) and quantitation of compounds present at low concentrations. The rapid development in these techniques has helped in the food chemistry, especially of food contamination, e.g. by pesticides, veterinary drugs, mycotoxins, etc. The chemical ionization process can be employed in the positive (PCI) or negative (NCI) ion mode, depending on the nature of the analyte, reagent gas and reaction conditions.

In CI, a reagent gas, e.g. methane, isobutane, ammonia or helium (electron capture), is introduced into a so-called chemical ionization chamber. Since this chamber is a nearly gas-tight modification of an EI source it permits the operation at a relatively high gas pressure of about 1 torr (133 Pa). Under these CI conditions gas phase ion–molecule reactions take place in contrast to EI ionization where collisions are avoided by maintaining low source pressure around 10^{-5} torr (1.33×10^{-3} Pa). Commonly, in CI, a high-energy electron beam (>200 eV) (Milberg & Cook, 1979) is used to ionize the reagent gas, e.g. methane, forming an array of primary ion species and low-energy secondary electrons as expressed in reaction 2.1:

$$CH_4+e^-(\text{high energy})\rightarrow CH_4^{+\bullet}, CH_3^+, CH_2^{+\bullet}, \ldots +2e^-(\text{low energy}) \qquad (2.1)$$

These ions yield in subsequent reactions with neutral methane molecules (reactions 2.2 to 2.6) secondary ions, CH_5^+ (48%), $C_2H_5^+$ (41%), $C_3H_5^+$ (6%) as major reactant species (Arsenault, 1972).

$$CH_4^{+\bullet}+CH_4\rightarrow CH_5^++CH_3^\bullet \qquad (2.2)$$
$$CH_3^++CH_4\rightarrow C_2H_5^++H_2 \qquad (2.3)$$
$$CH_2^{+\bullet}+CH_4\rightarrow C_2H_4^{+\bullet}+H_2 \qquad (2.4)$$
$$CH_2^{+\bullet}+CH_4\rightarrow C_2H_3^++H_2+H^\bullet \qquad (2.5)$$
$$C_2H_3^++CH_4\rightarrow C_3H_5^++H_2 \qquad (2.6)$$

The secondary ions can react then with the sample molecules (M) which are present at low concentrations (about 0.1%).

Among the variety of ionization reactions that are possible (proton transfer, hydride transfer, electron transfer, and ion association), a relatively common one is the protonation of sample molecules that have a higher proton affinity than the reagent ion (CH_5^+, etc.) which reacts as a Bronsted acid (in the gas phase):

$$CH_5^++M\rightarrow MH^++CH_4 \qquad (2.7)$$
$$C_2H_5^++M\rightarrow MH^++C_2H_4 \qquad (2.8)$$

acid base acid base

The reactant species (Bronsted acids) of the most commonly employed reagent gases (listed earlier) are CH_5^+, $C_2H_5^+$, t-$C_4H_9^+$, and NH_4^+ which decrease in their protonating ability from CH_5^+ to NH_4^+. The strongest acid is CH_5^+, which can protonate any organic compound containing a functional group, while NH_4^+,a weak acid, can react with a strong base but not a weak one. Of intermediate protonation ability is t-$C_4H_9^+$. The difference in proton affinities between the reactant gas (e.g. CH_4) and sample M correlates to the amount of excess energy transferred to the protonation product MH^+. Thus, a protonation by the strong reactant species CH_5^+ will subsequently favour fragmentation reactions while using t-$C_4H_9^+$, a much weaker protonating agent, leads mainly or exclusively to the formation of the MH^+ ion. It is apparent, significant diagnostic information can be obtained by the appropriate selection of the CI reagent gases and conditions. An example is the identification of food aroma components from fresh products such as fruits where many aroma components (alcohols, aldehydes, acetates, etc.) lack molecular ions. As discussed in detail by Cronin & Caplan (1987), in the identification of flavour constituents, chemical ionization is an important complementary technique to EI in providing the diagnostic information (e.g. $M^{+\bullet}$) needed. Depending on the nature of the sample molecules and the selected CI reagent gas, other reactions such as hydride abstraction and charge exchange can take place (Harrison, 1983).

Chemical ionization was initially developed as a technique with the main interest in the formation of positive ions. More recently (early 1970s) negative CI mass spectrometry evolved as a special recognizable discipline. In analytical chemistry, NCI has shown a very high selectivity and sensitivity for particular classes of molecules (Dougherty, 1981).

The formation of negative ions can be initiated by three different mechanisms:

$$AB + e^- \rightarrow AB^{-\bullet} \quad \text{resonance capture} \quad (2.9)$$
$$AB + e^- \rightarrow A + B^- \quad \text{dissociative resonance capture} \quad (2.10)$$
$$AB + e^- \rightarrow A^+ + B^- + e^- \quad \text{ion-pair production} \quad (2.11)$$

Generally, the ionization of the reagent gas molecules by the primary electrons is accompanied by the formation of low energy secondary electrons (2.1). In subsequent ionizing and non-ionizing collisions between electrons and neutral molecules (acting as a moderating gas), the energy level of the electrons decreases to near-thermal energy (0 to 0.1 eV)., In the negative-ion mode, the major reaction is the resonance capture (2.9) of these low-energy electrons by sample molecules of significant electron affinity forming an odd electron, negative molecular ion as primary product. The dissociative resonance capture (2.10), occurring at electron energies in the range of 0 to 15 eV, and ion-pair production (2.11), at energies above 10 eV, are two potential loss mechanisms for molecular weight information. A significant feature in electron-capture chemical ionization (ECCI) is that sample molecules having a strong electron affinity can be detected up to three orders of magnitude more sensitively than in PCI.

Brumley and Sphon (1987) discuss in a special treatise the principles and many applications of NCI in the analysis of food samples for contaminants (mycotoxins, pesticides, veterinary drugs, food additives, etc.) and food constituents (fatty acids/lipids, carbohydrates, steroids, etc.). The reader is referred to this excellent source of information on NCI mass spectrometry.

A comparison of the three different ionization modes, EI, PCI and NCI, is presented in Fig. 2.4 using the pentafluoropropionyl (PFP) derivatives of dopamine (DA) and 3,4-dihydroxyphenylacetic acid (DOPAC) as examples (Roboz, 1986). In the EI mode (top trace), a molecular ion ($M^{+\bullet}$) is not detectable for DA (*m*/*z* 591) but it is present for DOPAC (*m*/*z* 610). However, PCI (middle trace) produced the $(M+H)^+$ as the most intense ion (base peak) for both DA (*m*/*z* 592) and DOPAC (*m*/*z* 611). The high abundance of the $(M+H)^+$ ion, correlating to the molecular weight, makes it very suitable for selected-ion monitoring. In the NCI spectra (bottom trace), the predominant ions (base peaks), *m*/*z* 571 and *m*/*z* 463, resulted from the loss of HF (DA) and PFP (DOPAC); however, molecular ions did not appear.

DA and DOPAC have been involved in identification and quantification

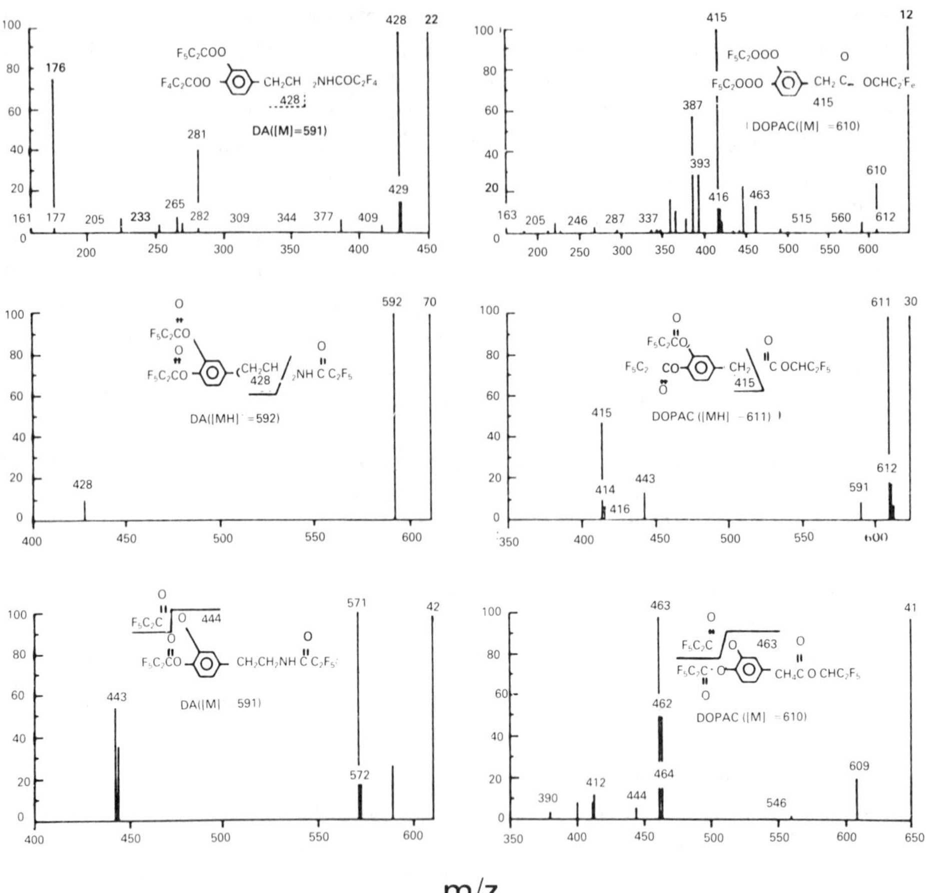

Fig. 2.4 — Electron ionization (EI) positive chemical ionization (PCI), and negative chemical ionization (NCI) mass spectra of the PFP derivaties of DA and DOPAC. Reprinted from Roboz (1986). Copyright 1986, Ellis Horwood Limited.

problems associated with the commonly used HPLC–ED (high-performance liquid chromatography–electrochemical detection) assays for catecholamines, their metabolites and salsolinol (1-methyl-6,7-dihydroxy-1,2,3,4-tetrahydroisoquinoline) in food and beverage samples (Duncan *et al.*, 1984). The identification and precise quantification of the above-listed compounds was accomplished by Duncan *et al.* (1984) using GC–MS and selected-ion monitoring (SIM).

2.3.3 Mass analyser

The ion beam accelerated out of the source passing through an exit slit is a mixture of ions differing in the mass-to-charge ratios (m/z). The purpose of the next stage of the mass spectrometer is to resolve the different masses from each other, which can be performed by a variety of techniques (magnetic deflection, quadrupole mass filter, time-of-flight, radio frequency, etc.). Among the commercial mass spectrometers, the two most common but different principles of mass analysis employed are magnetic deflection (magnetic sector instruments) and quadrupole mass filtration (quadrupole instruments). Recently, the ion-trap detector (ITD), a variation of the quadrupole geometry, has gained recognition as an inexpensive, highly sensitive gas chromatography detector (Brodbelt & Cooks, 1988). The ITD is derived in principle from the classical quadrupole mass filter.

2.3.3.1 Magnetic analyser

In the magnetic field, ions of the ion beam are deflected according to their mass-to-charge ratios. The motion of an ion of mass m and charge z extracted from the source with an accelerating voltage V in a magnetic field of strength B is in a circular path with radius r_m, where:

$$m/z = r_m^2 B/2V \tag{2.12}$$

Only those ions exiting the magnetic field with the correct radius of curvature will be detected by an ion collector (detector) placed behind a resolving slit. To collect ions over a range of masses (entire mass spectrum), one has to vary either the accelerating voltage or the magnetic field strength. Commonly, the accelerating voltage is kept constant and the magnetic field is continuously changed (scanned) to collect or focus sequentially ions of different m/z onto the detector.

A measure of the separating ability of the mass analyser, e.g. magnetic sector, is the resolution (R) or resolving power which is expressed as $R = m/\Delta m$, where m is the nominal mass of one of two ions (peaks) and Δm the actual mass difference between the two resolved peaks. The ions are considered resolved or separated when the overlap between the two peaks is 10% or less of their height (10% valley definition). Usually, such a single-focusing instrument provides only nominal mass determination (accuracy of 0.5 to 1.0 a.m.u.). Not all ions of the same m/z value entering the magnetic sector have exactly the same velocity or kinetic energy. This insufficient homogeneity of the ions causes some dispersion of the deflected ion beam and is the reason for the reduced resolution. By introducing an electrostatic sector (a pair of curved plates) for velocity or energy focusing in combination with the magnetic analyser this energy spread can be limited. The motion of an ion of mass m and charge z traversing the electric field E of the electrostatic analyser (ESA) is governed by the ion's kinetic energy, $\frac{1}{2}mv^2 = zV$, and the counterbalancing centrifugal force, $zE = mv^2/r_s$, where v is the velocity of the ion, r_s the radius of the curved ESA plates, and V the initial accelerating potential. The resulting equation is $r_s = 2V/E$, and if r_s and E are held constant, the ESA focuses the ions according to their translational energy. Therefore, the ESA functions as an energy-focusing device limiting the energy spread of the ion beam. With the addition of an electrostatic field or ESA to the magnetic field the resolution can be increased significantly. Such double-focusing instruments can achieve resolution in excess of 100 000, allowing exact mass measurements with an accuracy of 0.001 a.m.u. or better. Fig. 2.5 shows two double-focusing mass spectrometers. With the Mattauch–Herzog geometry, the focal points of all masses (m/z) are in a plane and a photographic plate can be used for simultaneous recording of every ion. The mass spectrum can also be recorded electrically by varying the magnetic field. Another common double-focusing

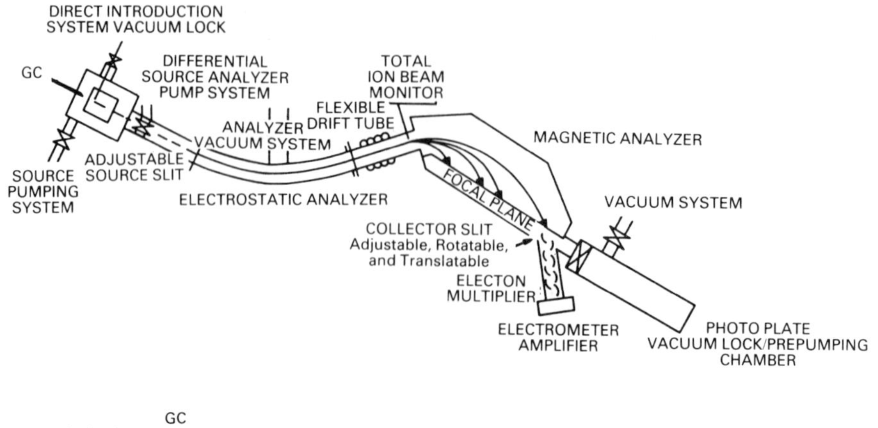

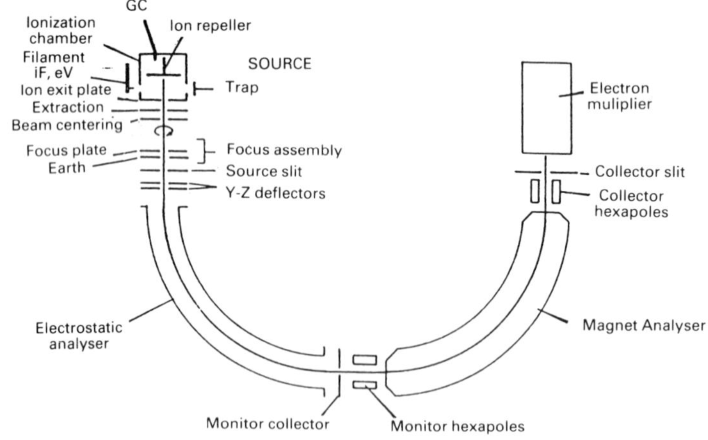

Fig. 2.5 — Schematic diagrams of double-focusing mass spectrometers. Mattauch–Herzog geometry (top) of a CEC.21-110B high-resolution instrument. Nier–Johnson geometry (bottom) of a Kratos MS25, 7500 resolution. Courtesy of Kratos Analytical.

arrangement is the Nier–Johnson geometry where electrostatic and magnetic deflections are in the same direction. By changing the magnetic field, ion masses are brought to a single focal point and recorded with an electric detector. However, the sequence of ESA and magnetic sectors is in some instrument designs reversed (reverse-geometry instruments).

Based on the exact mass measurements, the elemental composition of ions can be determined, thus obtaining information that aids significantly in the identification or confirmation of the sample molecules. However, increased resolving power is consequently accompanied by some loss of sensitivity. In addition, more expensive double-focusing instruments are needed for this technique.

The majority of GC–MS analyses are performed at low resolution (around 1000), yielding the nominal masses for scans of a chromatographic run. Most commonly, quadrupole mass filter instruments are used for low-resolution work. An interesting comparison of the quadrupole mass filter and the double-focusing MS is presented by Rosen *et al. (1988)*.

2.3.3.2 Quadrupole mass filter

The principle of the quadrupole mass filter was introduced by Paul and Steinwedel (1953). The advantage of this mass analyser is the small size and the simplicity of construction as compared with magnetic sector instruments. However, a disadvantage is the decrease in transmission efficiency for ions above mass 500, resulting in reduced sensitivity (Beynon & Brenton, 1982). For some time the fast repetitive scanning ability (cycle time) has made the quadrupole analyser a popular detector for capillary CG work. However, today's new magnet designs provide double-focusing mass spectrometers also with the capability of quick repetitive scanning, thus making these instruments quite competitive.

The quadrupole mass analyser (Fig. 2.6) consists of four precisely parallel aligned

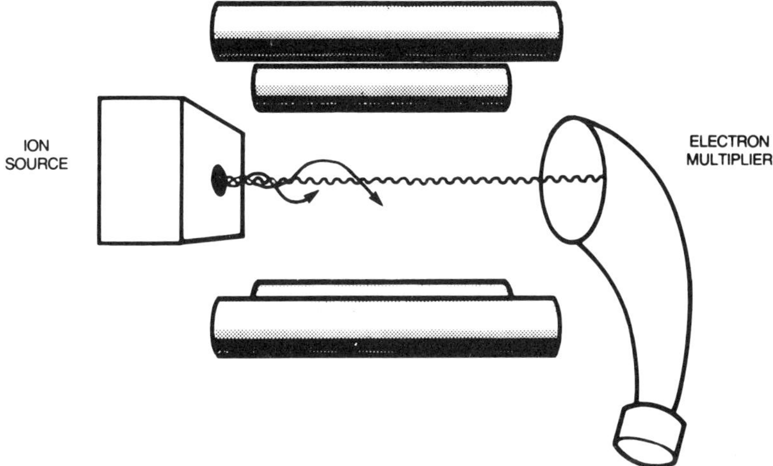

Fig. 2.6 — Schematic diagram of a quadrupole mass filter. Copyright 1983, Finnigan Corporation. Reprinted with permission.

metal rods of circular, or according to the theory, preferably hyperbolic, cross-section (Feser & Kögler, 1979). The ions extracted from the ion source are accelerated through a small potential difference entering along the principal axis, the rod array. A combination of direct-current (d.c.) voltage (*U*) and radio frequency (Rf) voltage (*V*) is applied to diagonally opposite rod pairs. Opposite rods are electrically connected. Under the influence of appropriately programmed d.c. and Rf fields, ions undergo oscillations, but only ions of a specific m/z ratio pass the mass filter and reach the detector while all others move too far away (with increasing amplitude) from the principal axis of the quadrupole assembly and strike the rods (filtered out). By rapidly and precisely changing the potentials on the rods, ions of different m/z values sequentially traverse the analyser and strike the detector.

2.3.3.3 Ion-trap detector

The ITD, developed by Finnigan MAT, has gained recognition as a low-cost, highly sensitive mass spectrometer for capillary gas chromatographs. A schematic of the GC–ITD assembly is shown in Fig. 2.7. The ITD can be operated both in the full-scan

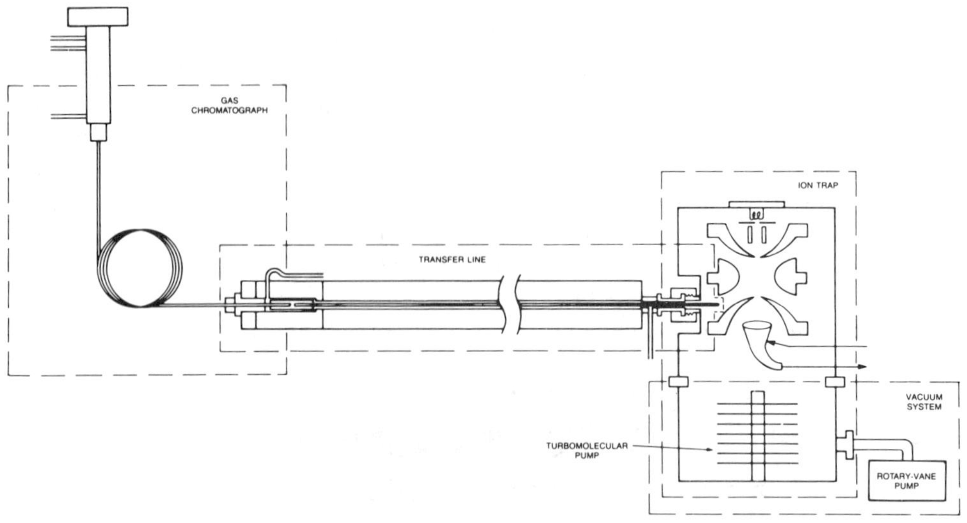

Fig. 2.7 — Schematic of GC–ITD assembly: GC, transfer line, ion trap and vacuum system. Copyright 1984, Finnigan Corporation. Reprinted with permission.

or multiple-ion (selected-ion) monitoring mode. The device is a three-electrode arrangement and a central cavity formed by two end-cap electrodes with cylindrical hyperbolic surfaces and a central ring electrode (Fig. 2.8). Application of a proper Rf voltage (storage voltage) causes all ions of interest to oscillate on stable trajectories. As the Rf voltage is increased in a precisely controlled way, ions of successively larger *m/z* ratios develop unstable oscillations (trajectories) and are ejected. A portion of these ions exit through holes in the end-cap (exit) electrode striking the electron multiplier (detector). In contrast to the quadrupole analyser, the ITD operates in the mass-selective instability mode. The sample molecules exiting the capillary GC column (Fig. 2.7) are ionized within the ion trap by electron impact, with thermal electrons emitted from a rhenium filament using a gated electron beam technique (Louris *et al.*, 1987), (Brodbelt & Cooks, 1988). A schematic of the operation and timing sequence of the ITD is shown in Fig. 2.9. The full-scan acquisition range is from 10 to 650 a.m.u. In addition, CI is also possible with the ITD, involving only a change in scan function (no change in source volume). An advantage over conventional CI is that instead of a mixture of reagent ions, selected ions, for example CH_5^+, $C_2H_5^+$ or $CH_5^+/C_2H_5^+$, may be selected. The ITD has been used in determining diethylene glycol in wine (Hübschmann & Katzlinger, 1985),the identification of essential oil components (Hübschmann & Schubert, 1986) and volatile organic compounds related to flavour and fragrance (Parliment, 1987). An application of the ITD for the quantification of the PCB-substitute Ugilec in fish has been described by Fürst (1988).

2.3.4 Ion detection

There are a number of means to detect the ion beam emerging after mass separation from the mass analyser, e.g. by photographic plate, Faraday cup, and electron

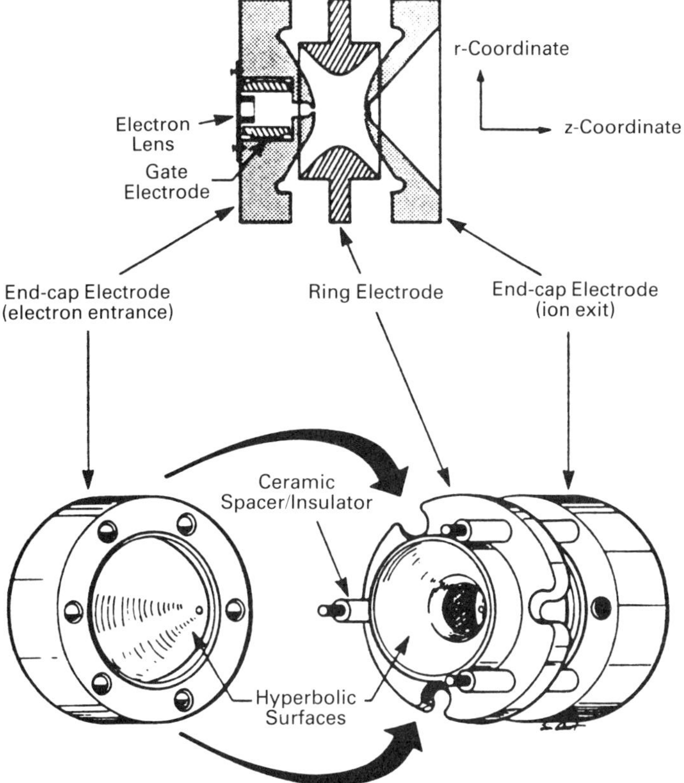

Fig. 2.8 — Schematic of ion-trap electrodes. An Rf field is established between the ring electrode and the two end-cap electrodes. The assembled structure is shown in cross-section. Reprinted with permission from Louris *et al.*, 1987, American Chemical Society.

multiplier. The most common device used in modern instruments is the electron multiplier, consisting of a series of electrodes (dynodes), usually a copper–beryllium alloy, to amplify the very small ion current (10^{-9} to 10^{-13} A) typical for a mass spectrum. The ion beam is focused first on a conversion dynode where the impinging ions liberate from the metal surface secondary electrons, which are directed and accelerated toward another electrode (dynode), producing more electrons, and are accelerated again dislodging additional electrons from another dynode. The accelerating voltage is increased successively at each new stage, resulting in a large build-up of the electron population, finally producing a measurable current. The continuous dynode electron multiplier (Fig. 2.10) operates similarly. The secondary electrons formed at the conversion dynode by the impact of the ion beam strike the curving inner wall of a 'horn-like' tube (lead oxide–glass). The inner wall has a coating of a thin semi-conducting film and a potential difference is applied across the ends of the tube. More secondary electrons are formed, accelerated again, and then passed further into the electron multiplier. This process is repeated along the tube toward

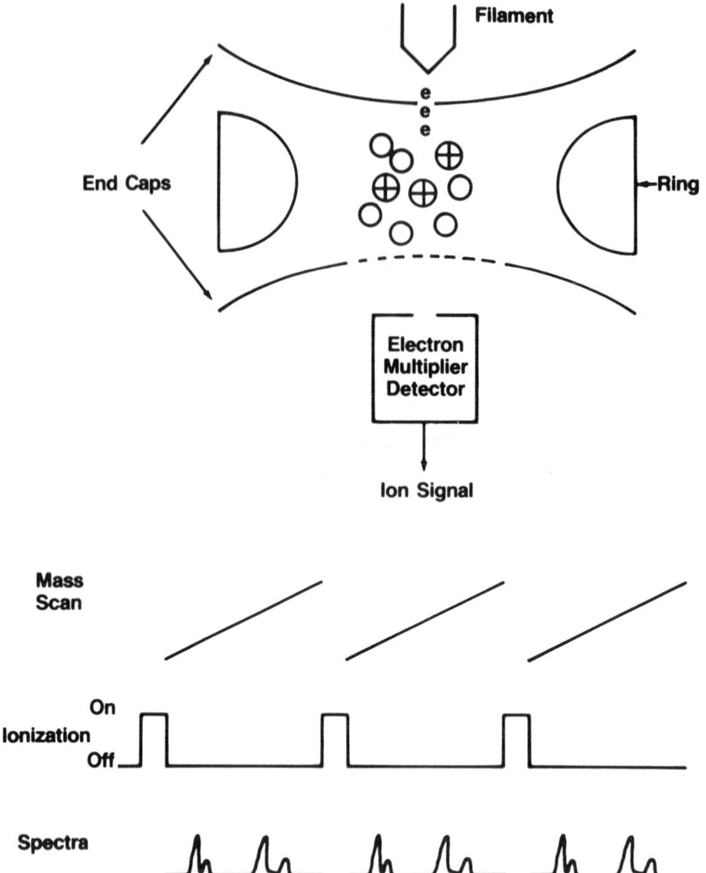

Fig. 2.9 — Schematic of the operation of the ion-trap detector. Copyright 1984, Finnigan Corporation. Reprinted with permission.

the anode, resulting in a large increase of electrons. For assessing the performance of the electron multiplier the reader is referred to Beynon and Brenton (1982).

2.4 SELECTED-ION MONITORING

The GC–MS system can be utilized in two general modes to obtain mass spectral information. One mode is to acquire complete low-resolution mass spectra displaying all ions formed in the fragmentation process of each component eluting from the chromatographic column. A full spectrum is essential for the identification of an unknown compound or the confirmation of derivatives as well as otherwise tentatively identified compounds (e.g. using retention time). In some cases (e.g. where there is weak or missing molecular ion), it may be necessary to acquire spectra in both the EI and CI modes to establish the identity of the compound of interest, and

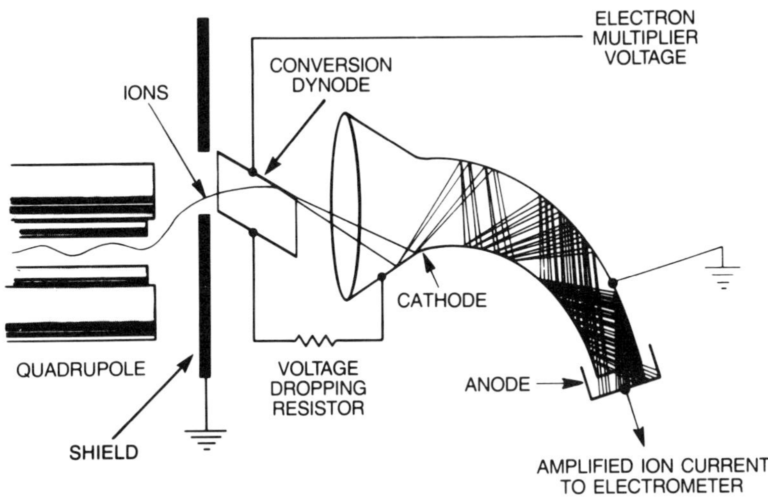

Fig. 2.10 — The standard electron multiplier, showing the conversion dynode, the voltage dropping resistor, and the cathode and anode of the electron multiplier. Copyright 1983, Finnigan Corporation. Reprinted with permission.

additional information may be attained from exact mass measurement. Commonly, a range of 1 to 10 ng of compound injected on a capillary GC column may be necessary to obtain a full spectrum.

The alternative mode is to use the mass spectrometer as a selective and very sensitive detector. Only a few pre-selected ions from the known fragmentation pattern of a compound to be analysed are monitored. This method is called selected-ion monitoring (SIM). It is apparent that in SIM mode significantly more time can be apportioned to measuring ion currents of a few selected ions, thus dramatically increasing the sensitivity of detection. For example, when scanning a mass range from 50 to 550 a.m.u. in 1 s (total time), only a very short time, 2 ms, can be spent recording each *m/z* value, including ions of no useful information, but much more time can be spent on monitoring the intensities of ions when only a few ions or one ion is chosen. Highest sensitivity could be achieved by single-ion monitoring, though at a loss of specificity. In the SIM mode the lower limit of detection attainable is in the picogram to femtogram range. The principal application of the SIM technique is in the quantification of compounds. The combination of specificity and sensitivity makes this method most suitable for residue analysis in the area of food safety (Stan, 1984).

In setting up a SIM experiment an important decision is the choice of either one or several ions. If sensitivity is of utmost concern single-ion monitoring will be the appropriate technique. However, the decreased selectivity may be improved by selecting an ion at high *m/z* value (at least above 200), e.g. a molecular ion. In some cases derivatization may provide better chromatographic properties as well as increased molecular weight for better selectivity.

The specificity increases with the selection of more ions (in many cases two or three may be sufficient). Ideally, the ions should be characteristic (e.g. the molecular ion), of high abundance (e.g. base peak) and preferably of high m/z values. Ions originating from column background as column bleed should be avoided. In some cases chromatographic retention data may be essential for distinguishing between two compounds that have practically the same set of characteristic ions, e.g. benz(a)anthracene and chrysene (Gilbert, 1987). In addition, monitoring the abundance ratios of the chosen ions can be used to increase the confidence in the measurements. A change in the ratios would indicate interfering material.

To ensure reliable quantitative results, it is strongly advisable to use an internal standard. The internal standard should be added to the sample as early as possible in the work-up procedure (extraction, derivatization, etc.). Any loss of analyte during work-up will be matched by loss of internal standard and, analogously, fluctuations in instrumental response (GC–MS), which could have a large effect on the reproducibility of data, will be compensated by the internal standard. The internal standard must be structurally as similar as possible to the analyte, ideally; for example an isotopically labelled deuterium or carbon-13 analogue would have the same chemical (derivatization) and chromatographic behaviour. Fig. 2.11 presents an example of SIM for the acetyltrifluoroacetyl derivatives of 3-methyl-4-hydroxyphenylglycol (MHPG), 3,4-dihydroxyphenylethyleneglycol (DHPG) and their deuterium-labelled analogues (Roboz, 1986) as internal standards. The ions monitored are fragment ions because the molecular ions were not detectable under EI conditions. As widely as this isotope dilution technique is used in biochemical and clinical research, its application in food science is relatively limited. Cost is a major factor in not using stable isotopes and in the case of natural products (e.g. toxins) a chemical synthesis may not be feasible. For practical reasons, homologues or isomers of the compound(s) of interest may serve as suitable substitutes. Commonly, quadrupole and double-focusing instruments are used for SIM. The recently introduced ITD can also be employed for SIM. However, quadrupole MS offers greater ease in carrying out SIM, because ions from any range of the mass spectrum can be monitored, and monitoring parameters can be changed more rapidly.

The majority of SIM is done in low-resolution mode (nominal masses); however, double-focusing instruments have the capability of high-resolution SIM, allowing exact mass measurements. Monitoring of exact masses may be essential in some cases to obtain confirmatory information, e.g. in distinguishing 2,3,7,8-tetrachloro-*p*-dioxin from other similar structures.

The discussion on SIM is by no means comprehensive and the reader is referred to two chapters by John Gilbert (1984, 1987). A wealth of information is provided on applications of SIM in the areas of pesticides, mycotoxins, veterinary drug residues, and miscellaneous applications (e.g. natural toxicants in food, food adulteration).

2.5 DATA HANDLING

A mass spectrometer generates an enormous amount of data. However, it was the union of the mass spectrometer and the computer that increased tremendously the utility of the GC–MS system for a wide range of analytical applications. In addition, the introduction of hard-disk technology expanded significantly the capacity for data storage. Data systems are standard components of modern commerical mass spectrometers.

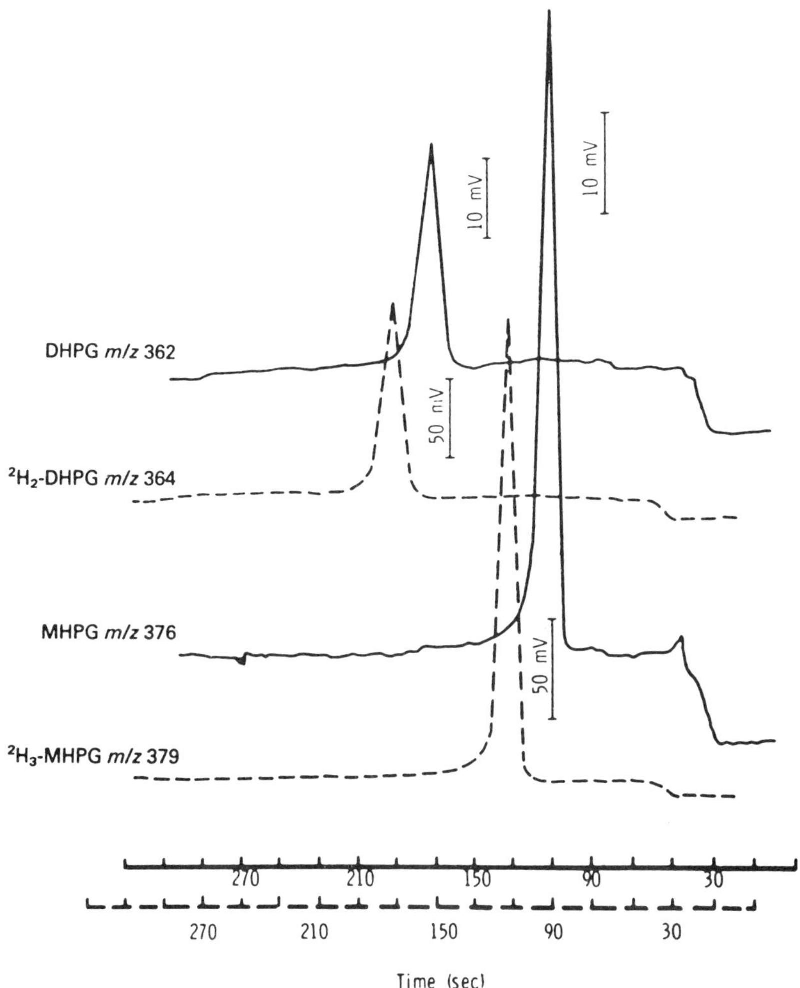

Fig. 2.11 — Selected-ion monitoring of MHPG and DHPG. Solid abscissa indicates the elution of the solid-line chromatograms. The dashed abscissa indicates the elution times of the dashed line chromatograms. Reprinted from Roboz (1986). Copyright 1986, Ellis Horwood Limited.

The basic operations of the data system include:

(a) control of the scanning of the mass spectrometer,
(b) data acquisition,
(c) data processing and manipulation.

However, the trend in modern mass spectrometer design is toward completely controlling, via the data system, all instrument control functions as well as data acquisition and processing, thus excluding manual operation of the mass spectrometer.

2.5.1 Data collection

During GC–MS analysis, modern instruments are capable of repetitively acquiring in very short time spectra over a wide mass range (*m/z* values) from nanogram amounts of hundreds of eluting peaks. The repetitive scanning of the mass spectrometer in the case of a GC–MS analysis may result in 2000 or more mass spectra. These spectra are stored on a magnetic disk and available for data processing to the analyst. For effective data acquisition, the mass range has to be selected wide enough to obtain the complete mass spectrum of each compound from GC–MS analysis, and the scan speed, in s/mass decade, has to be sufficiently fast to sample several spectra across each eluting peak. For narrow and fast eluting peaks from a capillary column rapid scan speeds of 0.1 to 0.5 s per decade may be necessary.

The data acquisition involves digitization of the analogue signals of each mass spectrometer scan and computation of the mass spectrum (mass/intensity files) from time/intensity files (based on reference files, e.g. perfluorokerosine as calibration compound). The multitude of data (spectra) is stored on a magnetic disk and accessible for further processing.

2.5.2 Data processing

A wide variety of manipulations are available with a modern data system to display, process, and manipulate the large amount of mass spectral information. The subtraction of background (e.g. column bleed, leaks) is an important feature for obtaining corrected or 'clean' spectra for subsequent computer-aided library search. The subtraction can be automatic or by manual operation on a peak-by-peak basis according to need. In the latter case the operator selects a spectrum (scan) which contains background only and subtracts this spectrum from the spectrum of interest.

Another feature of the data system is the facility to reconstruct a chromatogram from all scans during an entire GC–MS analysis, which is done by the summation of all ion abundances of each spectrum scanned, normalized to the largest of these values, and plotted against the scan number (time). This plot is called the total ion-current chromatogram (TIC). The peak areas of the TIC may be quantified if so desired. In addition to monitoring the TIC, the data system can be used to search all or selected scans (peaks) for the presence of ions e.g. $M^{+\bullet}$, unique for a specific compound or group of compounds. By plotting the abundance of the chosen ions as a function of the scan number (time), traces will be obtained which are called mass chromatograms or ion chromatograms (Fig. 2.12). The mass chromatograms will increase in abundance to a maximum together at the same scan number or retention time for ions (fragment ions and molecular ion) formed from the same compound. The maximum number of ions that could be selected at one time may be five (this number may vary depending on the data system). This display technique can be used very effectively to identify groups of compounds in complex mixtures by using characteristic ions (e.g. *m/z* 136 for monoterpenes, or the McLafferty rearrangement ion, *m/z* 74, which is typical for longer straight-chain fatty acid methyl esters) as well as to examine apparently homogeneous chromatographic peaks for unresolved components. If the peak is more than one compound the mass chromatograms will quickly reveal the exact position of the co-eluting components. Both background subtraction and mass chromatograms are powerful data manipulations for generat-

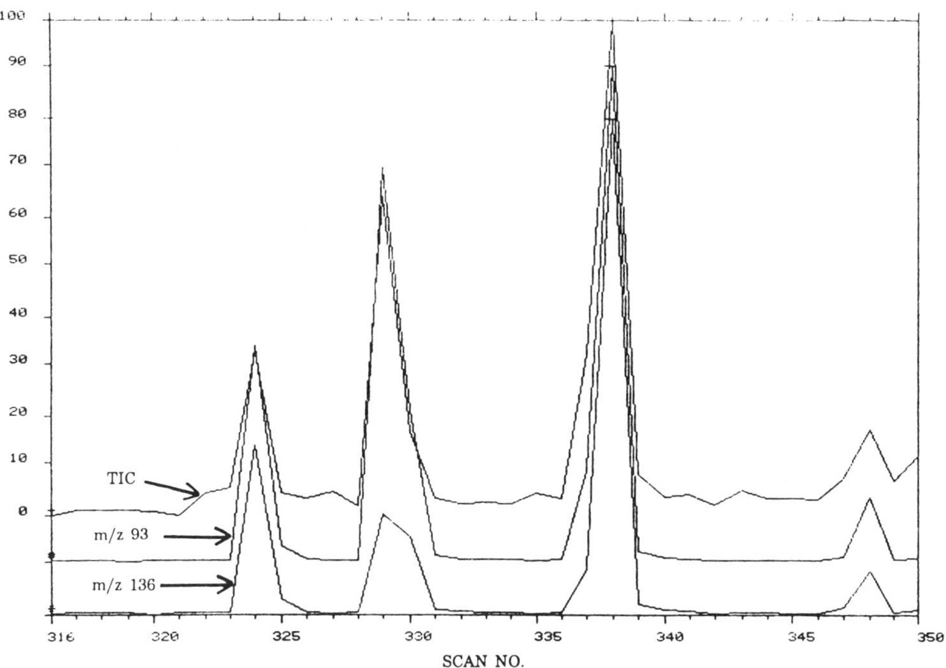

Fig. 2.12 — Mass chromatograms of m/z values 93 and 136 and the TIC. Unpublished results from the author's laboratory. Instrument: Kratos MS25.

ing representative mass spectra that can be used for a subsequent library search if the compounds analysed are unknown.

Mass spectral libraries are an integral component of current data systems. The number of spectra available may vary somewhat depending on the year of purchase of the GC–MS instrumentation. The most widely used database is the EPA/NIH library comprising about 44 000 spectra (some are duplicates). However, most recent GC–MS data systems (Fall, 1988) may have a spectra collection of more than 120 000. This database comprises a mass spectra registry by McLafferty and Abrahamsson, which is also available as the 1988 Registry of Mass Spectra Data (CD–ROM edition) of over 120 000 spectra from J. Wiley & Sons, New York.

With the data systems are provided search routines for the retrieval of spectra that match unknown compounds. A number of approaches have been devised for effective and rapid identifications of unknowns, e.g. PBM (probability-based matching), as discussed by McLafferty and Venkataraghavan (1979). A unique feature of library search routines is the so-called reverse searching. A reference spectrum (or several) is selected from the library and the data system compares it with each spectrum of all scans from an analysis by only considering the ions of the reference spectrum but ignoring the extraneous ions in the spectra to be searched. This technique aids in screening rapidly a GC–MS run for the presence of a specific compound.

Even if a matching spectrum cannot be found for tentative identification, some information may be deduced from a set of retrieved spectra, based on a number of characteristic ions that are indicative of certain types of compounds. Futhermore, some isomers (positional, double bond, *cis-trans,* etc.) are difficult to distinguish from another by MS; in this case chromatographic properties (retention time) may help in the identification process. Some attempts have been made at the automatic interpretation of mass spectra when no match was found for the identification of an unknown. The example is STIRS (self-training interpretive and retrieval system) which can aid the interpreter (McLafferty and Venkataraghavan, 1979), and another method is SISCOM, which has been developed by Henneberg and Weimann (1984).

2.6 USE OF MASS SPECTROMETRY FOR IDENTIFICATION OF UNKNOWN COMPOUNDS

Food materials can be considered among the most complex mixtures of chemicals. A most challenging task is the original identification of totally unknown compounds. The flavour research area is an example where application of GC–MS (Flath, 1981; Merritt and Robertson, 1982; Schomburg *et al.*, 1984; Cronin & Caplan, 1987), especially with the advent of high-performance fused-silica capillary GC columns, resulted in an outburst of activity in the identification of flavour volatiles. A list of studies of volatile constituents of foods and beverages presented by Horman (1984) illustrates this expanding activity very clearly. However, the increase in studies is not necessarily matched by an equal increase in the identification of 'new' compounds because certain groups of compounds are likely to reappear (e.g terpenoids, aldehydes). It is beyond the scope of this chapter to even enumerate the many applications of GC–MS for identification of unknowns (food safety, food flavour, etc.). A good example of the use of these technologies is the identification of flavour volatiles from ovine fat by Suzuki and Bailey (1985). Resulting chromatograms are shown in Fig. 2.13 and the names of compounds identified are listed in Table 2.1. The effect of forage (clover) on flavour volatiles of lamb compared to finishing on corn grain was studied. There was a significant increase for clover samples in the concentration of 2,3-octanedione, aldehydes and terpenes (neophytadiene, phyt-2-ene, etc.). The significant increase in the amounts of 2,3-octanedione, aldehydes and terpenes correlated with undesirable or 'grassy' flavour based on sensory panel scores. For experimental details of the studies, the reader is referred to Suzuki and Bailey (1985), Larick *et al.* (1987) and Dupuy *et al.* (1987).

2.7 CONCLUSION

Combined GC–MS interfaced with a modern data system presents a powerful analytical tool of high specificity and sensitivity essential in many areas of food chemistry. One of the most significant applications is in food safety based on the exceptional reliability of GC–MS to identify trace amounts of toxic contaminants. The availability of low-cost GC–MSD and GC–ITD instruments will help to broaden the use of this analytical methodology in food chemistry.

ACKNOWLEDGEMENTS

The preparation of the manuscript was in part supported by National Science Foundation Grant No. PCM–8 117 116.

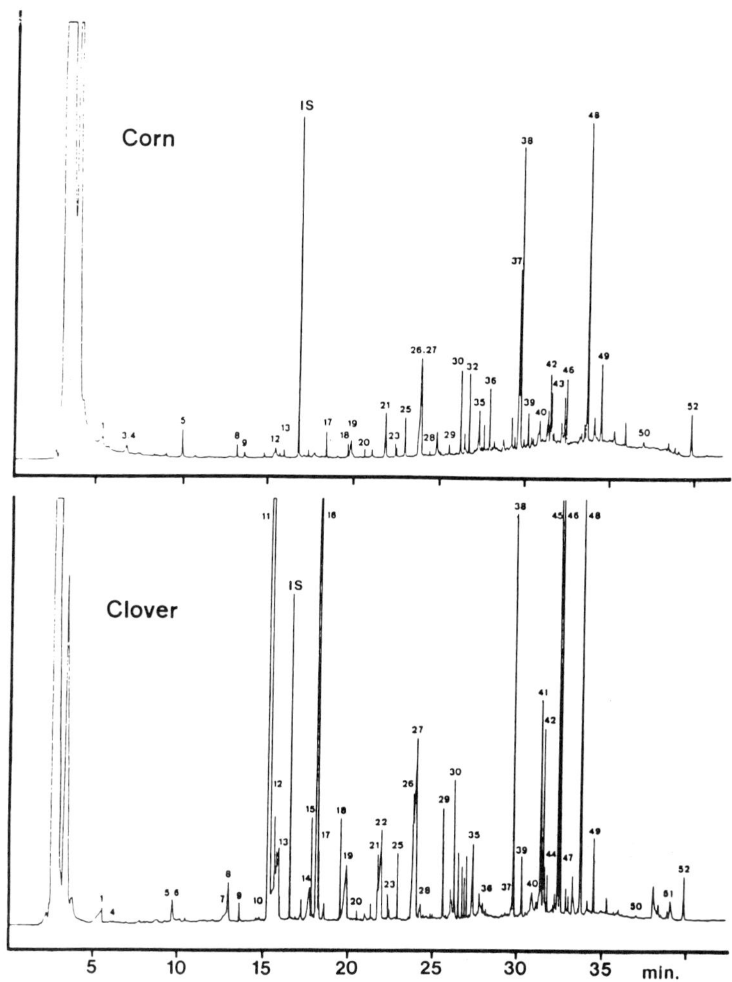

Fig. 2.13 — Separation of volatiles from fat of corn-fed and clover-fed lamb (New Zealand) by capillary GC. Identification by EI–MS (Kratos MS25). Courtesy of Dr. M. E. Bailey, University of Missouri, Columbia, MO 65211.

The author thanks Roy H. Rice for helpful discussions, Karen G. Sayre for expert technical assistance, and Suzanne Black of Finnigan MAT for providing a selection of illustrations.

REFERENCES

Arsenault, G. P. (1979). Chemical ionization mass spectrometry. In *Biochemical Applications of Mass Spectrometry*, Waller, G. R. (ed.), Wiley-Interscience, New York, p. 817.

Beynon, J. H. and Brenton, A. G. (1982). *An Introduction to Mass Spectrometry*. University of Wales Press, Cardiff, p. 30.

Table 2.1 — Volatile compounds found in lamb fat by GC–MS analysis (Suzuki and Bailey, 1985)

1	Acetic acid	27	2-Undecenal
2	Propionic acid	28	Tetradecane
3	Pentanal	29	Unknown 1
4	Heptane	30	2-Tridecanone
5	Hexanal	31	δ-Decalactone
6	Octane	32	BHT
7	Pentanoic acid	33	Unknown 2
8	Heptanal	34	Unknown 3
9	2,4-Pentadienal	35	Undecanoic acid
10	Sabinene	36	Hexadecane
11	2,3-Octanedione	37	γ-Dodecalactone
12	Hexanoic acid	38	2-Pentadecanone
13	Octanal	39	δ-Dodecalactone
IS	2-Methyl-3-octanone	40	Dodecanoic acid
14	Heptanoic acid	41	Phyt-1-ene
15	2,6-Dimethyl-4-heptanone	42	Octadecane
16	3-Hydroxy-2-octanone	43	2-Hexadecanone
17	Nonanal	44	Phytane
18	2-Nonenal	45	Neophytadiene
19	Octanoic acid	46	Phyt-2-ene
20	Dodecane	47	Phytadiene
21	Nonanoic acid	48	2-Heptadecanone
22	2-Decenal	49	δ-Tetradecalactone
23	2-Undecanone	50	δ-Pentadecalactone
24	Tridecane	51	Phytol
25	2,4-Decadienal	52	Hexadecalactone
26	Decanoic acid		

Brodbelt, J.S. and Cooks, R. G. (1988). Ion trap tandem mass spectrometry, *Spectra —A Finnigan MAT Publication*, **11**, 33.

Brumley, W. C. and Sphon, J. A. (1987). Application of negative ion chemical ionization. In *Applications of Mass Spectrometry in Food Science*, Gilbert, J. (ed.), Elsevier Applied Science, New York, p. 141.

Cramers, C. A., Scherpenzeel, G. J. and Leclercq, P. A. (1981). Increased speed of analysis in direct coupled chromatography–mass spectrometry systems, *J. Chromatogr.*, **203**, 207–216.

Cronin, D. A. and Caplan, P. J. (1987). Application of GC/MS to identification of flavour compounds in food. In *Applications of Mass Spectrometry in Food Science*, Gilbert, J. (ed.), Elsevier Applied Science, New York, p. 1.

Dougherty, R. C. (1981). Negative chemical ionization mass spectrometry, *Anal. Chem.* **53**, 625A.

Duncan, M. W., Smythe, G. A., Nicholson, M.V. and Clezy, P. S. (1984).

Comparison of high-performance liquid chromatography with electrochemical detection and gas chromatography–mass fragmentography for the assay of salsolinol, dopamine and dopamine metabolites in food and beverage samples, *J. Chromatogr.*, **336,** 199–209.

Dupuy, H. P., Bailey, M. E., St. Angelo, A. J., Vercellotti, F. R. and Legendre, M. G. (1987). Instrumental analyses of volatiles related to warmed-over flavor of cooked meats. In *Warmed-Over Flavor of Meat,* St. Angelo, A. J. and Bailey, M. E., (eds.), Academic Press, Orlando, p. 165.

Feser, K. and Kögler, W. (1979). The quadrupole mass filter for GC/MS applications. *J., Chromatogr. Sci.*, **13,** 57.

Flath, R. A. (1981). Identification in flavor research. In *Flavor Research — Recent Advances,* Teranishi, R., Flath, R. A. and Sugisawa, H. (eds.), Marcel Dekker, New York, p. 83.

Friedli, F. (1981). Fused silica capillary GC/MS coupling: a new, innovative approach, *J. HRC & CC,* **4,** 495–499.

Fürst, P. (1988). Determination of the PCB-substitute Uglic in fish with the ion trap detector, *Spectra — A Finnigan MAT Publication,* **11**(2), 26–29.

Giang, B. Y. (1984). Improved capillary GC/MS interface, *J. HRC & CC,* **7,** 137–139.

Gilbert, J. (1984). Confirmation and quantification of trace organic food contaminants by mass spectrometry–selected ion monitoring. In *Analysis of Food Contaminants,* Gilbert, J. (ed.), Elsevier Applied Science, New York, p. 265.

Gilbert, J. (1987). Application of quantitative mass spectrometry in food science. In *Applications of Mass Spectrometry in Food Science,* Gilbert, J. (ed.), Elsevier Applied Science, New York, p. 73.

Grayson, M. A. (1986). The mass spectrometer as a detector for gas chromatography, *J. Chromatogr. Sci.*, **24,** 529–542.

Gudzinowicz, B. J., Gudzinowicz, M. J. and Martin, H. F. (1977). *Fundamentals of Integrated GC–MS.* Parts II and III, Marcel Dekker, New York.

Harrison, A. G. (1983). *Chemical Ionization Mass Spectrometry,* CRC Press, Boca Raton, FL, p. 63.

Hatch, F. W. and Parrish, M. E. (1978). Vacuum gas chromatography using short glass-capillary columns combined with mass spectrometry, *Anal. Chem.*, **50,** 1164–1168.

Henneberg, D. and Weimann, B. (1984). Search for identical and similar compounds in mass spectral data bases, *Spectra — A Finnigan MAT Publication,* **10**(1), 11–14.

Henneberg, D., Henrichs, U. and Schomburg, G. (1975). Open split connection of glass capillary columns to mass spectrometers, *Chromatographia,* **8,** 449–451.

Horman, I. (1984). Mass spectrometry. In *Analysis of Foods and Beverages: Modern Techniques,* Charalambous, G. (Ed.), Academic Press, Orlando, p. 141.

Hübschmann, H. J. and Katzlinger, H. (1985). Bestimmung von Diethylenglycol in Wein mit dem Ion Trap Detector, *Lebenss. Biotechn.*, **4,** 148.

Hübschmann, H. J. and Schubert, R. (1986). Ion trap detector: the techniques and its application. In *Progress in Essential Oil Research,* Brunke, E.-J. (ed.), Walter de Gruyter, Berlin, p. 643.

Hunt, D. F., Stafford, Jr., G. C., Crow, F. W. and Russel, J. W. (1976). Pulsed positive negative ion chemical ionization mass spectrometry, *Anal. Chem.*, **48**, 2098–2105.

Koller, D. W. and Tressl, G. (1980). Simple GC/MS interface with transfer-line of fused silica for open or direct coupling, *J. HRC & CC*, **3**, 359.

Larick, D. K., Hedrick, H. B., Bailey, M. E., Williams, J. W., Hancock, D. L., Garner, G. B. and Morrow, R. E. (1987). Flavor constituents of beef as influenced by forage- and grain-feeding, *J. Food Sci.*, **52**(2), 245–251.

Ligon, Jr., W. V. (1979). Molecular Analysis by Mass Spectrometry, *Science*, **205**, 151–159.

Louris, J. N., Cooks, R. G., Syka, J. E. P., Kelley, P. E., Stafford, Jr., G. C. and Todd, J. F. J. (1987). Instrumentation, applications and energy deposition in quadrupole ion-trap tandem mass spectrometry, *Anal. Chem.*, **59**, 1677.

McFadden, W. H. (1973). *Techniques of Combined Gas Chromatography/Mass Spectrometry: Applications in Organic Analysis*. Wiley-Interscience, New York.

McFadden, W. H. (1979). Interfacing chromatography and mass spectrometry, *J. Chromatogr. Sci.*, **17**, 2–16.

McLafferty, F. W. and Venkataraghavan, R. (1979). Computer techniques for mass spectral identification, *J. Chromatogr. Sci.*, **17**, 24–29.

Merritt, Jr., C. and Robertson, D. H. (1982). Techniques of analysis of flavours, gas chromatography and mass spectrometry. In *Food Flavours, Part A. Introduction*, Morton, I. D. and MacLeod, A. J. (eds.), Elsevier Scientific, New York, p. 49.

Milberg, R. M. and Cook, Jr., J. C. (1979). Design considerations of MS sources: EI, CI, FI, FD and API, *J. Chromatogr. Sci.*, **17**, 17–23.

Munson, M. S. B. and Field, F. H. (1966). Chemical ionization mass spectrometry. I. General introduction, *J. Am. Chem. Soc.*, **88**, 2621.

Parliment, T. H. (1987). Sample analysis in flavor and fragrance research, *American Laboratory*, **19**(1), 51–57.

Paul, W. and Steinwedel, H. (1953). Ein Neues Massenspektrometer Ohne Magnetfeld. *Z. Naturforschung*, **8a**, 448.

Peltonen, K., Lakkisto, U.-M. and Rosenberg, C. (1988). Use of a deactivated capillary between the GC column and the mass spectrometer to facilitate column changing, *LC-GC*, **6**(6), 524.

Roboz, J. (1986). Gas–liquid chromatography/mass spectrometry. In *Quantitative Analysis of Catecholamines and Related Compounds*, Krstulovic, A. M. (ed.), Ellis Horwood, Chichester (UK), p. 51–55.

Rosen, R. T., Hartman, T. G. and Lech, J. (1988). Why not purchase a high resolution magnetic mass spectrometer?, *Mass Spec. Source*, **XI**(3), 13.

Schomburg, G., Husmann, H., Podmaniczky, L., Weeke, F. and Rapp, A. (1984). Coupled gas chromatographic methods for separation identification and quantitative analysis of complex mixtures: MDGC, GC–MS, GC–IR, LC–GC. In *Analysis of Volatiles: Methods and Applications*, Schreier, P. (ed.), Walter de Gruyter, Berlin, p. 121.

Stan, H.-J. (1984). Contribution of mass spectrometry to food safety. In *Chromato-*

graphy and Mass Spectrometry in Nutrition Science and Safety, Frigerio, A. and Milon, H. (eds.), Elsevier Science, Amsterdam, p. 91.

Stan, H.-J. and Abraham, B. (1978). All-glass open-split interface for gas chromatography–mass spectrometry, *Anal. Chem.*, **50**(14), 2161–2164.

Suzuki, J. and Bailey, M. E. (1985). Direct sampling capillary GLC analysis of flavor volatiles from ovine fat, *J. Agric. Food. Chem.*, **33,** 343–347.

Vangaever, F., Sandra, P. and Verzele, M. (1979). Influence of the vacuum on the separation efficiency in coupled $(GC)^2$-MS, *Chromatographia,* **12,** 153–154.

Wetzel, E., Kuster, T. and Curtius, H.-C. (1982). A split system applicable as a gas chromatographic–mass spectrometric interface and as effluent splitter for specific gas chromatographic detectors, *J. Chromatogr.*, **239,** 107–114.

3

Analysis of aroma volatiles

Joseph A. Maga

3.1 INTRODUCTION

The flavour chemist who has been charged with separating and identifying the impact level of compounds in a food system that have a characteristic aroma has indeed a difficult task. For centuries, man has relied almost exclusively on his subjective organoleptic senses to solve such a problem. With the introduction of simple wet chemistry techniques his task became somewhat easier, but it was not until the introduction of gas chromatography (GC) in 1952 that he had a reliable objective ally.

Today the compounding of relatively simple aromas is still guided by the imaginative and artistic talents of an individual and his nose, but since most food aromas are quite complex, man has learned to rely heavily on GC as a necessary and important tool. In fact, some feel that the modern flavour chemist may rely too heavily on GC. In this light, we must remember that man is still more sensitive in detecting aromas than any GC designed to date. Also, many food aromas are quite complex and composed of compounds of limited chemical and thermal stability. Thus, one must consider the influence of compound isolation and separation when one attempts to use GC. Some still contend that the application of GC is 50% art and 50% science, and thus a great deal of training and experience are required of an individual who expects to make significant contributions to the flavour area.

It should also be noted that, although volatile compounds contribute to food aroma, non-volatile compounds contribute to taste, with both aroma and taste forming food flavour. Therefore, one can become fascinated with attempting to identify every one of 700 volatile compounds isolated from a food using GC only to find that no combination of the compounds has the same characteristic flavour as the food in question. However, for foods where volatile compounds are indeed important, we must realize that, since the mode of human odour perception and the interrelationships of compound structure versus odour type and potency have not been resolved, it is indeed fortunate that a tool like GC is available to assist us in better understanding aroma chemistry.

In addition, it has become quite obvious that food aroma is due to a vast array of compounds of dramatically different structures that cannot be effectively separated using the same or simple chromatographic conditions. Therefore, the field of GC, with now over 200000 units in operation worldwide, is far from static, and special thanks must go to engineers and specialists in electronics who are continually making strides in improving man's ability to separate more effectively and efficiently potent and important aroma compounds.

We should also note that originally GC utilized packed columns, which for the modern flavour chemist are almost a thing of the past, except for rather routine analyses. Today various forms of capillary columns are almost a necessity to the flavour chemist. Also, it is almost impossible to speak of experimental aroma chemistry utilizing only GC. Granted, aroma chemistry has made great strides with GC, but today the GC–MS analytical set-up is almost routine among flavour chemists.

3.2 AROMA COMPOUND EXTRACTION AND CONCENTRATION

Choosing the proper extraction technique, since it is the first step in most GC analyses, is quite critical and can significantly influence the results obtained. One must first consider the form of the food from which the compounds of interest are to be extracted. For example, if it is in solid form, the food usually has to be ground or processed in some manner to make extraction easier and more efficient. However, with raw food systems this step may initiate enzymatic activity resulting in artefact formation.

Considering the facts that the total quantity of aroma compounds normally present in natural systems can range from only a few p.p.b. to perhaps a maximum of several hundred p.p.m., and that this fraction usually contains 100–700 individual compounds, some of which are thermally unstable, have various structures and reactive groups, and vary in volatility and molecular weight over a boiling point range of 20–300°C, it becomes quite evident that efficient and yet gentle methods are required to extract aroma compounds.

From a quantitative standpoint it is imperative that the extraction procedure chosen be effective considering the above physical and chemical properties of the compounds in question. Also to be considered is the fact that most aroma compounds are lipophilic. This latter factor along with compound volatility are primarily utilized in most separation schemes. In addition, it should be noted that the primary objective of any extraction is to remove as completely, and as true to the original odour as possible, the aroma fraction of the food in question. Thus, if one were interested in a particular raw fruit aroma, extraction systems that require any amount of heating would not be appropriate. For some aromas it is essential that the extraction be conducted at low temperature, under inert gas or in the absence of light. In some cases, it is also necessary to inactivate enzymes which lead to changes in the volatile components during the isolation process.

In addition, for most forms of GC analysis a concurrent or subsequent concentration step usually representing at least a 5000-fold concentration is required. During this stage, compound volatility becomes especially important if true quantitative values are required. Ideally extraction and concentration should be accomplished in one step, but this is not always practical and thus artefact formation is a major concern with multi-step procedures.

There are no uniform methods of extracting and concentrating aromas that work for all food systems and, as a result, a wide variety of techniques are available. These can generally be classified into physical, solubility, adsorption and chemical methods. (Cronin, 1982; Jennings, 1980a; Peyron, 1982; Sugisawa, 1981; Teranishi, 1981).

3.2.1 Physical methods

Historically the most popular manner of separating volatiles from a food was to disperse the food in water, heat to the boiling point and collect the resulting condensate. Then a second separation was required to remove the water from the volatiles. This is classical steam distillation at atmospheric pressure and the resulting initial condensate usually is then extracted with solvents and concentrated. Obviously, prolonged heating at an elevated temperature is not an ideal system for all foods and one has to question the degree of artefact formation using this technique.

An improvement on the above system is distillation under reduced pressure, which in effect lowers the boiling point of the system and thus reduces both the time required and artefact formation. Normally the distillation is performed at 10–20 torr (1330–2660 Pa) although some workers use pressures as low as 1 torr (133 Pa). Many researchers prefer to use this technique in conjunction with a Likens–Nickerson apparatus, which combines distillation, extraction and concentration in one continuous step. With this system simultaneous condensation of the steam distillate in an immiscible extracting solvent occurs. Since the extracting solvent is recycled, a relatively small amount of solvent is required and thus artefacts that may be present in the solvent are kept to a minimum.

A technique called flash distillation can also be employed, at either atmospheric or reduced pressures. Instead of an external heat source being applied to obtain the appropriate boiling point, steam is injected directly into the liquefied food system to be distilled, thereby significantly reducing the time at which the food is at elevated temperatures. The method is very effective for extremely volatile compounds but cannot be considered to be quantitative and can result in significant amounts of artefact formation due to localized thermal reactions.

Fractional distillation at atmospheric or reduced pressures can also be used. Usually a Vigreux-type distillation column 30–40 cm in length is used in conjunction with traps operating at −20° and −40°C. Also, a column packed with Raschig rings or wire gauze can be used in conjunction with a reflux condenser to achieve a similar effect.

High vacuum or molecular distillation conducted at 10^{-2} to 10^{-5} torr (1.33 to 1.33×10^{-3} Pa) is also useful. Normally the liquid to be distilled passes over as a thin film on a spinning band that is heated from the inside by the circulation of an appropriate liquid.

Cold-finger distillation is also carried out in a vacuum of approximately 10^{-4} torr (1.33×10^{-2} Pa) by means of a diffusion pump. The resulting volatiles are condensed in traps cooled with liquid nitrogen.

A gentle but usually not quantitative method involves passing an inert gas, such as nitrogen, through the food system and condensing the resulting vapour in a cold trap or by dissolving it in a suitable solvent. Its major advantages are that it is effective for very volatile compounds and the food state does not have to be altered. For example, a powdered food would not have to be reconstituted and could be evaluated in its natural state.

3.2.2 Methods based on solubility

If differences in solubility are large enough between the food mass and a specific individual or combination of solvents, aroma compounds can be effectively separated. Historically the solvent had to be in a liquid state, but recent developments have been directed to compound separation using supercritical carbon dioxide or liquefied gases, such as carbon dioxide, butane and freons. These systems are extremely efficient, quantitative and result in little if any artefact formation since cold temperatures are used. Their major limitation is the fact that highly sophisticated equipment capable of withstanding high pressures is required.

Actually the simplest method of extracting aroma compounds based on solubility is the liquid extraction of compounds that are either water soluble or insoluble in the presence of an appropriate solvent. By simply shaking the food in question in a water/solvent system one can separate a vast number of important aroma compounds. Solvents commonly used include ether, hexane, pentane and isooctane. The procedure can be done in a separating funnel at room temperature in a batch or continuous process. Also, temperature can be elevated and the extraction carried out in a Soxhlet extractor.

If common solvents are used with fat-containing foods, one extracts not only aroma compounds but also unwanted lipids. With these systems the food can be dry-mixed with Celite 545, packed into a liquid chromatography column and eluted with acetonitrile to minimize lipid interference.

As mentioned earlier, when distillation systems result in aroma compounds and water, these are usually subjected to a subsequent solvent extraction and/or concentration step. Historically the rotary evaporator has been used to selectively remove solvent in such a situation. Heating and a partial reduction in pressure are usually employed, but, as with most systems, heating can result in artefacts.

3.2.3 Adsorption methods

The adsorption of very dilute concentrations of aroma compounds present in the airspace above a food sample has become a popular separation technique (Nunez *et al.*, 1984). This technique, which is sometimes termed dynamic headspace analysis or the purge and trap method, represents an improvement over direct analysis of the vapours in equilibrium above a sample of food, which is commonly called (static) headspace analysis. Static headspace analysis is insensitive since it involves injecting a dilute sample, and the headspace also usually contains water vapour that can cause havoc with GC stationary phases.

However, if the volatiles in a headspace sample are swept onto an activated carbon column or a column composed of other adsorbents, one can effectively concentrate volatiles and can even minimize the negative aspects of water vapour. For example, an activated charcoal trap strongly adsorbs organic compounds but has little affinity for water and thus it is not deactivated by the presence of moisture.

Major advantages of this general technique are that the apparatus design is simple, the adsorbed volatiles can easily be transported in a stable form which makes on-site sampling quite easy, and the volatiles can easily be desorbed from the trap, either by applying heat for direct thermal desorption onto the chromatographic column or by dissolving them in a small volume of solvent for injection of a liquid sample.

Recently, the use of various porous polymers has been advanced. Their major advantages are that they have the ability to retain most organic compounds even at ambient temperatures, while having little if any affinity for water. If the actual trapping is done on chromatographic-type columns, they can serve as a pre-column or the actual GC column that can be installed into the GC unit.

Several types of these polymers are commercially available with the most popular being Porapak Q, which is a non-polar copolymer of ethylvinylbenzene and divinylbenzene. Also popular are the Chromosorb Century polymers, which have low susceptibility to both oxidative and thermal decomposition. A Tenax series of polymers varying in surface areas and polarities is also commonly used. The retention characteristics of a wide range of volatile compounds on Tenax TA have recently been reported (Maier & Fieber, 1988).

In practice, up to 500 mg of polymer is packed into a small column and conditioned by heating at temperatures up to 225°C, depending on the specific polymer, and purged with purified oxygen-free nitrogen. The column is then placed as a trap above a food system which is then purged with purified air or nitrogen at a flow rate of up to 300 ml/min. Volatiles can be collected from headspace volumes of up to 250 l. In order to minimize contamination of the volatiles by trace impurities in the purging gas, Grob (1973) introduced a closed-loop stripping apparatus in which purging of the sample is performed by continuous recirculation of a limited volume of gas. The apparatus (Fig. 3.1) consists of a sample container placed in a thermostatted water bath, an all glass and metal circuit which is gas tight, and a stainless-steel bellows pump. The gas stream leaving the sample is always heated about 10°C above the bath temperature to prevent water condensation and to optimize the adsorption of organic components on the charcoal microfilter.

After trapping the volatiles, the trap is then disconnected and purged with purified nitrogen to remove most of the water vapour. The resulting column can then be extracted with an appropriate solvent or can be placed in line in a GC unit and through the use of an auxiliary heating block and switching valve, it can be made to release the volatiles onto the main GC column for separation.

Isolation of aroma volatiles by headspace concentration using an adsorbent is a useful alternative to vacuum distillation and solvent extraction, or the use of a Likens–Nickerson apparatus. Either of the last two techniques tends to give better recovery of the less volatile components in a food aroma, but headspace concentration usually gives good recovery of the more volatile components. The chromatograms corresponding to volatiles isolated from a food by two different techniques often differ markedly. An example of this is shown in Fig. 3.2, which shows volatile components in the headspace from the Australian mango (Bartley & Schwede, 1987). Loss of esters in the chromatogram following isolation by the Likens–Nickerson technique was ascribed to evaporation and hydrolysis.

A similar difference is evident from the work of Takeoka *et al.* (1988). This study reported that a total of 62 compounds including 21 alcohols, 15 esters, 10 carbonyls,

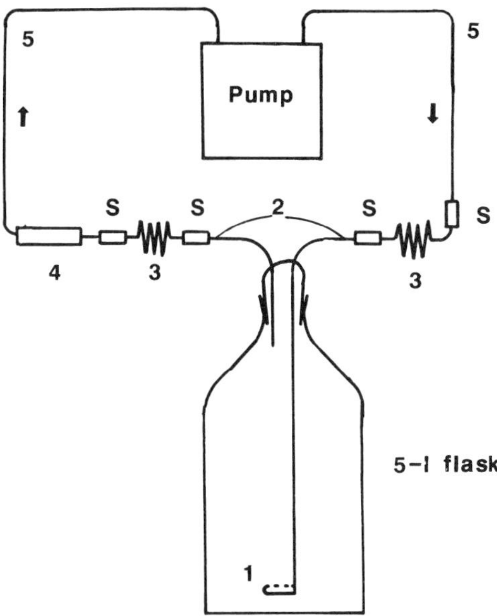

Fig. 3.1 — Closed-loop stripping apparatus. 1, coarse glass frit; 2, fused glass–metal connections; 3, coiled steel tubing; 4, filter holder; 5, stainless-steel tubing (3.2/2.0 mm); S, Swagelok fittings. Redrawn with permission from Grob (1973).

two acids, one hydrocarbon and 13 miscellaneous compounds were isolated from nectarines by vacuum distillation followed by solvent extraction. However, a total of 99 components including 43 esters, 33 hydrocarbons, 10 alcohols, six carbonyls and seven miscellaneous components were isolated by headspace concentration using an adsorbent followed by thermal desorption onto the GC column.

Headspace concentration has found application in following the changes in volatile composition as influenced by the ripening process in fruits, for example, whereby a chamber can be constructed around a ripening fruit and the headspace volatiles analysed without physically disrupting the fruit. However one major limitation must be remembered and that is that the human nose characterizes headspace aroma in an equilibrium state, which is dependent upon individual compound concentrations and vapour pressures, and that, when large volumes of a gas are passed through a headspace, these equilibrium conditions are no longer present. Therefore, the composite volatiles trapped in one of these columns may not necessarily smell like the natural food aroma. Because of this it is quite appropriate to compare subjectively the composite aroma obtained from a trap with that of the food aroma being evaluated. If the two aromas are significantly different, the exclusive use of this technique should be questioned.

Volatile compounds possessing different functional groups can be separated using water-deactivated silica gel and a combination of solvents. When Freon 11 is used alone, the early fractions contain non-polar hydrocarbons, and these are

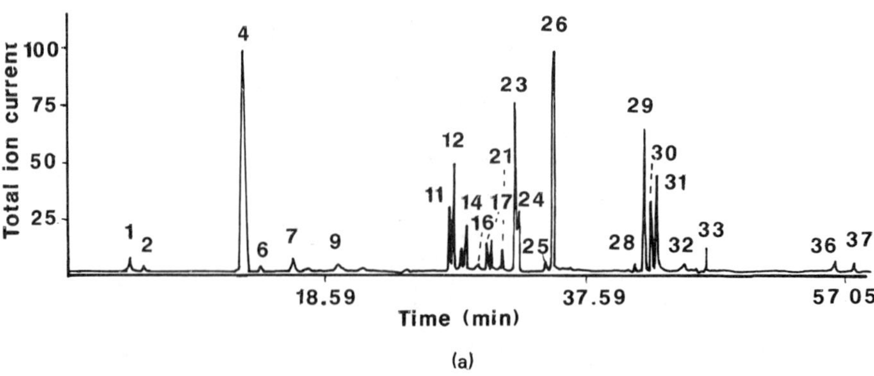

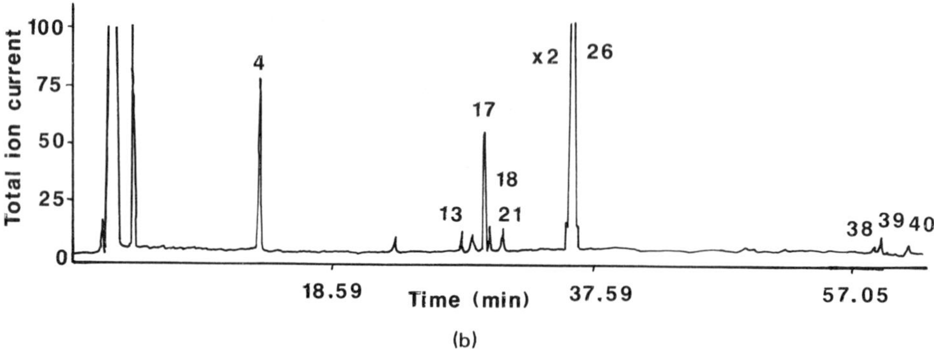

Fig. 3.2 — Gas chromatograms of 'Bowen' mango constituents. (a) Fresh, ripe fruit volatiles isolated by headspace concentration; (b) Fresh, ripe fruit volatiles isolated by simultaneous distillation–extraction (Likens–Nickerson). Redrawn with permission from Bartley and Schwede (1987).

followed by a fraction containing moderately polar compounds, such as esters and carbonyls. When 20% ether and Freon 11 is used, the highly polar alcoholic fraction can be effectively separated.

3.2.4 Chemical procedures

In many cases the utilization of an entire volatile sample for GC analysis may not be practical because a poor separation will result due to the complex mixture present, or the presence of minor but important aroma compounds will be masked by the presence of major compounds. In these situations it may be desirable to isolate or fractionate compounds or classes of importance by various chemical means. However, major limitations include the facts that usually relatively large amounts of volatiles are required, the method is time-consuming, and resulting quantitative data can be questioned. Many traditional wet chemistry separations exist but usually they are not sensitive or quantitative enough for the quantities normally utilized in conjunction with GC.

In theory, because of the wide range of functional groups associated with the volatiles being evaluated, compound class separations can often be achieved through the use of appropriate reagents and/or pH adjustments. For example, the total aqueous volatile phase from a food can be rather simply fractionated into acidic, basic, and neutral fractions by just adjusting the pH of the solution. In this case, compounds like pyrazines would be present in the basic fraction and their GC analysis can be performed without interference from acidic and neutral volatiles.

A typical successive separation could include isolation of the acids present by treating the aroma extract with 6% aqueous sodium carbonate. The mixture is then acidified with 2 M HCl and the resulting organic acids removed from the aqueous system with ether. The now acid-free solution is then treated with 6% NaOH, and after acidification of the alkaline solution with 2 M HCl, the phenols and the lactones of hydroxy acids are extracted with ether. Organic bases, such as pyrazines, pyridines, pyrroles and thiazoles, can be removed from the original extract by first treating with aqueous HCl, followed by 10% aqueous KOH, and extraction with ether.

A volatile solution containing carbonyl compounds can be reacted with 2,4-dinitrophenylhydrazine to form stable hydrazone derivatives that can then be flashed onto a GC column for analysis.

The concept of on-line chemical abstractors should also be considered for certain aroma research situations. A chemical abstractor is an organic or inorganic compound that has the ability to chemically react with a specific type of functional group thus forming non-volatile derivatives. The compound is placed in line either before or after the GC analytical column on a suitable solid support material.

Numerous compounds have been evaluated and it appears that boric acid and 3-nitrophthalic anhydride are effective in complexing alcohols, and zinc oxide binds most carboxylic acids. The compounds benzidine, sodium bisulphite and semicarbazide complex aldehydes and ketones, and phosphoric acid is effective in eliminating nitrogen bases.

The concept has only received limited attention and usually only with rather simple volatile mixtures. Problems encountered include the fact that the abstraction of compounds may not be complete due to temperature of operation, and that other functional groups may also be partially affected. In any event, if used properly, it can be used to provide information on the presence of certain compound classes.

Aside from providing information on the presence of certain compound classes the utilization of specific abstractors can provide interesting information on the sensory contribution of removed or abstracted compound classes if a series of abstracting compounds is used. From these data conclusions can be drawn as to which compound class contributes the most to a specific food aroma.

One can also use on-line GC catalytic hydrogenation as an important chemical tool. In this system hydrogen is used as the carrier gas and unsaturated compounds are converted to their saturated form via hydrogenation of double bonds. By running a sample without on-line hydrogenation and another with, it can rather easily be determined which peaks represent unsaturated aroma compounds.

3.2.5 Miscellaneous concentration techniques

Some of the separation procedures already discussed inherently include a concentration step, but many occasions arise when one wishes to concentrate a sample containing volatiles to improve GC sensitivity. First a word of caution about sample concentration. One can attempt to concentrate a sample too much and some of the more volatile compounds may be lost or compounds of limited solubility will not be injected into the GC. Obviously both situations result in erroneous quantification.

When large quantities of water are associated with volatiles of interest, some researchers prefer to use the freeze concentration technique instead of extracting the volatiles from the aqueous system with solvents. Concentration of volatiles using solvents can present inherent problems such as loss of volatiles and accumulation and concentration of solvent impurities.

In this technique, batches of solution are placed in beakers that are continuously stirred at below freezing. However, the rate of temperature decline has to be carefully controlled since, if freezing is too rapid, some organic compounds may become trapped in the ice phase. The rate of stirring can also influence the efficiency of the process. The ice phase is physically separated and thus the volatiles are concentrated in the unfrozen phase. If the beakers are covered, evaporation of volatiles is not a problem. Also, heat-labile compounds are not affected and in general artefact formation is minimal and limited to oxygen-labile compounds owing to the stirring.

A similar method is zone melting whereby a frozen sample is gradually thawed, with the organic compounds being present in the liquid rather than frozen phase owing to their distribution coefficients. The technique also minimizes loss of volatiles and artefact formation but requires highly purified solvents and complex equipment, and can only be performed with small amounts of sample.

Lyophilization is a useful tool for concentrating volatiles from solid or semisolid foods that cannot be effectively treated with traditional distillation techniques. However, the method is far from quantitative since many of the aroma compounds are retained in part in the food. Another limitation is that water is also removed along with the volatiles and thus the volatiles have to be separated from the water in a subsequent step.

3.3 GC IN AROMA ANALYSIS

GC was first introduced in 1952 and since that time major improvements have been made in the important components of successful GC, namely injection systems, column types and supports, and detection systems, but the basic principles essentially remain the same today as 35 years ago. Current state-of-the-art GC units can detect compounds in the 10^{-13} g range thus making them a valuable tool for the flavour chemist (Dal Nogare & Juvet, 1962; Jennings, 1980b).

In order to avoid band-spreading, it is of the utmost importance that a sample be quickly and completely transferred to the GC unit. In most situations the aroma compounds are injected in a solvent carrier in microlitre or smaller quantities and thus the sample-to-solvent ratio becomes quite important since large amounts of solvent, relative to compound concentration, may not permit the effective separation of the most volatile compounds present in the mixture. They may be eluted in the broad solvent band. This effect is most severe with narrow-bore WCOT capillary columns, but may be reduced by the use of wide-bore or SCOT capillary columns.

Also, if compound concentration is insufficient, certain peaks may not be detectable, and thus key aroma compounds may be missed.

Traditionally, a liquid sample was injected through a septum into a heated chamber, which causes the volatiles to vaporize onto the column. However, several precautions need to be taken. Ideally the vaporization chamber should be as small as feasible so that dead volume, which results in band broadening, can be kept to a minimum. Most heated injector blocks can be modified to accommodate a glass liner insert, thereby decreasing dead volume and improving separation.

Pieces of septum and a dirty injector in general can be a major source of band-broadening and the appearance of artefactual peaks. If pieces of septum are inside the injection port, they can absorb some of the volatiles injected and then these compounds are gradually desorbed, causing major band-broadening.

If acidic residues are present in the injection port, certain aroma compounds can undergo rearrangement, and conversion of *cis* and *trans* isomers can occur. If a basic residue is present, compound epimerization is possible. Undoubtedly the major concern is the possibility for thermal decomposition due to high injector-port temperature. The concern often arises as to whether all of the compounds observed in a chromatogram are naturally present in the original sample, or whether some were thermally generated via injection conditions.

The capillary column has become a favourite of the flavour chemist and thus has necessitated unique injection systems due to the relatively small amount of sample that can normally be injected. Initially sample vaporization onto the column in both split and splitless modes was common. The Grob splitless technique (Grob, 1973) provides a way of partially condensing solvent after injection, thus increasing the actual amount of true sample that can be placed on the column (Schreier & Idstein, 1985).

Also available is cold injection into an ambient-temperature heating block, which is then temperature programmed. A similar technique is direct cold on-column injection. Both of these techniques provide reproducible introduction of known sample volume and provide gentle conditions for thermally unstable compounds.

Mention was made earlier of utilizing polymer traps that can be placed in line, thus serving as the injection mechanism. Also, attempts have been made to directly apply aqueous aroma extracts using fused silica capillary columns coated with a non-polar liquid phase so as to minimize bleeding and degradation normally associated with most columns to which water is applied.

3.4 COLUMN CONSIDERATIONS

The commercial choice of coatings and/or packings for GC columns approaches several hundred, and thus the GC operator must be somewhat knowledgeable in this area if he is to obtain a column that is ideal for his specific separation requirements.

Traditionally the flavour chemist has found two types of packings useful for aroma analysis using packed columns. Carbowax 20M, which is a polyethylene glycol, is the most popular polar phase material and SE-30, which is a methyl silicone, is the standard non-polar liquid phase of choice. Either of these two phases is applied at levels of 10% on the stationary phase of choice, which is usually 80–100-mesh Chromosorb W-HA. In the case of Carbowax 20M lighter loads of around 2%, coated on a finer mesh support (100–120 mesh), usually reduce overall compound elution time and give better resolution of late-eluting compounds. Using loads of less than 10% SE-30 usually results in increased compound-tailing and thus is not normally recommended. Today most flavour chemists working with packed columns prefer glass columns since they exhibit greater efficiency and produce less tailing and/or decomposition than stainless steel columns.

Packed columns are also used by the flavour chemist as preparative columns which are usually in the neighbourhood of 0.5 inches (13 mm) in inside diameter and are more heavily loaded with liquid phase, around 20%, than normal packed columns. As a result, they can handle total sample sizes of several millilitres representing up to 100 mg of material per peak. If a column splitter is used in conjunction with this type of column, it represents an effective means of isolating individual compounds for further flavour evaluation or collecting fractions that can then be rechromatographed on other columns for more detailed separation.

Packed columns are limited to a maximum of 20 000 theoretical plates while it is quite common for the WCOT capillary or Golay column to have 100 000–200 000 theoretical plates, and therefore most flavour analyses utilize capillary columns.

Various forms of capillary columns are available to the flavour chemist, including wide-bore, porous-layer, multichannel, micropacked and fused silica (Boyko *et al.*, 1978; Jennings, 1980c, 1985; Sandra *et al.*, 1985, Schreier & Idstein, 1985; Teranishi, 1981; Wyllie *et al.*, 1978).

3.5 COLUMN SELECTION

Selection of a column type and liquid phase can be a frustrating experience with one usually depending upon word of mouth or the scientific literature for clues. Also, with experience one can learn to predict which column would be most suitable for the situation at hand. However, when working with an aroma extract of unknown composition, one usually just evaluates separations derived from several columns before deciding which column finally to use to attempt to optimize a separation.

Another source of limited assistance is commercial GC column suppliers, but for the most part their catalogues depict separations of model composite volatiles and when biological volatile extracts are evaluated using the same column and conditions, the presence of interfering compounds or a significantly larger number of compounds makes the separation far from ideal.

The normal initial step in evaluating an aroma mixture of unknown composition is to observe its separation on a short (3–10 m) apolar (SE-30 or OV-101) high-temperature column. Two general pieces of information can be obtained by using this technique. First, the degree of component dispersion can be observed, and secondly, the presence of high-boiling components can be ascertained. By temperature programming from 50°C to 280°C, and holding for 30 min, the full range of compounds present can be seen. In most GC separations the presence of unexpected high-boiling compounds is always of special concern, since if they take an especially long time to elute, they can interfere with repeated short-time injections.

When a sample does not appear to have any high-boiling compounds present, optimization of separation can be achieved on the original apolar short column by altering carrier gas velocity or column temperature. If, on the other hand, high-boiling compounds are present, a longer column can be used but usually it is more practical to change to a column that has a more polar liquid phase. However, it should be noted that the polarity of a liquid phase is inversely related to column stability. A highly polar liquid phase usually has low temperature limits, is subject to high bleed rates, has low coating efficiency, and is susceptible to oxygen and water interactions. Because of all these potential problems, most flavour chemists would prefer to go to only a slightly polar liquid phase such as Carbowax 20M.

In theory, no single liquid phase will completely separate a complex volatile mixture and thus some compromise must be made. This has led some researchers to blend liquid phases to maximize compound separation. For example, one can determine what volume fraction of two, or possibly more, liquid phases will result in the highest relative retentions for all compounds in a mixture, how many individual peaks will result, and how elution order is influenced.

Some researchers have promoted thin-film liquid phase columns for the separation of high-boiling or thermally labile compounds in order to combine high separation efficiency with short analysis times. However, it must be remembered that resolution is directly proportional to the partition ratio and thus at any specific temperature thin-film columns require more theoretical plates to obtain the same degree of resolution as a thick-film column which has fewer plates.

The properties of various liquid phase substances that are used by the flavour chemist are compared in Table 3.1. It can be seen from this comparison that if all properties need to be considered for a certain separation, Carbowax 20M represents a good middle-of-the-road choice. In the case of SE-30, all properties except polarity are in the high category, but for certain separations, as discussed above, if high polarity is a major prerequisite, SE-30 is not appropriate.

Table 3.1 — Properties of liquid phases commonly used by the flavour chemist

	Low	*Medium*	*High*
Upper temperature limit	Polyester	Carbowax 20M	SE-30
Polarity	SE-30	Carbowax 20M	Polyester
Coating efficiency	Polyester	Carbowax 20M	SE-30
Resistance to water	Polyester	Carbowax 20M	SE-30
Resistance to oxygen	Polyester	Carbowax 20M	SE-30

Proper column selection will manifest itself in the type of separation obtained and much of the art of GC centres around observing a separation and knowing if it is ideal. If it is not, the next question is which factor or factors are not optimum and how they should be changed. For the moment, let us assume that in general the types of compounds present are known and that according to the literature the proper column has been chosen relative to packing and/or coating, and length/diameter. Other variables that can interact include sample injection size, carrier flow rate, temperatures of injection, column and detector, extraneous dirt, and electrical impulses.

For example, in Fig. 3.3 which represents a separation on a capillary column,

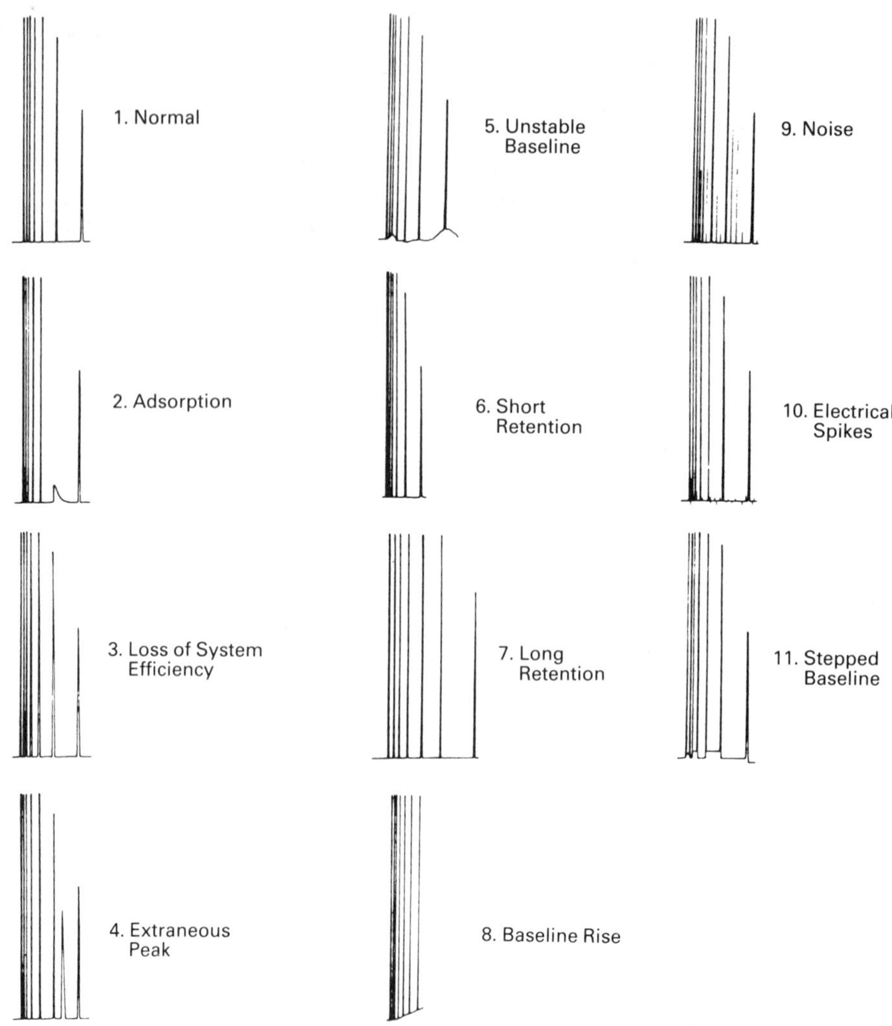

Fig. 3.3 — Typical problems with GC separations.

separation 1 is an example of an ideal separation, whereas in 2, there is an obvious tailing of one compound, probably an alcohol, which would indicate that some active sites on the column have been exposed. If latter peaks tend to have leading tails, it is usually an indication that too large a sample has been injected. This obviously can be corrected by decreasing sample size or, if possible, raising column temperature.

Separation 3 is a typical example where all of the peaks are too broad at their base, thus demonstrating overall loss of separation efficiency. Separation 4 shows the appearance of an extra peak which in all probability does not belong with the current separation due to its shape relative to the rest of the compounds. It represents a peak that had a relatively long retention time from a previous injection. A very common separation defect is typified in separation 5, which represents a normal separation, but there is drift and an unstable baseline. Most times this is caused by high-boiling compounds that have spent days on the column and are finally being eluted. This problem can usually be resolved by replacing in-line filters or by conditioning the column at a high temperature to drive off lingering compounds. Separation 6 shows a loss in partition ratios for all compounds, probably due to a loss of liquid phase. This column has a limited useful life. The other extreme is shown in separation 7, where all partition ratios have increased. In all probability the carrier-gas velocity has decreased, probably owing to a septum or column-connector leak. Separation 8 is typical of a temperature-programmed run, showing sharp baseline rise due to loss of liquid phase or deactivation compounds leaving the column. Separations 9 and 10 demonstrate spiking and noise problems. With 9 the detector is receiving extra signals, probably from a dirty detector or from the use of contaminated gases. Separation 10 is in all probability caused by an electronic noise problem. The last chromatogram in Fig. 3.3, number 11, shows a stepping problem. This usually represents a recorder problem which may be due to a sluggish or sticking pen drive or an electrical grounding problem within the recorder.

With care and attention a GC column can have a long and useful life, but when such a column is being used in quality control, such as headspace shelf-life analysis, special attention should be paid to gradual changes that can occur with time. In these situations columns should be routinely checked with standard reference compounds and their partition ratios and response relative to concentration verified.

3.6 MULTIDIMENSIONAL GC (MDGC)

MDGC has had some interesting aroma applications. Two GC columns are operated in tandem with the first column, which can either be a packed or capillary column, serving to preseparate compounds, all of which or only a portion of which can then be introduced into the second column, which is usually a capillary column. Thus one can select which parts of a chromatogram are introduced into the second column, thereby eliminating potential interfering compounds such as solvents. Obviously both columns can be operated at different temperature profiles and degrees of polarity, thus resulting in a rather sophisticated final separation. However, if two capillary columns are used, sample size is a limiting factor in that for some of the compounds not enough material will be available for detection after passing through the second column. Also, it may be necessary to put a trap just before the second capillary column to cryofocus the compounds going into the second column thus minimizing band-spread. Through the use of splitters on both columns, both sniffing and compound-trapping can be performed. Thus the major advantages of MDGC are that it can effectively separate extremely complex mixtures, and it can be used to enrich trace amounts of volatiles that normally are associated with the solvent portion of a sample (Jennings, 1980d; Schreier & Idstein, 1985).

3.7 GC ANALYSIS OF THERMALLY LABILE COMPOUNDS

As mentioned at the beginning of this chapter, a compound must have some degree of volatility to be evaluated by GC. For aroma compounds that have limited volatility the easiest solution is to perform the analysis at high inlet, column, and detector temperatures. However, since significant decomposition can result with a high inlet temperature, it is preferable to use splitless or on-column injection techniques for these compounds.

Another technique that can influence the separation of larger molecular weight compounds possessing limited volatility is change of carrier-gas velocity and type. Within reason, the flow rate can be increased, but with nitrogen and argon efficiency decreases with increasing velocity. However, with hydrogen efficiency does not decrease with increased velocity to the same extent and thus it is the preferred carrier gas in these situations. Analysis of aroma compounds that are thermally labile may require them to be derivatized in order to increase volatility or stability.

It is a well-accepted fact that the presence of certain functional groups will result in greater compound volatility. The key is that the compound in question has to possess a readily reactive group, such as active hydrogen atoms, or amino, hydroxyl or carbonyl groups that can interact with the derivatizing reagent. Obviously not all aroma compounds have such reactive groups, so only certain classes of flavour compounds, such as phenolic acids, can be evaluated using this technique. It can also be argued that if you preferentially react these groups via derivatization, there is less chance for the compounds to chemically react within the GC.

Historically the formation of methyl through propyl esters has been quite popular but these have little application for the flavour chemist. Instead, trialkylsilyl ether derivatives are preferred. In most reactions, an active hydrogen atom is replaced with a trimethylsilyl (TMS) group. In compounds with more than one active hydrogen, care must be exhibited in permitting enough time for all active sites of the compound to derivatize. If not enough time is permitted before injection a single compound may appear as several different peaks dependent upon the number of TMS groups that have become attached to the compound. The reaction is also dramatically reduced in the presence of water. In addition, if water comes into contact with a derivatized compound, the reaction is reversible. Also, since there is usually an excess of silylating reagent present at time of injection, liquid phases with reactive hydrogens are not suggested.

3.8 COMPOUND IDENTIFICATION BASED ON RETENTION INDEX

Using the theoretical assumption that if conditions are ideal with respect to temperature, carrier-gas flow, etc., each volatile compound in a mixture will exit a column at a different time, Kovats devised a retention index system for characterizing compounds separated by GC (see section 1.5).

Numerous data were originally generated for packed columns using this technique, but note that various isothermal conditions covering a wide range of temperatures is required to generate enough retention index ranges to cover the wide spectrum of compounds normally found in an aroma sample. In general these indices are most helpful in providing tentative identification of mono-functional compounds of low molecular weight (Teranishi *et al.*, 1971).

The Kovats system was later generalized for application to temperature programming, which dramatically shortens the amount of time and number of GC runs needed to obtain retention indices. Also, instead of aliphatic hydrocarbons for standard reference compounds, some of which because of their chain length take a long time to be eluted, the ethyl esters of normal aliphatic fatty acids can be used to generate reference indices. The number of published Kovats indices for individual flavour and fragrance compounds now exceeds 1000.

Recently data of this type have been generated for capillary columns. The technique involves the use of both polar (Carbowax 20M) and non-polar columns (OV-1 or SE-54) so that the data can be cross-correlated. Currently data on over 2000 compounds using this approach are available and the indices are highly reproducible.

The key to the successful application of the Kovats system is that all conditions of separation must be identical and usually, if both separations are not performed back-to-back, it is difficult to ensure that the conditions are indeed identical. Temperature is critical, especially for larger molecular weight compounds, and usually with older GC units temperatures are very difficult to duplicate exactly.

Instead of using hydrocarbons or fatty acids, two classes of compounds that are not that important from a sensory standpoint, other researchers have suggested using retention indices based on methyl ketones, but essentially the same logic as the Kovats method is employed. Since several systems are available; care should be taken in reporting and using retention indices to note conditions and which reference compound class is being used.

There are those who postulate that there is a possibility of two or more compounds having the same or very close indices, thus making the system somewhat suspect. Few if any flavour chemists would rely on retention index data as the sole means of identification and therefore auxiliary techniques will be discussed in a later section.

The possibility that two compounds may have the same retention index can be dealt with on most occasions by using both of two relatively simple techniques. The first involves spiking or adding to your sample the suspected compound and observing if an enlarged peak appears at the retention index under question. The second involves performing the separation using a different column.

As can be seen in Table 3.2, compounds having the same functional groups can be nicely classified according to their retention indices, and these indices are dependent on column polarity. Thus if identical conditions as those for which compound retention indices are available are utilized, one can usually predict with a good deal of certainty what the compound type is, and with further effort and an extensive retention index database, one can actually match the retention index of the unknown compound with that of a known compound.

3.9 DETECTOR CONSIDERATIONS

The last important component of a GC system is its detector. Some general considerations important for a detector that is to be used in conjunction with aroma analysis include sensitivity and limits of detection, selectivity, stability, versatility, and linear response. Another factor, which becomes especially important in the

Table 3.2 — Comparison of Kovats retention indices with respect to compound and column type

Compound Class	Column	
	Carbowax 20M	SE-30
Alcohols (MW 74)	1040–1150	620–720
Alcohols (MW 88)	1140–1280	720–820
Esters (MW 102)	910–1010	670–730
Esters (MW 116)	940–1110	730–820
Esters (MW 130)	1040–1210	830–920
Esters (MW 144)	1080–1290	880–1010
Esters (MW 158)	1180–1280	980–1040

analysis of trace amounts of unknown volatiles, is whether the detector is either destructive or non-destructive to the compounds being evaluated.

Ideally the flavour chemist wants a detector that is quite sensitive to all types of compounds and shows good linear response both in the p.p.b. range to detect trace compounds and to compounds that are the major components present in a mixture. Also, since the flavour chemist now knows that many types of compounds, some containing nitrogen and sulfur, contribute to flavour, the ideal detector should address this situation.

In view of the above requirements, it should come as no surprise that there is no all-purpose GC detector that can meet all the needs of the flavour chemist. The flame ionization detector (FID) is widely used because it is sensitive, has a linear response over a wide concentration range and is relatively stable if properly cleaned and maintained, with its limitations being that it is destructive and non-selective. However, through the use of a splitter its destructive nature ceases to be a significant factor, since a portion of the effluent can still be sniffed as each compound is being detected.

The flame photometric detector (FPD) is a common alternative to the FID because it is sensitive, but it has a rather narrow linear response range. However, the FPD can preferentially detect sulphur compounds, some of which make important organoleptic contributions to food aromas.

The use of GC alone is not a good tool in providing structural identification. However, the flavour chemist is indeed fortunate that the combination of GC–MS is available. More informative aroma data have been obtained by combined GC–MS analysis than by any other means (Durr, 1983: Flath, 1981; Merritt & Robertson, 1982).

By combining a gas chromatogram equipped with a capillary column and a modern mass spectrometer that is sensitive enough to evaluate effluent directly from the GC column and fast enough to process many closely separated peaks, one can find out a great deal regarding to compound structure. The mass spectrometer supplies a characteristic pattern of molecular fragments, which if analysed via a computer or by an experienced individual can provide information on actual chemical structure. However, care should be taken in the interpretation of MS data since often the information is not specific enough to make a positive identification. In some cases a combination of published MS data and Kovats indices can be used to identify compounds in a mixture. However, in experimental flavour chemistry one usually has the problem that one or both sets of data are not available.

Of course the purist can argue that these tools only provide relative proof of identification and that the true identity can only be verified by actual synthesis and analytical reconfirmation of the compound in question. However, in some cases synthetic routes are not known as yet and thus it is a fairly safe assumption that important flavour compounds below an effective concentration of 10 p.p.b. will continue to be identified by these techniques for a long period of time. The easy ones have been resolved and now the more elusive ones require investigation.

Since most mass spectrometers are capable of scanning at high speeds, one can scan a particular GC peak at its beginning, at the apex, and on the back side to obtain an estimate of the purity of the compound. If conflicting mass spectra are obtained at these different points, the observed peak probably represents a mixture of compounds that require further GC separation.

Tandem mass spectrometry (MS–MS) has also been proposed for aroma compound identification. The compound or chromatogram area of interest trapped from a traditional GC system is manually administered into one mass spectrometer where traditional molecular fragmentation occurs. These ions are then permitted to enter a second mass spectrometer where mass and amount are recorded. However, the system does not provide meaningful data if compound isomers are present.

Because of limitations among all detectors, the concept of multidetector analysis has been proposed. By connecting various types of detectors in series or parallel, some interesting results can be obtained. For example, an FID to determine total volatiles, an FPD for sulphur-containing compounds and a thermionic detector (TD) for nitrogen-containing components can be simultaneously run. In addition, with the aid of a splitter the effluent can be checked by the human nose, which for the flavour chemist is the ultimate detector. Therefore, in an area of a FID chromatogram there may not appear to be significant amounts of compounds, but when the same elution time is evaluated with the other detection systems in conjunction with the human nose, nitrogen- or sulphur-containing compounds may become apparent. Similar FID–ECD–FPD and FID–ECD–NPD (nitrogen, phosphorus detector) systems have been evaluated. Multidetector analysis is invaluable in both detecting and identifying trace volatiles of unique structure.

Earlier it was mentioned that the human nose can play an important role when used in conjunction with a GC detector/splitter system. Perhaps a few words should be written about this. In working with experimental aroma samples nothing can be more fascinating and rewarding than attempting to correlate peaks on a chromatogram with specific or characteristic descriptive terms obtained by the technique of sniffing column effluent. Undoubtedly, many complex mixtures have been separated and their individual components characterized in this manner. However, a major word of caution should be expressed. The long-term health implications associated with the continued inhalation of numerous relatively pure unknown compounds with various structures should be considered (Dal Nogare & Juvet, 1962; Durr, 1983; Gagliardi & Verga, 1982; Teranishi, 1981).

The practice of sniffing, although potentially harmful, seems to be especially useful in attempting to isolate compounds or areas on chromatograms where objectionable food aromas appear. In this type of situation the objectionable aroma can often be attributed to a single compound. However, when GC sniffing is directed towards attempting to find an area or areas on a chromatogram that possess the same characteristic aroma as the food being evaluated, less success usually results since the unique aroma property of most foods is not due to one compound, but usually to many compounds of diverse volatilities and diverse structures that are naturally blended in a specific ratio.

Assuming that all has gone well and an efficient separation is evident, the next question asked by most flavour chemists is what the identity is of each important sensory-related compound. This is where sniffing can again come into play. For instance, Drawert & Christoph (1984) described seven components with high odour intensities by sniffing the effluent from an apple aroma concentrate analysed by capillary GC (Fig. 3.4). Considering the fact that an efficient capillary column may

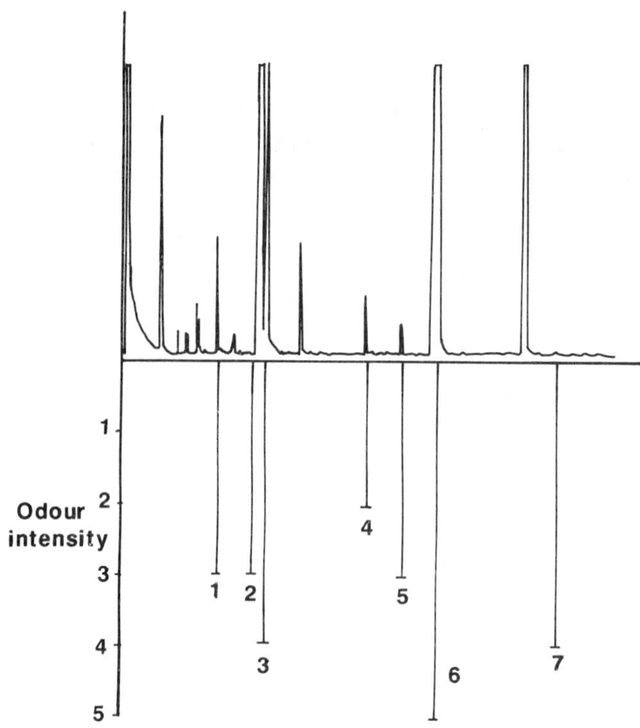

Fig. 3.4 — Part of a sniffing chromatogram of an apple aroma concentrate. Classification of odour intensities on a 0–5 scale. Redrawn with permission from Drawert and Christoph (1984). 1, isobutylacetate: fruity; 2, ?: black currant; 3, 1-butylacetate: fruity, pear; 4, isopentylacetate: fruity, apple; 5, *cis*-2-hexenal: like grass; 6, *trans*-2-hexenal: green; 7, ?: mushroom.

show a separation representing literally hundreds of compounds, one may not want to put much time and effort into identifying each and every compound detected. Thus an initial column/nose evaluation can be used to target compounds or chromatogram areas that are of sensory interest. These specific areas can then be subjected to further subjective and objective analyses.

3.10 ENANTIOMER CONSIDERATIONS

It is a well-accepted fact that the specific odour associated with an aroma substance is closely related to compound geometry, conformation, and optical isomerism. Thus if one considers the various forms of a complex aroma compound that can exist, it almost becomes demoralizing when one realizes that the isolation and GC analysis procedure can either form or decompose active forms that are primarily responsible for aroma properties.

Also, when synthesizing such compounds one wants to verify the aroma property exhibited by each possible form of the compound and then choose the form that is most potent or characteristic. For example, let us assume that one wants to synthesize the compound responsible for characteristic anise aroma. Research has shown that the compound responsible is the E-isomer of anethol, while the Z-isomer has a woody, minty odour that is nothing like anise. Thus in the synthesis, isolation and/or quality control of anethol, it is essential that the amount of each isomer be known. The accurate and complete separation of enantiomers is also an important consideration in being able to distinguish between natural and synthetic aroma compounds. If the normal ratio of isomers is known for the compound derived from a natural source, one can postulate as to the authenticity of an unknown source sample.

We should realize that a GC cannot be expected to do every type of separation, and an enantiomeric separation normally is difficult to perform using GC. However, GC techniques for the enantiomeric separation of important aroma compounds have been developed. GC separation of such compounds can be achieved in two ways. The first involves the formation of diastereomeric derivatives which is then followed by separation on a non-chiral phase. Depending on the compounds in question, various derivatives can be used. In the case of the compounds *Z*- and *E*-ethyl-3-methyl-3-phenylglycidate, for example, diastereomeric (−) methyl-glycidic esters can be used to derivatize. In the case of some secondary alcohols, esters of 2-, 3-, and 4-hydroxyacids and corresponding lactones can be helpful in separating structurally similar compounds. Another derivatizing compound of interest is R(+)-α-methoxy-α-trifluoromethylphenylacetic acid.

The other approach is to utilize an optically active stationary phase that is operating at high temperature. For example, the (2R,4R) (*trans*), (2S, 4S) (*trans*), (2S, 4R) (*cis*), and (2R,4S) (*cis*) isomeric forms of the aroma compound 2-methyl-4-propyl-1, 3-oxathiane can be effectively separated using Ni(II)-bis[3-heptafluorobutyl-(1*R*)-camphorate] as the column stationary phase. Of the two methods, the latter is preferred since, with the former, problems can be encountered of incomplete derivatization, and the formation of by-products or partial racemization can result. Normally when either of the above two procedures is used, it is not performed on the composite aroma sample. Usually the separation is first done on a standard capillary column with the chromatogram area of interest being trapped, then derivatized, and separated again on a specialized column (Schreier & Idstein, 1985).

3.11 CORRELATION OF SENSORY AND GC DATA

Since the human nose and mind control the final decision as to the acceptable or unacceptable nature of a specific food aroma, data obtained by GC analysis and the senses obviously have to correlate closely or the GC effort has gone in vain.

Earlier, the role of sniffing GC effluents was discussed, but considering the fact that many compounds present in a GC effluent are only present for not more than several seconds, a complete aroma profile using this technique is time-consuming and difficult to obtain. Another limiting factor is that most humans have a difficult time in quantifying eluting compounds. Also, the human nose, it has to be assumed, has the unique ability of only detecting important aroma-contributing compounds, whereas GC does not have this capability in that if the compound is present at a large enough concentration, it will be detected regardless of its overall contribution to human-perceived aroma.

Several approaches have been attempted to circumvent this dilemma. Historically the concepts of 'aroma values', 'odour units' and 'odour values' have been suggested whereby GC peak area is recalculated relative to the odour-detection threshold of each compound, thus giving a truer picture of the relative aroma impact of each compound. However the limitation to this approach has been primarily centred around the collection of threshold data, since data of this type are not available for a wide range of aroma compounds.

A recent approach purported to alleviate some of the above limitations is the 'charm' technique. Basically, effluent from a GC is sniffed and via a video terminal the human odour response is recorded relative to the time the response began and ended. In addition, the sensation is characterized by using descriptor terminology. Based on response duration measured in seconds, reaction time ($s \times 10^{-1}$), which is defined as the time it takes for the subject to choose a descriptor after initial odour detection, is calculated. Then a mixture of paraffin standards is chromatographed under identical conditions and their retention times used to convert human times to retention indices via linear interpolation. The sniffing procedure is repeated several times, usually at different concentrations, to allow the subjects to completely perform an evaluation and a resulting coincident response chromatogram is constructed. Therefore, areas in the GC chromatogram which contribute the most to an aroma property can be pinpointed. For example, with methyljasmonate isomers the methylepijasmonate isomer was found to only represent 5% of the total mass as compared to 95% for methyljasmonate as separated by GC, but according to the 'charm' technique the epijasmonate form contributed over 95% of the aroma (Acree *et al.*, 1984).

An alternative approach introduced by Grosch and coworkers involves diluting the sample successively with solvent prior to GC and determining the flavour dilution factor, which corresponds to the highest dilution at which a component is still detectable by sniffing at the end of the GC (Schmid & Grosch, 1986; Schieberle & Grosch, 1987; Ullrich & Grosch, 1987). The aromagram obtained in this way represents the plot of dilution factor against Kovats Retention Index (see Fig. 3.5).

The concept that the role of individual compounds in contributing to composite aroma is not as important as the overall aroma sensation itself means that GC and sensory data lend themselves to treatment by multidimensional scaling (MDS), whereby data based on total odour perception are treated on the differences between odours (Aishima, 1983; Christoph & Drawert, 1985; Derde & Massart, 1985; Martens & van der Burg, 1985; Powers, 1982; Schaefer *et al.*, 1983).

The use of a multivariate approach seems to be quite logical, since most food aromas are due to some degree of interaction among numerous compounds. If all food flavours were due to one specific but different compound that varies with food source, the flavour chemist would have been out of a job a long time ago. For example, through interaction, only 12 compounds can theoretically produce over 4000 aroma combinations. Obviously not all of these combinations actually naturally occur, and via data reduction based on factor analysis and GC peak area ratios, for example, the number of practical combinations can be significantly reduced. By dimensionally plotting resulting data, one can then begin to predict that a sample having certain peak combinations will have certain sensory properties. The advent of computer technology coupled with sophisticated plotting programs make this approach one of the most promising and exciting areas of flavour chemistry.

3.12 CONCLUSIONS

Depending upon whether qualitative or semi-quantitative results are desired, the actual extraction and concentration of volatiles from complex food systems with a minimum of artefact formation prior to actual GC analysis are major considerations.

Most detailed GC aroma analyses are performed on various types of capillary columns coupled to an FID detector. However, for certain classes of aroma compounds, other detectors (FPD, TD) are more appropriate. GC–MS systems are quite useful for compound identification.

Although GC is not as sensitive as the trained human nose, advances in column, electronics, detector, and data-processing technologies will soon result in GC systems that are comparable to human sensitivity, thus continuing to make GC the major analytical instrument utilized by the flavour chemist.

REFERENCES

Acree, T. E., Barnard, J. and Cunningham, D. G. (1984). A procedure for the sensory analysis of gas chromatographic effluents, *Food Chem.*, **14**, 273–286.

Aishima, T. (1983). Relationships between gas chromatographic profiles of soy sauce volatiles and organoleptic characteristics based on multivariate analysis. In *Instrumental Analysis of Foods: Recent Progress*, Vol. 1, Charalambous, G. and Inglett, G. (eds,), Academic Press, New York, pp. 37–56.

Bartley, J. P. and Schwede, A. (1987). Volatile flavor components in the headspace of the Australian or 'Bowen' mango. *J. Food Sci.*, **52**(2), 353–360

Boyko, A. L., Morgan, M. E. and Libbey, L. M. (1978). Porous polymer trapping for GC/MS analysis of vegetable flavors. In *Analysis of Foods and Beverages: Headspace Techniques*, Charalambous, G. (ed.), Academic Press, New York, pp. 57–79.

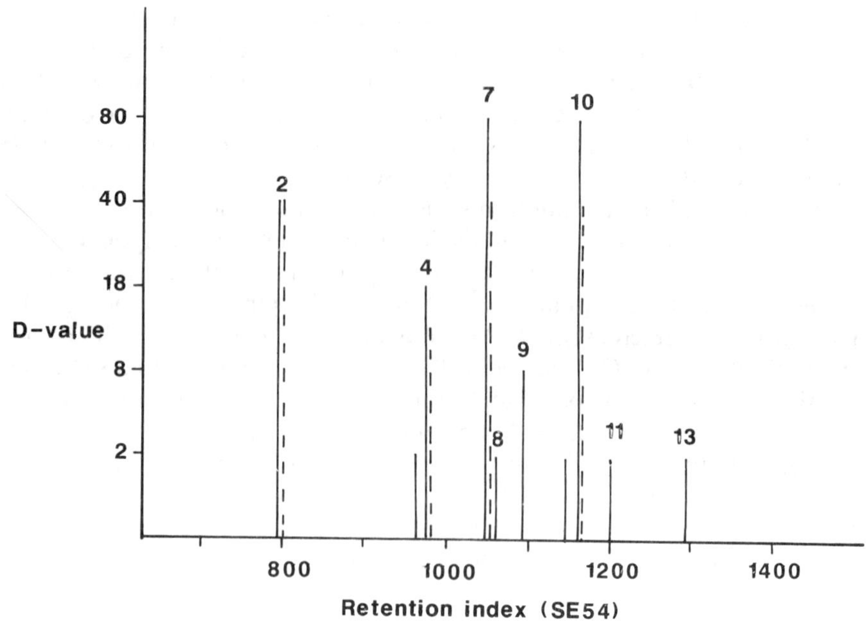

Fig. 3.5 — Aromagram of the flavour compounds formed after autoxidation of methyl linoleate at 22–24°C for 72 h. Redrawn with permission from Ullrich and Grosch (1987). 2, hexanal; 4, 1-octen-3-one; 7, 2(Z)-octenal; 8, 2(E)-octenal; 9, 3(Z)-nonenal; 10, 2(E)-nonenal; 11, 2,4-nonadienal; 13, 2,4-decadienal. —— D-value; --- D-value corrected on the basis of hexanal.

Christoph, N. and Drawert, F. (1985). Olfactory thresholds of odour stimuli determined by gas chromatographic sniffing technique: structure–activity relationships. In *Topics in Flavour Research,* Berger, R. G., Nitz, S. and Schreier, F. (eds), H. Eichhorn, Marzling-Hangehan, West Germany.

Cronin, D. A. (1982). Techniques of analysis of flavours: chemical methods including sample preparation. In *Developments of Food Science 3A-Food Flavours, Part A—Introduction,* Morton, I. D. and Macleod, A. J. (eds), Elsevier, Amsterdam, pp. 15–48.

Dal Nogare, S. and Juvet, R. S. (1962). *Gas–Liquid Chromatography: Theory and Practice,* Interscience, New York, pp 180–240.

Derde, M. P. and Massart, D. L. (1985). Introduction to multivariate statistical techniques. In *Progress in Flavour Research-1984,* Adda, J. (ed.), Elsevier, Amsterdam, pp. 113–130.

Drawert, F. and Christoph, N. (1984). Significance of the sniffing technique for the determination of odour thresholds and detection of aroma impacts of trace volatiles. In *Analysis of Volatiles* Schreier, P. (ed.) Walter de Gruyter, Berlin, pp 269–292.

Durr, P. (1983). Measuring of sensory quality: assessing aroma by gas chromatography. In *Sensory Quality in Foods and Beverages,* Williams, A. A. and Atkin, R. K. (eds.), Ellis Horwood, Chichester, England, pp. 188–195.

Flath, R. A. (1981). Identification in flavor research. In: *Flavor Research: Recent Advances,* Teranishi, R., Flath, R. A. and Sugisawa, H. (eds), Marcel Dekker, New York, pp. 83–123.

Gagliardi, P. and Verga, G. R. (1982). Automatic direct headspace GC analysis of flavors with capillary column and multidetector systems. In *Chemistry of Foods and Beverages: Recent Developments,* Charalambous, G. and Inglett, G. (eds), Academic Press, New York, pp. 49–72.

Grob, K. (1973). Organic substances in potable water and its precursor. Part 1. Methods for their determination by gas–liquid chromatography. *J. Chromatogr.,* **84**, 255–273.

Jennings, W. (1980a). *Gas Chromatography with Glass Capillary Columns,* Academic Press, New York, pp. 183–200.

Jennings, W. (1980b). *Gas Chromatography with Glass Capillary Columns,* Academic Press, New York, pp. 2–13.

Jennings, W. (1980c). *Gas Chromatography with Glass Capillary Columns,* Academic Press, New York, pp. 39–51, 161–171, 173–182.

Jennings, W. (1980d). Recent Advances in the Separation of Volatile Components. In *The Analysis and Control of Less Desirable Flavors in Foods and Beverages,* Charalambous, G. (ed.), Academic Press, New York, pp. 3–16.

Jennings, W. (1985). Developments in Analytical Gas Chromatography. In *Topics in Flavour Research,* Berger, R. G., Nitz, S. and Schreier, P. (eds), H. Eichhorn, Marzling-Hangenham, West Germany, pp. 3–25.

Maier, I. and Fieber, M. (1988). Retention characteristics of volatile compounds on Tenax TA. *J. High Resolut. Chromatogr. Chromatogr. Commun.,* **11**(8), 566–576.

Martens, M. and van der Burg, E. (1985). Relating sensory and instrumental data from vegetables using different multivariate techniques. In *Progress in Flavor Research-1984,* Adda, J. (ed.), Elsevier, Amsterdam, pp. 131–148.

Merritt, C. and Robertson, D. H. (1982). Techniques of analysis of flavours. Gas chromatography and mass spectrometry. In *Developments in Food Science 3A Food Flavours, Part A—Introduction,* Morton, I. D. and MacLeod, A. J. (eds.), Elsevier, Amsterdam, pp. 49–78.

Nunez, A. J., Gonzalez, L. F. and Janak, J. (1984). Pre-concentration of headspace volatiles for trace organic analysis by gas chromatography. *J. Chromatogr.,* **300**, 127–162.

Peyron, L. (1982). Recent techniques in the analysis of heterocyclic aroma compounds in foods. In *The Chemistry of Heterocyclic Flavouring and Aroma Compounds,* Vernin, G. (ed.), Ellis Horwood, Chichester, England, pp. 262–304.

Powers, J. J. (1982). Techniques of analysis of flavours: integration of sensory and instrumental methods. In *Developments in Food Science 3A-Food Flavours — Part A—Introduction,* Morton, I. D. and MacLeod, A. J., Elsevier, Amsterdam, pp. 121–168.

Sandra, P., Bicchi, C. and Belliardo, F. (1985). Possibilities and limitations of off-line HPLC–CGC and on-line selective sampling in flavor research. In *Topics in Flavour Research,* Berger, R. G., Nitz, S. and Schreier, P., H. Eichhorn, Marzling-Hangenham, West Germany, pp. 27–57.

Schaefer, J., Tas, A. C., Velisek, J., Maarse, H., ten Noever de Brauw, M. C. and Slump, P. (1983). Application of pattern recognition techniques in the differentiation of wines. In *Instrumental Analysis of Foods: Recent Progress,* Vol. 2, Charalambous, G. and Inglett, G. (eds), Academic Press, New York, pp. 335–355.

Schieberle, P. and Grosch, W. (1987). Evaluation of the flavor of wheat and rye bread crusts by aroma extract dilution analysis. *Z. Lebensm. Unters, Forsch.,* **178**, 479–483.

Schmid, W. and Grosch, W. (1986). Identifizierung fluchtiger Aromastoffe mit hohen Aromawerten in Sauerkirschen (Prunus cerases L). *Z. Lebensm. Unters. Forsch.,* **182**, 407–412.

Schreier, P. and Idstein, H. (1985). Advances in the instrumental analysis of food flavours, *Z. Lebensm. Unters. Forsch.,* **180**, 1–14.

Sugisawa, H. (1981). Sample preparation: isolation and concentration. In *Flavor Research: Recent Advances,* Teranishi, R., Flath, R. A. and Sugisawa, H. (eds), Marcel Dekker, New York, pp. 11–51.

Takeoka, G. R., Flath, R. A., Guntert, M., Jennings, W. (1988). Nectarine volatiles: vacuum steam distillation versus headspace sampling. *J. Agric. Food Chem.,* **36**, 553–560.

Teranishi, R. (1981). Separations. In *Flavor Research: Recent Advances,* Teranishi, R., Flath, R. A. and Sugisawa, H., Marcel Dekker, New York, pp. 53–82.

Teranishi, R., Issenberg, P., Hornstein, I. and Wick, E. L:. (1971). Gas chromatography separations. In *Flavor Research: Principles and Techniques,* Marcel Dekker, New York, pp. 77–106.

Ullrich, F. and Grosch, W. (1987). Identification of the most intense volatile flavour compounds formed during autoxidation of linoleic acid. *Z. Lebensm. Unters. Forsch.,* **184**, 277–282.

Wyllie, S. G., Alves, S., Filsoof, M. and Jennings, W. G. (1978). Headspace sampling: use and abuse. In *Analysis of Foods and Beverages: Headspace Techniques,* Charalambous, G., Academic Press, New York, pp. 1–15.

4

Gas chromatography of carbohydrates in food

Allan G. W. Bradbury

4.1 INTRODUCTION

The first report on the analysis of carbohydrates by GC (gas chromatography) appeared in the literature in 1958 but it was the application of the trimethylsilylation technique to carbohydrates by Sweeley and co-workers (1963) that initiated the rapid development of work in this field. GC became the leading technique for the analysis of carbohydrates in foods until the mid 1970s when HPLC (high-performance liquid chromatography) began to dominate. Although the convenience and rapidity of most HPLC-based procedures presently make this technique the preferred one for the determination of carbohydrates in foods, GC still offers advantages for a number of applications and should continue to serve as a valuable analytical tool for those food technologists in carbohydrates.

Carbohydrates in foodstuffs can be classified as simple (mono- and disaccharides), oligosaccharides and polysaccharides. Commonly encountered examples of the 'simple carbohydrate' fraction in foods are: sucrose, glucose and fructose (fruits and suger-sweetened products), galactose and lactose (dairy products), and sorbitol (humectant and non-cariogenic sweetener). The oligosaccharide fraction incorporates the maltodextrins (commercial starch hydrolysis products), 'honeydextrins' and the oligosaccharides of the raffinose family which occur in legumes. Polysaccharides are the principal components of dietary fibre and constitute the gums or thickeners which are used in a number of food applications.

Analysis of the various and often complex carbohydrates in foodstuffs requires a range of analytical techniques with respect to sample work-up, derivatization procedures, and chromatography equipment and conditions. The recently developed high-resolution capillary columns are particularly suitable for separating complex mixtures such as those found in polysaccharide hydrolysates, and examples of their application will be given.

The analysis of carbohydrates in food has been previously reviewed by Birch (1973) and more recently by Folkes (1985). The major emphasis of these reviews was on the analysis of low molecular weight carbohydrates, i.e. sugar alcohols and mono-, di- and oligosaccharides. This review will include recent applications in this area and also deal with the analysis of the higher molecular weight carbohydrates, the polysaccharides.

4.2 COMPARISON OF GC AND HPLC

The analyst faced with determining the carbohydrates in a foodstuff needs to consider a number of factors in deciding whether to use GC or HPLC. HPLC is particularly suitable for rapid analyses of large sample quantities of simple mixtures where ease of sample preparation (no derivatization required) and speed of analysis are advantageous. GC is generally favoured for complex samples or for those containing unknown or low concentrations of carbohydrates.

Some researchers have compared GC- and HPLC-based procedures for the analysis of carbohydrates and have outlined certain advantages of the former. Schaeffler and Morel du Boil (1984) stressed the higher column efficiencies and detector sensitivities of GC, which allow better resolution of complex mixtures and detection of low component concentrations. A GC using a flame ionization detector (FID) is typically a factor of 10 times more sensitive than HPLC using refractive index (RI) detection. Date *et al.* (1982), in their analyses of a series of foods by GC and HPLC, claimed that the former was more effective for samples with low carbohydrate content and for those containing sorbitol, as they were unable to separate sorbitol from glucose by HPLC. Iverson and Bueno (1981) showed that the analysis of simple sugar mixtures in processed foods was more rapid and more accurate for HPLC than for GC. Reyes *et al.* (1982) showed in the analysis of sugars in strawberries that, although analysis times were longer for GC than HPLC, GC offered greater sensitivity and better resolution. The analytical capability of GC can be extended by interfacing with a mass spectrometer (GC–MS). This technique allows confirmation of peak identities and can assist in the identification of unknown components.

The extensive clean-up required for some materials prior to HPLC analysis is not always necessary for GC analysis. For example, Brobst and Scobell (1981) described how samples that require removal of protein and salt before carbohydrate analysis, such as soy products, are often more readily analysed by GC than by HPLC. The HMDS/TMCS silylation procedure (see later) gives a precipitate of ammonium chloride as a byproduct, and Schaeffler and Morel du Boil (1984) have shown how this can also assist sample clean-up by co-precipitating suspended particulate materials. Derivatization, in particular silylation procedure, is specific to compounds with 'active hydrogens'. Other compounds are not derivatized and are subsequently not monitored. In contrast, HPLC, which does not require derivatization and usually, for carbohydrate analyses, utilizes refractive index detection, is less specific.

Zuercher and Hadorn (1976) also compared sugar determination by GC with enzymatic and reducing procedures and concluded that, for a series of syrup samples, the GC method was at least as accurate as the others and, in addition, provided the distribution of sugar types.

4.3 DERIVATIVES

Carbohydrates are non-volatile and in order to make them suitable for analysis by gas chromatography, it is necessary to form chemical derivatives that have enhanced volatility. This is usually done by applying derivatization procedures that convert the hydroxyl groups of the carbohydrate molecules to ether or ester groups. The most widely used derivatives are the trimethylsilyl (TMS) ethers first described for the analysis of carbohydrates by Sweeley *et al.* (1963). GC-based procedures for the separation of simple mixtures of the common sugars often use direct trimethylsilylation or silylation, as it is more often termed. Silylation is convenient and easy to perform.

For more complex mixtures, a modified derivatization procedure is usually more appropriate. This is because more carbohydrates exist in solution as a mixture of anomeric and acyclic forms, and direct ether formation or esterification leads to a number of stereochemical isomers, each of which gives a peak on the GC chromatogram. Although direct silylation offers a convenient means of monitoring the anomeric forms of monosaccharides in solution, for practical purposes, the potentially large number of peaks from a food sample (usually two to four for each reducing sugar) can cause problems of resolution. Fructose exists in solution in at least five isomeric forms. As their TMS derivatives, these forms elute close to the two glucose peaks and any other hexoses that may be present, causing difficulties with accurate quantification, particularly with packed columns.

In order to reduce the number of isomers produced, an extra derivatization step, involving modification of the reducing function at the anomeric carbon, is introduced. Conversion to oximes, methyloximes or reduction are the modifications that are usually preferred. Oxime and methyloxime derivatives exist in the two closely related *syn-* and *anti*-isomeric forms. As their TMS derivatives, these two forms have similar retention times and usually elute as a single peak on packed columns. Li *et al.* (1983) showed that, of the sugars present in commercial yoghurts (fructose, galactose, glucose, sucrose, lactose and maltose) only the oxime–TMS derivatives of galactose gave two peaks on an SP-2250 packed column. Non-reducing carbohydrates are not converted to oximes, so samples that contain non-reducing and reducing carbohydrates, such as honey (Mateo *et al.*, 1987), are actually analysed as a mixture of TMS and oxime–TMS derivatives.

Reduction of the reducing group gives the corresponding alcohol or alditol of the parent sugar. Silylation of the alditols is rarely used because of the poor resolution between the hexitols: mannitol, glucitol and galactitol. In addition, the reduction products of some monosaccharides such as glucose (glucitol or sorbitol) and xylose (xylitol) may be present in some samples and these would not be distinguished from the free monosaccharides if this procedure were used. Also, ketose sugars form two alcohols on reduction; for example, reduction of fructose yields a mixture of mannitol and glucitol.

Alditols are more frequently analysed as their acetates. These derivatives are widely used for the monosaccharides produced by acid hydrolysis in the determination of natural polysaccharides. The alditol acetate derivative gives a single peak for each monosaccharide and those derivatives prepared from the common natural polysaccharides are usually readily separated on columns of medium to high polarity.

Acetylation of the oxime and methyloxime derivatives of monosaccharides leads to the aldononitrile and methyloxime acetate derivatives, both of which are gaining in popularity for the determination of polysaccharides. Excellent separation is readily obtained on packed and capillary columns with the former (Fig. 4.1) and sample

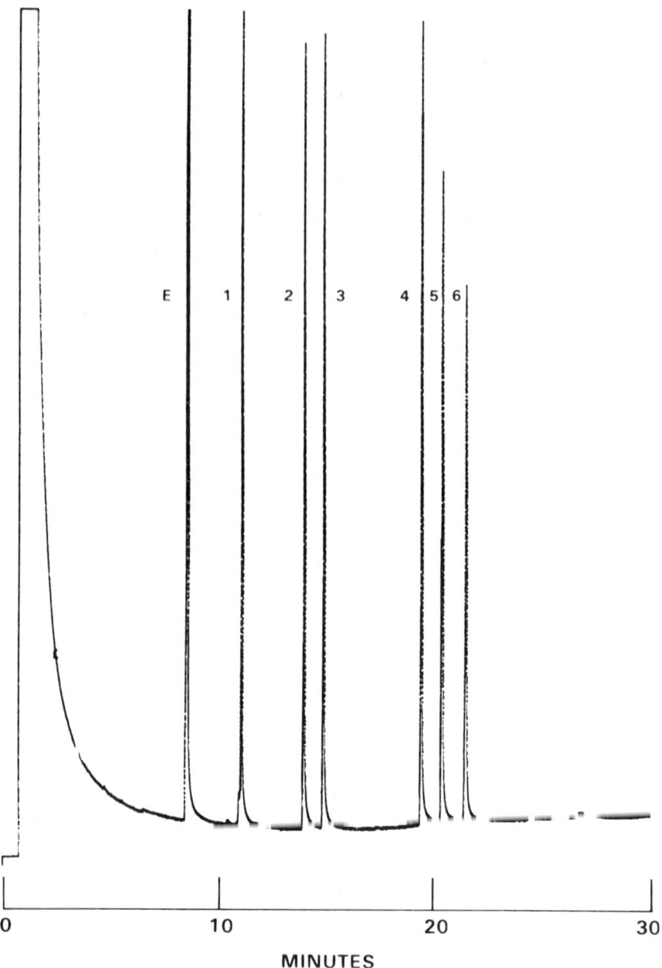

Fig. 4.1 — Analysis of monosaccharides as aldononitrile acetate derivatives. Column: 30-m DB-Wax carbowax bonded fused-silica capillary column. Temperature-programmed: 180°C to 210°C at 2°C/min, hold at 210°C. Peaks: E, erythritol; 1, rhamnose; 2, arabinose; 3, xylose; 4, mannose; 5, galactose; 6, glucose.

preparation is easy. A drawback of this procedure, however, is that dehydration of the oxime to the aldononitrile is not quantitative and cyclic byproducts, at a yield of about 30% for monosaccharides, are formed (Furneaux, 1983). The methyloxime acetate derivatives are readily prepared, but the derivative from each monosaccharide exists, as mentioned above, in the *syn*- and *anti*- forms. The peak pairs for the common hydrolysate products, mannose, galactose and glucose, all elute closely together, but, as shown in Fig. 4.2, efficient separation of all the isomers of the derivatives of these monosaccharides is possible on a capillary column.

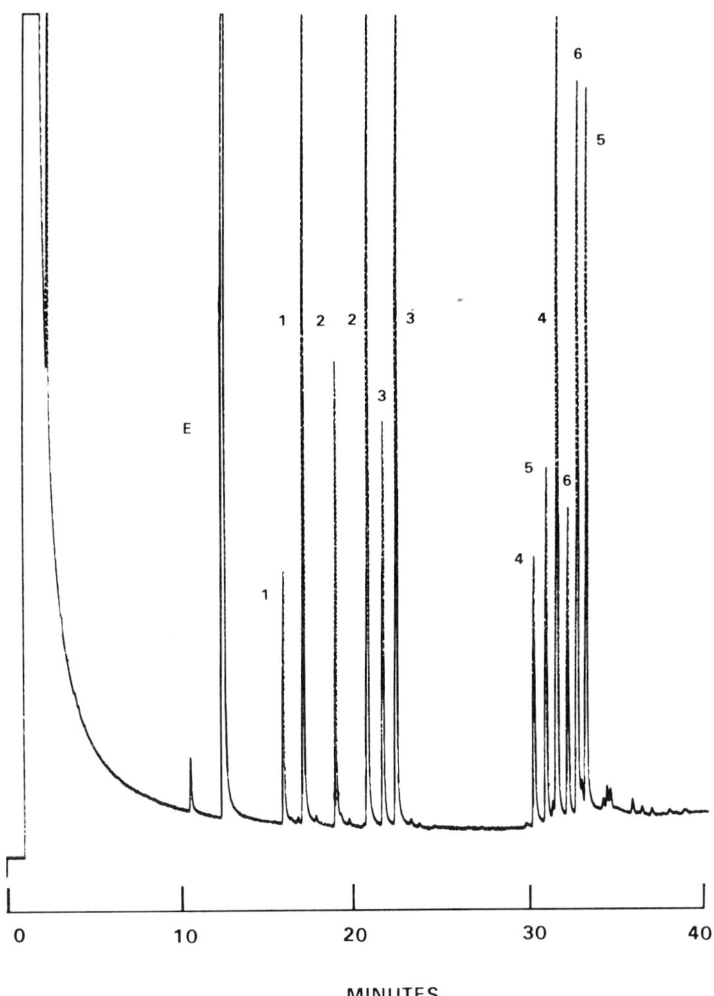

Fig. 4.2 — Analysis of monosaccharides as methyloxime acetate derivatives. Column: 30-m DB-Wax carbowax bonded fused-silica capillary column. Temperature-programmed: hold at 190°C for 3 min, 3°C/min to 220°C, hold at 220°C. Peaks: E, erythritol; 1, rhamnose; 2, arabinose; 3, xylose; 4, mannose; 5, galactose; 6, glucose.

For substituted sugars, such as the uronic acids produced by the hydrolysis of pectin, GC analysis is more difficult, mainly because of the problems associated with quantitative derivatization as well as those of elution due to interactions with the GC columns. Several workers have used GC-based procedures to analyse polysaccharides containing acidic groups and most of these utilize methanolysis as the polysaccharide degradative procedure. Methanolysis converts the component monosaccharide residues to their corresponding methyl glycosides, which are then silylated. Chromatograms are complex owing to the presence of the various anomeric forms, but the procedure has the advantage that the uronic acids are converted to the methyl glycosides. Excellent separation can be obtained on low-polarity, high-resolution capillary columns (Fig. 4.3).

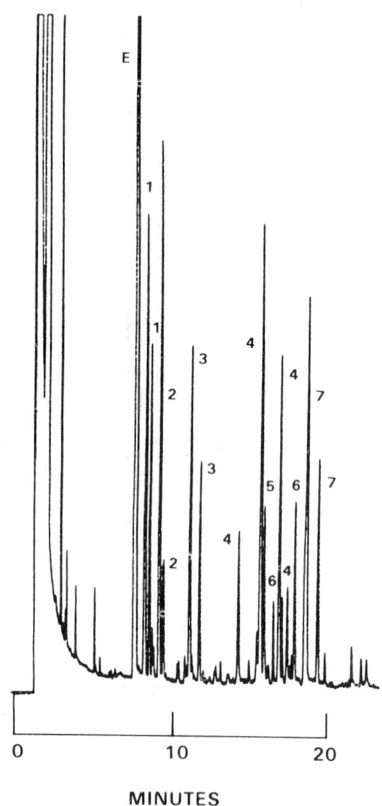

Fig. 4.3 — Analysis of monosaccharides as methyl glycoside–TMS derivatives. Column: 30-m DB 1 bonded phase fused-silica capillary column. Temperature-programmed: 135°C to 250°C at 3°C/min, hold at 250°C. Peaks: E, erythritol; 1, arabinose; 2, rhamnose; 3, xylose; 4, galacturonic acid; 5, mannose; 6, galactose; 7, glucose.

Trifluoroacetate derivatives of carbohydrates are among the most volatile known and are particularly effective for the determination of oligosaccharides. Although some workers have reported difficulties with quantitative conversion (Molnar-Perl and Szakacs-Pinter, 1981), they have been shown to be more stable than TMS derivatives (Englmaier, 1985).

There are a number of other derivatives which have found application for the GC analysis of carbohydrates. These include methyl ethers and acetate esters. The former are tedious to prepare, while the acetates offer no real advantages over the silyl derivatives and are less conveniently prepared. Dimethylsilyl derivatives have retention times up to 50% shorter than the corresponding TMS derivatives, but their extreme sensitivity to hydrolysis hinders their application to the analysis of food carbohydrates (Mueller and Goeke, 1972).

4.4 DERIVATIZATION PROCEDURES

Trimethylsilylation or silylation, either performed directly or subsequent to an additional reaction step, is the derivatization method most used for the analysis of carbohydrates in foodstuffs. A book issued by the Pierce Chemical Co. (1979) describes theoretical and practical aspects of silylation in detail. The Pierce catalogue (1989) lists silylation reagents and gives procedures for their application to carbohydrates. Many researchers have studied the silylation of carbohydrates and have discussed the advantages and disadvantages of the procedures and reagents used.

The original silylation procedure for carbohydrates of Sweeley *et al.* (1963) is still one of the most widely used. Dry sample containing up to 5 mg carbohydrate is dissolved in 1 ml pyridine; 0.2 ml hexamethyldisilazane (HMDS) and 0.1 ml trimethylchlorosilane (TMCS) are then added and the sample heated at 80°C for 15 minutes. The fine precipitate of ammonium chloride formed does not interfere with quantification and the sample can be injected directly into the GC. The particulate material can, however, cause blockages in syringes and it is recommended that the precipitate is separated by means of a bench-top centrifuge. Schaeffler and Morel du Boil (1984) have also referred to the difficulty of storing TMCS under anhydrous conditions.

A silylation reagent which is second in popularity to the HMDS/TMCS combination for carbohydrates is trimethylsilylimidazole (TMSIM), which is marketed by Pierce Chemical Co. as a 21% solution in pyridine (Tri-Sil 'Z', Pierce, 1989). It is normally used as is, although some workers have added other reagents. For example, Tsuda and Nakanishi (1983) obtained rapid silylation of carbohydrates by combining TMSIM with TMCS. Advantages of TMSIM include: ease of use (direct addition), no precipitation and rapid rate of quantitative ether formation. It can also be more readily used with an auto sampler, whereas the HMDS/TMCS reagent can cause corrosion at metal surfaces. Other workers, for example Schaeffler and Morel du Boil (1984), have criticized the use of TMSIM because of problems caused by imidazole, a byproduct of silylation: its elution can lead to interference, it can give peak tailing and it solidifies at room temperature, which can cause blockages. They favoured derivatization with HMDS and trifluoroacetic acid (TFA) by means of the Brobst procedure (Brobst and Lott, 1966). This method calls for dissolution of sample (size as above) in pyridine (1.0 ml) and then addition of HMDS (0.9 ml) and TFA (0.1 ml). After vigorous shaking and standing at room temperature for 15 min, the sample is ready for GC injection. Brobst and Lott mention that this method also has economic advantages for large numbers of samples as, on a silyl donor basis, HMDS is eight times cheaper than TMSIM. Another silylating reagent sometimes used for carbohydrates is *N,O*-bistrimethylsilyltrifluoroacetamide (BSTFA), either alone in pyridine (Pelletier and Cadieux, 1982) or together with TMCS where silylation is quantitative at room temperature (Adam, 1981; Mueller and Goeke, 1972). Zuercher and Hadorn (1975) carefully monitored the effect of reaction conditions when they used *N,O*-bis(trimethylsilyl)acetamide (BSA), HMDS and TFA to silylate carbohydrates. They found that reaction was quantitative if the reagents were added to a pyridine solution of sample in the above order and that heating was unnecessary.

Some workers prefer to silylate under strictly anhydrous conditions. Kline *et al.* (1970) added Drierite (anhydrous calcium sulphate) before silylating and Mateo *et al.* (1987) used anhydrous magnesium sulphate to pretreat pyridine solutions of honey before silylation. Others have shown that, by appropriate choice of silylating reagents and conditions, derivatization of carbohydrates in aqueous solution can be quantitative. Weiss and Tambawala (1972) used excess 2:1 HMDS/TMCS reagent in pyridine to derivatize sugars in aqueous solutions of 10% strength. Also, Gianotto *et al.* (1986) were able to derivatize lactose in dilute aqueous solution by using direct addition of TMSIM before adding pyridine.

It has been claimed that the rate of silylation can be accelerated by the addition of cyclohexane to the reagent mixture (Leblanc and Ball, 1978). Silylation can also be performed in alternative solvents to pyridine, which has an unpleasant odour. For example, Davis (1988) silylated sugars extracted from potato tubers with BSTFA in dimethylformamide.

Reducing sugars are converted to oximes by reacting the sample with hydroxylamine hydrochloride in pyridine at, typically, 75°C for 30 min. Silylation is then performed directly on the reactant mixture. Pierce markets 'Stox' reagent, which is a pyridine solution containing 25 mg/ml of hydroxylamine hydrochloride plus 6 mg/ml phenyl-β-D-glucopyranoside as an internal standard (Pierce, 1989). Care must be taken when carbohydrate mixtures containing sucrose are converted to oximes as Schaeffler and Morel du Boil (1981) noticed that some hydrolysis of the disaccharide to glucose and fructose can occur. They attributed this to the acidity of the derivatizing reagent. By using a reagent which incorporated dimethylaminoethanol as an organic buffer, they were able to prevent inversion without inhibiting the rate of oxime formation of the reducing carbohydrates. Methyloxime–TMS derivatives are prepared in the same way as the oxime–TMS derivatives, except methoxylamine hydrochloride is used instead of the hydroxylamine salt (Laine and Sweeley, 1973).

Alditol acetates are prepared by treating reducing carbohydrates with sodium borohydride to give their corresponding alditols and then acetylating with a mixture of acetic anhydride and pyridine, or acetic anhydride and sodium acetate. With these procedures, it is necessary to remove sodium borate from the reaction mixture prior to acetylation, usually by treatment with cation exchange resin (to remove sodium ions) and then by a series of evaporations in the presence of methanol, so as to remove boric acid as its volatile methyl ester. This procedure is time-consuming and Blakeney *et al.* (1983) have developed a simplified 'one-pot' procedure which does not involve the transfer of solutions. Their method utilizes dimethyl sulphoxide as solvent and methylimidazole as an acetylation catalyst. The latter allows acetylation to proceed without the necessity of removing borate. This method is gaining general acceptance, although the two above-named reagents may lead to the appearance of some extraneous GC peaks as well as solvent tailing.

Aldononitrile acetate derivatives are easy to prepare: following oxime formation as for the oxime–TMS derivatives described above, sample is acetylated by heating in acetic anhydride and pyridine. The solution can then be directly injected into the GC, but solvent tailing is significantly reduced if the reagents are evaporated in a stream of dry air and the derivatives extracted into chloroform and washed with water. Methyloxime acetates are prepared similarly, using methoxylamine hydrochloride instead of hydroxylamine hydrochloride (Neeser and Schweizer, 1984).

Of the two reagents, *N*-methylbis(trifluoroacetamide) (MBTFA) and bis(trifluoroacetamide) (BTFA) suggested for the trifluoroacetylation of carbohydrates (Donike, 1973), the former has been recommended by Sullivan and Schewe (1977) and Englmaier (1985) due to its higher reactivity. Sample is dissolved in pyridine, MBTFA added in excess and the sample heated for 10 min at 75°C. The resulting mixture can be injected directly into the GC.

4.5 GC EQUIPMENT

4.5.1 GC columns

Both packed and capillary columns are used for the analysis of food carbohydrates. The former has traditionally been preferred for most applications, particularly in the analysis of low molecular weight carbohydrates. Capillary columns are now being produced that are less fragile and more thermally stable and they are rapidly gaining in popularity.

Packed columns are suitable for routine, straightforward analyses because they are robust and easy to use. The importance of inert supports for the stationary phase in the separation of carbohydrate derivatives has been stressed by Folkes (1985). Suitable commercial supports are available which have been acid-washed, treated with silanizing reagent and carefully sieved to a narrow particle size distribution. Most packed columns are made out of glass, but carbohydrate analyses which use copper and steel columns have also been reported. Schaeffler and Morel du Boil (1984), in their work on the analysis of sucrose processing streams, have considered in detail the advantages and disadvantages of using capillary columns and these may be summarized:

Advantages:

- more efficient, higher-resolution separation;
- shorter analysis times;
- low pressure drop, thus leaks have less effect on performance;
- narrower peaks lead to easier integration;
- lower sample loads lead to less detector contamination and longer trouble-free operation.

Disadvantages:

an inlet splitter is necessary, care must be taken that the splitter does not discriminate and that the vented sample portion does not lead to clogging in the ventline, back-pressure regulator or needle valves;
higher sensitivities require stable FID and electrometer back-up;
make-up gas is required;
capillary columns are more expensive.

Capillary columns were, until recently, constructed exclusively from glass and thus required careful handling in order to avoid breakage. However, the development of the more durable fused silica type has facilitated installation and operation. Fused silica columns with internal diameters down to 0.25 mm are available. The liquid phase, which typically has a maximum thickness of 1 μm, can be simply coated on the column surface or, as is now more usual, be chemically bonded. The bonded phases are more thermally stable and can be subjected to washing treatments.

For the relatively simple analysis of the common sugars, i.e. sucrose, glucose, and fructose, a short packed column will suffice. However more complex mixtures, or samples containing interfering components, will often require the higher resolution provided by the capillary columns. The resolution of the capillary columns is increased with decreasing diameter, although the sample capacity is reduced. Typically, sample component capacity of a 0.25-mm-i.d. (internal diameter) column is of the order of 20 to 50 μg, and this increases by a factor of about 20 as the internal diameter is doubled. Sample splitters, which will typically split the sample in a 100:1 ratio, are necessary when using the 0.25-mm-i.d. columns.

A variety of stationary phases are available, ranging from the lower polarity methyl silicone gums, which are suitable for the separation of TMS derivatives, to the higher polarity phases such as cyanopropylphenylated silicones, which can be used for acetylated and trifluoroacetylated derivatives. The former phase types are more thermally stable and can be operated at higher temperatures, thereby speeding analysis. The stability of the higher polarity capillary columns is continually being improved and thermally stable, silica-bonded stationary phase columns are now available.

Tables 4.1, 4.2 and 4.3 list some stationary phases used for packed and capillary columns in the analysis of carbohydrates in foods. Derivative types compatible with the stationary phases are also listed. Detailed tabular data on derivatives and GC conditions for the analysis of carbohydrates has been published (Churms, 1982).

4.5.2 Injection systems

Generally a hot injection system (200–300°C) is necessary for the volatilization of derivatized carbohydrates. Loosely packed plugs of silanized and phosphoric acid-treated glass wool in the injection port glass liner aid in the removal of contaminants, such as fats, that may be present in the sample. Traitler *et al.* (1984) describe how the hot injection system is limiting for higher molecular weight oligosaccharide derivatives, which degrade or are insufficiently volatile. They recommend an on-column injection system, with which they demonstrate improved separation efficiency and shorter retention times for oligosaccharides up to DP 6 as compared to the traditional

Table 4.1 — GC of sugars and sugar alcohols in foods

Analysis	Derivatives	GC column	Reference
Sugars in fruit	Oxime–TMS	2% OV-17 on 100/200 Chromosorb W	Long and Chism (1987)
Sugars in honey	Oxime–TMS, TMS	OV-101-coated fused-silica capillary	Mateo *et al.* (1987)
Sugars in royal jelly	TMS	SE-52-coated fused-silica capillary	Lercker (1986)
Sugars in milk	TMS	2% OV-17 on 120/140 Volaspher A-2	Olano (1983)
Sugars in honey	Oxime–TMS	4% SE-52 on 80/100 Varaport 30	Deifel (1985)
Sugars, sugar alcohols in vegetables	TMS	OV-17 coated glass capillary	Sorensen and Goldberg (1984)
Sugars in almonds	TMS	3% SE-30 on 80/100 Supelcoport	Saura-Calixto *et al.* (1984)
Sugars in sucrose processing streams	TMS	OV-1701 coated glass capillary	Preuss and Thier (1983)
Sugars in yoghurt	TMS	3% SP-2250 on 80/100 Supelcoport	Li *et al.* (1983)
Sugar alcohols in mints	Acetates	10% Silar 10C on 60/80 Chromosorb W/AW	Daniels *et al.* (1982)
Sugars in soft drinks	TMS	Dimethylsilicone-bonded glass capillary column	Martin-Villa *et al.* (1981)
Sugars in cane molasses	Oxime–TMS	SP-2250 coated glass capillary	Schaeffler and Morel du Boil (1981)
Sugars in cheese	TMS	3% OV-1 on 80/100 Supelcoport	Harvey *et al.* (1981)
Sugar alcohols in chewing gum	Acetates	OV-17 coated capillary	Oesterhelt *et al.* (1980)
Sugars in pears	TMS	5% SE-52 on Chromosorb W HP	Akhaven *et al.* (1980)
Sugars in dates	TMS	3% SE-30 on 100/120 Diatomite CQ	Jaddou and Hakim (1980)
Sugars in confectionary	TMS	2% OV-17 on 100/200 Supelcoport	Prager and Miskiewicz (1979)

Table 4.2 — GC analysis of oligosaccharides in foods

Analysis	Derivatives	GC Column	Reference
Mono-, di-, trisach. in honey	Alditol–TMS	DB-5-coated fused-silica capillary	Low and Sporns (1988)
Fructans to DP 6	TFA	3% Dexsil 410 on 80/100 Chromosorb W-HP	Englmaier (1985)
Raffinosaccharides in soy beans	TMS	3% SP-2550 on 80/100 Supelcoport	Molnar-Perl *et al.* (1984)
DP 1–3 in flour	TMS	4% OV-1 on 100/120 Gas Chrom A	Mbugua *et al.* (1983)
Raffinosaccharides in legumes	TMS	3% Dexsil on 80/100 Chromosorb W	Sosulski *et al.* (1982)
DP 1–3 in maltitol syrups	TMS	15% Dexsil GC 300 on 100/120 Chromosorb W AW-DMCS	Molnar-Perl *et al.* (1981)
Dextrins to DP 8	TMS	3% JXR on 80/100 Chromosorb W	Beadle (1969)

Table 4.3 — GC of polysaccharides in foods

Analysis	Derivatives	GC column	Reference
Plant cell walls	Alditol acetates	DB-225-coated fused-silica capillary	Hoebler *et al.* (1989)
Hydrocolloids	Me–Glycoside –TMS	DB-5-coated fused-silica capillary	Ha and Thomas (1988)
Dietary fibre	Alditol acetates	OV-225-coated glass capillary	Theander and Westerlund (1986)
Orange juice	Me–Glycoside –TMS	SE-30-coated glass capillary	Kauscus and Thier (1985)
Gums in foods	Aldononitrile acetates	3% NGPS on 100/200 Gas Chrom Q	Lawrence and Younger (1985)
Non-starch polysaccharides in cereal grains	Alditol acetates	Silar 10C glass capillary	Henry (1985)
Plant cell walls	*O*-Methoxime acetates	Carbowax 20M fused silica capillary	Nesser and Schweizer (1984)
Gums in foods	Me–Glycoside TMS	SE-30 glass capillary	Preuss and Thier (1983)
Fruits and vegetables	Alditol acetates	10% SP-2330 on Supelcoport W-AW	Bittner *et al.* (1982)
Dietary fibre	Alditol acetates	3% SP-2330 on 100/120 Supelcoport	Englyst *et al.* (1982)
Hydrocolloids	Aldononitrile acetates	3% PNGS on 100/120 Chromosorb W-AW	Mergenthaler and Scherz (1976)

technique. Davies (1988) preferred the hot injection system for his analysis of sugars in potato tubers owing to low response for sucrose with his on-column injection system.

4.5.3 Detectors

The most-favoured detector for carbohydrates is the flame ionization detector, which allows high-sensitivity detection down to the nanogram level. For organic substances, the response with this detector is approximately proportional to the molecular weight and thus introduction of the bulky trimethylsilyl group, which triples the molecular weight of a typical carbohydrate, gives high sensitivity. A drawback of the TMS derivative is the accumulation of silica, which is deposited at the detector electrode. Although this is less of a problem with newer designs, some loss in sensitivity can occur with time. Zuercher and Hadorn (1975) added heptafluorobutyramide to their silylated samples to prevent silica deposition at the electrode.

Even higher sensitivities can be achieved via the use of alternative detectors such as the nitrogen-selective FID or the electron capture detector. By means of the former, sensitivity for monosaccharides can be significantly increased by using the oxime derivatives (Morita and Montgomery, 1978). The electron capture detector is effective with halogen-containing compounds and Eklund *et al.* (1977) demonstrated its application to the high-sensitivity detection of trifluoroacetate derivatives of monosaccharides.

An alternative detector type that is finding more use in special applications is the mass detector (Fox *et al.*, 1989). This is essentially a bench-top mass spectrometer which is interfaced via a capillary direct inlet to a gas chromatograph equipped with a capillary column. The detector can be used to monitor total ion current or selected ions in the mass spectra of the eluting components. This has the advantage of facilitating the determination of components that elute at similar retention times, particularly if the mass spectra are dissimilar due to structural differences. Pelletier and Cadieux (1982) described application of the technique to the monitoring of methyloxime–TMS derivatives of monosaccharides.

4.6 QUANTITATIVE ANALYSIS

For purpose of quantification, sample components are usually calibrated against a suitable reference material, an internal standard, which is added to the sample to be analysed. It is advantageous to use a carbohydrate as an internal standard as this will also undergo derivatization. It should obviously not be present in the original sample and should elute well clear of the components being analysed. The response of the standard against the sample should be checked for linearity over a range of concentrations and the amount used should be no more than that of the main sample component. The internal standard should also be added prior to work-up in order to minimize errors due to sample preparation. In the case of polysaccharides, some workers prefer to add the internal standard subsequent to hydrolysis because of possible degradation. This is particularly the case with the Saeman hydrolysis procedure, which utilizes concentrated sulphuric acid (Theander and Westerlund, 1986).

The majority of the internal standards used are carbohydrate based. The most commonly encountered in the determination of low molecular weight carbohydrates is phenyl-β-D-glucopyranoside as, for example, in the analysis of sugars in breakfast cereals (Li and Schuhmann, 1980). This compound is also present in 'Stox' reagent (Pierce, 1989) for use as an internal standard for sugars determined as their oxime-TMS derivatives. The cyclic alcohol, *myo*-inositol, is also popular. Other standards include linear alcohols, e.g. sorbitol (Kallio *et al.*, 1985) and erythritol; monosaccharides, e.g. arabinose (Jaddou and Al-Hakim, 1980) and glucoheptose (Kline *et al.*, 1970). Others prefer to use inert hydrocarbons, e.g. *n*-docosane (Date *et al.*, 1982).

For polysaccharide analyses, *myo*-inositol is frequently used. Harris *et al.*, (1988) recommend allose as an internal standard for the analysis of plant cell wall polysaccharide hydrolysates as their alditol acetates. Its advantages include: allose is a monosaccharide and will thus monitor the reduction step; neither allitol or psicose (which yields allitol on reduction) are encountered in natural polysaccharides; and allitol acetate is eluted between the pentitol and hexitol acetates and is thus a better retention time marker than inositol, which is eluted after the hexitol acetates. Mateo *et al.* (1987) identified the sugars in honey by determining their retention times relative to sucrose and isomaltose. These disaccharides gave characteristic peaks, were present in all honey samples and did not interfere with other peaks.

4.7 ANALYSIS OF CARBOHYDRATES IN FOOD PRODUCTS AND INGREDIENTS

The large variety of food products and the different types of carbohydrate that they contain require a wide range of GC-based analytical procedures. The analysis of food carbohydrates is considered here in three sections: simple sugars and sugar alcohols, oligosaccharides and polysaccharides.

4.7.1 Sample preparation in the analysis of simple sugars and sugar alcohols

There are many examples in the literature of the analysis of the sugars and sugar alcohols in foodstuffs by GC analysis. As mentioned above, the majority of these analyses use trimethylsilyl derivatives, either with or without a pre-derivatization step, which is usually oxime formation. Samples with high carbohydrate contents may be analyzed directly; other samples are often subjected to work-up conditions, which involves removal of interfering substances, extraction and enrichment. Consideration will be be given first to work-up conditions.

4.7.1.1 Removal of interfering substances

In the GC analysis of sugars in fruit juices, organic acids can interfere and are often removed as insoluble salts. The sugar profiles of 28 fruit juice samples were determined by GC of the trimethylsilyl ethers following removal of organic acids with saturated lead acetate solution (Kline *et al.*, 1970). In their method, fruit juice diluted with 80% ethanol was treated with saturated lead acetate solution to precipitate polybasic acids. A small aliquot was then taken to dryness (vacuum, 30°C), internal standard added and the sample trimethylsilylated for GC analysis. Some workers have used similar procedures for fruit juices, whereas others have extended this approach to other foodstuffs. Rumpf (1969) used barium acetate to precipitate citric acid from potato extracts where this interfered with the GC determination of fructose.

Acids may also be removed by passing food extracts through ion exchange columns. Procedures have been developed, based on ion exchangers, which allow the analyses of both acids and sugars to be combined into a general method. Reyes *et al.* (1982) passed a centrifuged, ethanol-treated aqueous extract from fruit through a cationic resin (Bio-Rad AG 1-X8, 200–400 mesh, acetate form) and an acidic ion exchange column (Bio-Rad AG 1-X8, 200–400 mesh, acetate form) before drying the sample by rotary evaporation and analysing the TMS derivatives. The acid fraction was analysed after elution from the acidic column. Selective fractionation was obtained by passing wines and beers through a polyvinylpyrrolidone column (bound phenolics), a medium-strength basic ion exchange column (organic acids) and a strong acidic ion exchange column (amino acids). The neutral carbohydrates, which passed through all of the resins, and the other isolated fractions were analysed as their TMS derivatives by GC (Drawert *et al.* 1976).

Akhavan *et al.* (1980) compared the lead salt precipitation method with the ion exchange method for the analysis of sugars and acids in pears and concluded that, for sugars, both methods were equally effective, but for acids, the latter method was preferable.

Interference is also reduced by removal of the lipids from food samples and this is often the first step of an analytical work-up procedure for low molecular weight carbohydrates. In the analysis of sugars in breakfast cereals (Li and Schuhmann, 1980), *n*-hexane was used to extract the fat and other non-polar material prior to polar solvent treatment to extract the sugars for GC analysis. Other solvents used to remove fat include ether and petroleum ether in the analysis of cocoa products (Luke, 1971). Alternatively, the initial fat extraction can be omitted and any fat associated with the aqueous ethanol extracts removed by shaking with chloroform (Mueller and Goeke, 1972).

4.7.1.2 Extraction of sugars

Sugars for GC analysis are usually extracted from foods with aqueous alcohol (normally 80–95% v/v methanol or ethanol) or water. Jaddou and Al-Hakim (1980) used soxhlet extraction with 95% ethanol to extract the sugars from Iraqi dates. Martin-Villa *et al.* (1981) used hot 80% ethanol to extract the sugars from neutralized commercial soft drinks. Saura-Calixto *et al.* (1984) used 80% ethanol at 50°C to extract the sugars from almonds. Fuchs *et al.* (1974) showed that extraction with 80% ethanol at room temperature overnight was effective for the quantitative determination of monosaccharides and disaccharides in a variety of foods. Verwack *et al.* (1978) analysed mono-, di-, and trisaccharides in various foods, using water for extraction from samples with high carbohydrate content and ethanol from those with low contents. Date *et al.* (1982) used water (maceration at room temperature in a blender) for foods containing large quantities of water-soluble polysaccharides or protein, and ethanol (reflux in 80% ethanol for 15 min) for the other samples. Li and Schuhmann (1980) have warned against invertase activity with some foods on extraction with water. They observed extensive hydrolysis of sucrose during extraction of raisin-containing cereal. They obtained no invertase activity when they extracted the cereal with 80% methanol (room temperature, vigorous stirring, two hours). This method also gave less extraneous material and required less drying before extraction, although they observed significantly lower extraction yields of maltose compared with water extraction. Care must also be taken to avoid sucrose inversion when extracting acidic substrates such as fruits.

4.7.1.3 Direct derivatization

The development of silylating reagents and procedures that are insensitive to water has allowed the use of GC analyses performed on directly derivatized samples without prior extraction of the sugar fraction. Weiss and Tambawala (1972) used excess silylating reagent (HMDS and TMCS) to derivatize sugars in dilute aqueous solution, and Long and Chism (1987) showed how freeze-dried fruit tissue could be directly converted to oximes (STOX reagent) before silylation to determine the sugars present. Li *et al.* (1983) compared a typical work-up procedure (de-fat and aqueous alcohol extraction) with direct derivatization (oxime formation from freeze-dried sample followed by silylation) for yoghurt and found the approaches gave comparable results.

Direct derivatization is convenient for foods and ingredients that contain high levels of sugars and has been successfully applied to a variety of sample types. Examples are candy (Daniels *et al.*, 1982), molasses (Schaeffler and Morel du Boil, 1984), syrups (Sennello, 1971) and honey (Mateo, 1987).

4.7.2 Analysis of simple sugars and sugar alcohols in selected foods

4.7.2.1 Food profiles

The versatility of the GC technique allows the adaption of analytical procedures for carbohydrates in a wide range of foodstuffs. Fuchs *et al.* (1974) established conditions for the quantification of monosaccharides and disaccharides by GC in a variety of foods such as bread, sausage and oranges and Birkhed *et al.* (1980) determined sugars and sugar alcohols as their TMS derivatives in 228 commercial food products. Examples of the analysis of low molecular weight carbohydrates in various food types with the derivatives and column conditions used are given in Table 4.1.

4.7.2.2 Milk products

The total amounts of the three principal milk sugars, glucose, galactose and lactose, were determined by GC of the TMS ethers prepared using the Sweeley procedure (Reineccius *et al.*, 1970). They showed that for a range of milks and cheeses with various fat contents, the relative proportions of the sugars were similar. Martinez-Castro and Olano (1978) used GC of the TMS ether to determine the low levels of the disaccharide, lactose, in milk. Mouillet *et al.* (1977) used GC of the TMS derivatives to study lactose hydrolysis during the manufacture of yoghurt. They were able to show that approximately 35% of the lactose was hydrolyzed to glucose and galactose, and thereafter established the mechanism of action of the yoghurt culture by monitoring glucose and galactose concentrations of a wide range of commercial yoghurt samples. In a detailed study of yoghurts sold in the U.S., Li *et al.* (1983) showed that total sugars (fructose, galactose and lactose, but no glucose) ranged from 4 to 6% in plain and 12 to 18% in flavoured brands (which also contained glucose, sucrose and maltose). Corzo *et al.* (1986) used GC to show the presence of lactulose and epilactulose in sterilized milk. Harvey *et al.* (1981) found lactose, galactose and glucose (determined as their TMS ethers) in a range of cheddar cheeses. α-Lactose and sucrose are not separated as TMS derivatives on the low-polarity columns normally used, and Larson *et al.* (1974) achieved the separation by using a moderately polar phase such as XE-60, a cyanomethylpolysiloxane.

4.7.2.3 Alcoholic beverages

Olano (1983) used GC of TMS derivatives to determine trehalose and inositol in wines and GC of the acetates to determine sorbitol, mannitol and inositol in sherries. In the analysis of wines and beers, Drawert *et al.* (1976) used an ion exchange-based separation method to isolate 11 neutral carbohydrates (nine monosaccharides and the sugar alcohols, sorbitol and inositol) which were then analysed as their TMS derivatives. Uronic acids were collected from a basic ion exchange resin and also analysed directly as their TMS derivatives. Jamieson (1976) used oxime–TMS derivatives to determine the sugars in brewing materials. Samples with high sugar contents, such as syrups and wort, were derivatized directly, whereas beer required freeze-drying.

4.7.2.4 Syrups/sugar products

Schaeffler and Morel du Boil (1984) published a detailed account of the application of gas chromatography to the analysis of sugar-cane-derived products, such as molasses, and processing streams. They emphasized the suitability of GC analysis of TMS derivatives on capillary columns as the preferred method for accurately analysing the high concentrations of sucrose and the lower quantities of the other low molecular weight carbohydrate components. Preuss *et al.* (1984) also demonstrated the suitability of capillary GC for the analysis of the intermediate and end-products of sucrose manufacture. For example, they showed that low concentrations of byproducts in raw sugar could be determined by direct analysis (Fig. 4.4). According to Nurok and Reardon (1977), sucrose can be accurately analysed (standard deviation of only 0.1%) in 1 min on an OV-17 coated capillary column.

4.7.2.5 Chewing gum

Oesterhelt *et al.* (1980) screened both silyl and acetate derivatives for the determination of the sugar alcohols xylitol, mannitol and sorbitol in chewing gum. Although separation of the TMS derivatives was more difficult and required the application of a 40-m OV-17 capillary column, these derivatives gave more-reproducible results than the acetate. In contrast, Daniels *et al.* (1982) preferred the acetate derivatives for the GC determination of xylitol, mannitol and sorbitol in sugar-free chewing gum and mints.

4.7.2.6 Honey

The large number of low molecular weight carbohydrate components in honey make it a suitable system for analysis by GC. Sixteen mono-, di-, and trisaccharide components were identified and quantified both as their TMS ethers and as their TMS oximes using capillary GC (Figs. 4.5); (Mateo *et al.*, 1987). Deifel (1985) used two different SE 52 support-coated packed columns to analyse honey carbohydrates.

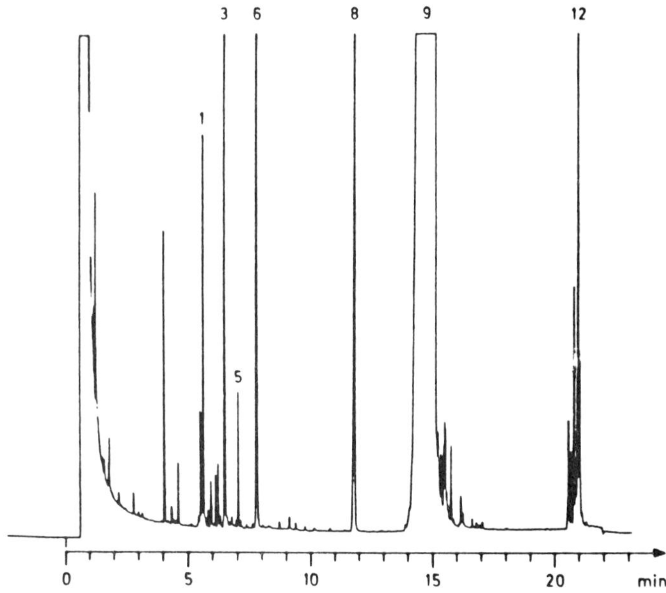

Fig. 4.4 — Analysis of TMS derivatives of carbohydrates in raw sugar. Column: 15-m OV-1701 coated glass capillary column. Temperature-programmed: 100° to 170°C at 8°C/min. Peaks: 1,5, fructose; 3,6, glucose; 8, phenyl-β-D-glucopyranoside (Internal standard), 9, sucrose; 12, raffinose. (From Preuss, A. *et al.* (1983)). *Lebensm.-Wiss. Technol.* **1**, 163–166; with permission.)

He used a 2-m-long, 4% SE 52-coated column to separate the important components, melezitose and erlose, and a shorter 60-cm, 0.5% SE 52-coated column to separate oligosaccharides. Lercker *et al.* (1986) showed that capillary GC chromatograms of the TMS derivatives of the neutral components of bee larvae food and honeys can be used to detect adulteration by commercial starch syrups. GC patterns of the latter are characteristic and reproducible, which allows their ready detection.

4.7.2.7 Fruits

The principal sugars in fruits are glucose, fructose and sucrose, and these can be readily separated and accurately quantified by GC as their TMS or oxime–TMS derivatives. Kline *et al.* (1970) determined the contents of these three sugars and sorbitol as their TMS derivatives in 28 different fruits. Akhaven *et al.* (1980) used GC and GC–MS to confirm the presence of fructose, glucose, sorbitol, sucrose, xylose and inositol in pears. Fruit sugars are preferably analysed as their oxime–TMS derivatives, where glucose and fructose each give one peak on packed columns and two closely eluting peaks on capillary columns, for example, in orange juice (Fig. 4.6). Long and Chism (1987) used the oxime–TMS derivatization procedure to determine the sugars in apples, bananas, nectarines and tomatoes. Li and Schuhmann (1983) also used this procedure to determine the fructose, glucose and sucrose

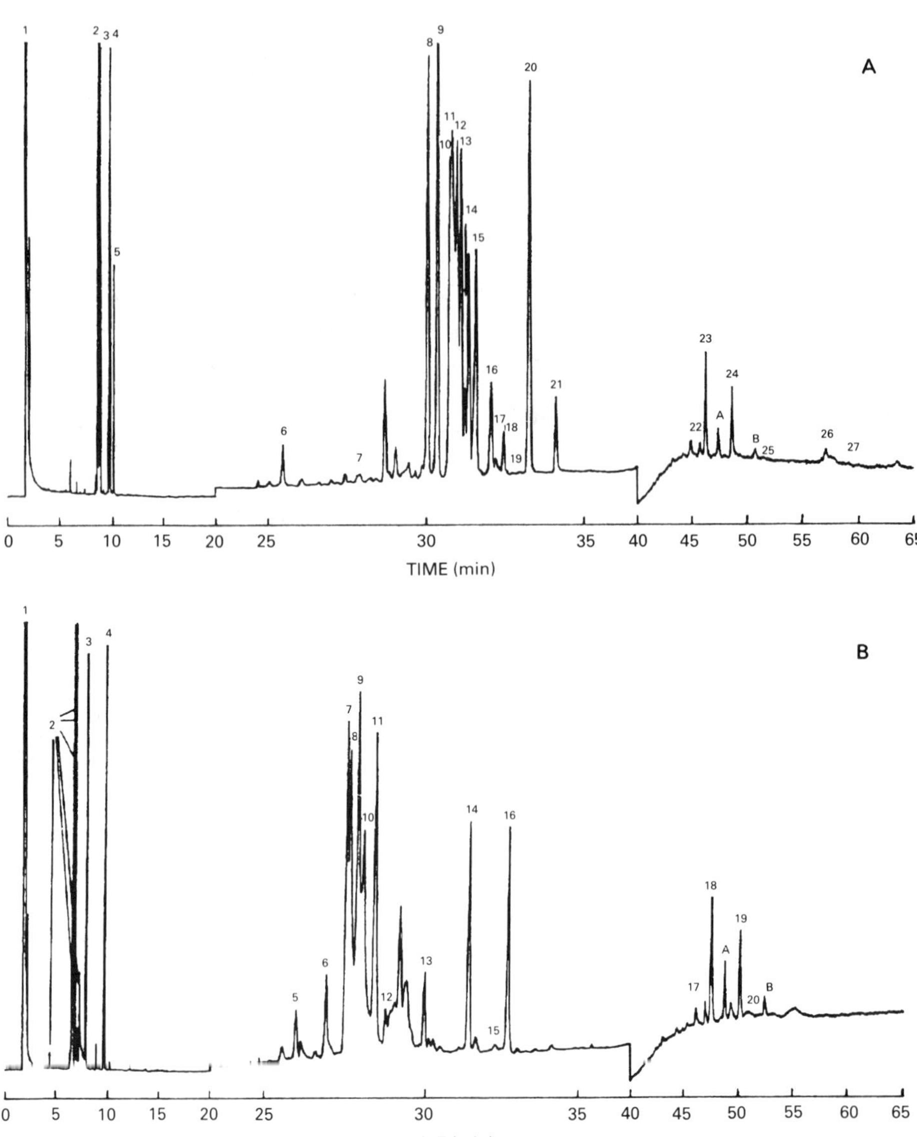

Fig. 4.5(A) — Analysis of honeydew honey sugars following oxime formation and silylation. Column: 25-m×0.23-mm i.d. OV-101 fused-silica capillary column. Temperature-programmed: 180°C to 280°C at 3°C/min, hold 4 min, 2°C/min till 290°C, held at 290°C. Peaks: 1, solvent; 2, 3, fructose; 4, 5, glucose; 6, sucrose; 7, α,α-trehalose; 8, 9, maltulose; 10, nigerose; 11, turanose and maltose; 12, turanose; 13, kojibiose; 14, maltose and isomaltose; 15, nigerose; 16, kojibiose and palatinose; 17, gentiobiose; 18, palatinose; 19, gentiobiose; 20, 21, isomaltose; 22, raffinose; 23, 1-kestose; 24, melezitose, 25, 26, 27, maltotriose. (B) — Analysis of honeydew sugars as their TMS derivatives. Conditions as in 4.5(A). Peaks: 1, solvent; 2, fructose (several isomeric forms); 3, α-glucose; 4, β-glucose; 5, sucrose; 6, α-maltose; 7, 8, maltulose; 9, turanose and nigerose; 10, β-maltose; 11, β-kojibiose; nigerose and α,α-trehalose; 12, palatinose, 13, α-kojibiose; 14, α-isomaltose; 15, gentiobiose; 16, β-isomaltose; 17, raffinose; 18, 1-kestose; 19, melezitose; 20, maltotriose. (From Mateo, R. *et al.* (1987), *J. Chromatogr.* **410**, 319–328, with permission.)

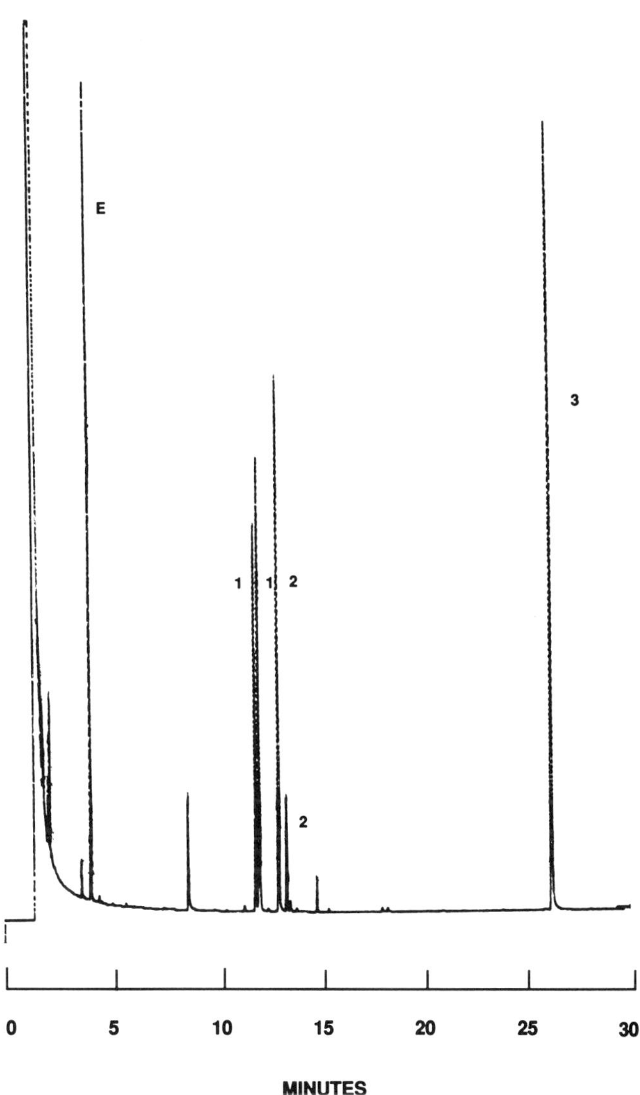

Fig. 4.6 — Analysis of sugars in orange juice following oxime formation and silylation. Column: 30-m DB-1 bonded phase fused-silica capillary column. Temperature-programmed: 170°C for 2 min, 4°C/min to 275°C, held at 275°C.

contents of 11 different types of fruit juice. Tarrach and Herrman (1987) used capillary GC of oxime–TMS derivatives to demonstrate the formation of esters from the acids and sugars present in fruit concentrates.

4.7.2.8 Other foods

GC of TMS derivatives has shown that sucrose (content ranges from 4 to 8%) is the principal low molecular weight carbohydrate in green coffee beans. Reducing sugars are only present at a very low concentration (total less than 0.5%). Blanc *et al.* (1989) used GC of TMS derivatives to determine the contents of sucrose and maltose in soluble coffee; the presence of elevated levels of the latter observed in some samples was an indication of added maltodextrins.

In a comprehensive study of the sugar profiles in ready-to-eat breakfast cereals, Li and Schuhmann (1980), detected fructose, glucose, lactose, maltose (oxime–TMS derivatives) and sucrose (TMS derivatives), with total sugars in the cereals varying from less than 0.5% to as much as 56%.

Davies (1988) obtained baseline separation of the TMS ethers of fructose, glucose and sucrose extracted from potatoes in only 10 min on a 5-m HP-1 (SE-30 equivalent)-coated fused silica capillary column (Fig. 4.7).

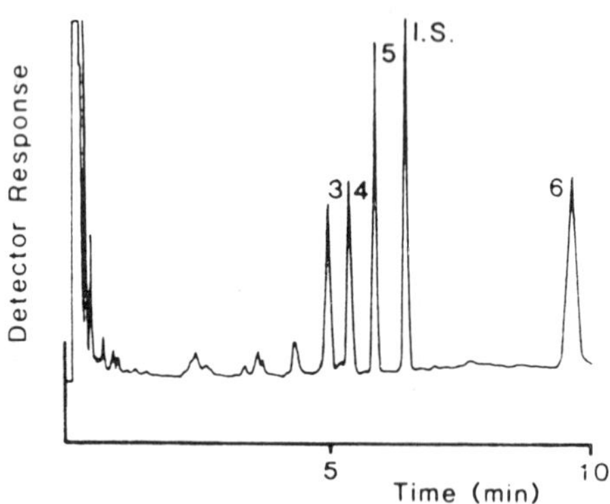

Fig. 4.7 — Analysis of sugars in potato tubers as TMS derivatives. Column: 5 m×530 μm i.d. HP-1(equivalent to SE-30)-coated fused-silica capillary column. Temperature programmed: 140°C for 2 min, rising to 250°C at 20°C/min. Peaks: 3, fructose; 4, α-glucose; 5, β-glucose; 6, sucrose. (From Davis, H. V. (1988), *Potato Res.* **31**, 569–572, with permission.)

Chepurnoi *et al.* (1986) found by GC analysis that glucose and fructose were the main components of crystals that separated from jam, syrup and honey samples. They were also able to determine the anomeric form of the sugars in the crystals.

Brobst and Scobell (1981) were able to observe difructose anhydride compounds in chicory syrup by GC analysis of their TMS ethers.

4.7.3 Oligosaccharides

GC analyses of oligosaccharides usually utilize either the TMS or trifluoroacetate derivatives. Englmaier (1985) favoured trifluoroacetates for the analysis of oligosaccharides, citing their higher volatility, good separation on high polarity columns and their thermal stability, which is superior to TMS derivatives. In addition, they can be detected at low concentrations using an electron capture detector, which is at least 100 times more sensitive than an FID detector. Molnar-Perl and Szsakas-Pinter

(1981) compared the use of trifluoroacetate and TMS derivatives for the analysis of maltitol syrups. They showed that the method was more sensitive to the TMS derivatives, where the response of the FID detector was two to three times greater. Other disadvantages they claimed for the trifluoroacetates can be related to the trifluoroacetic anhydride used: this reagent was more caustic than the silylation reagents, trifluoroacetylation was more laborious and silylation of the alditols gave the expected one peak, whereas trifluoroacetylation led to several peaks for each alcohol, indicating incomplete derivatization. Englmaier (1985) has recommended MBTFA for the trifluoroacetylation of oligosaccharides for the following reasons: the reagent is easy to use; it gives quantitative trifluoroacetylation; and it does not result in any depolymerization, which he had observed on use of the anhydride.

GC of trifluoroacetates was used by Englmaier (1985) to analyse fructan oligosaccharides up to DP 6 and by Sullivan and Schewe (1977) to analyse raffinose (DP 3) and stachyose (DP 4). Kamiyana and Sakai (1974) quantified xylo-oligosaccharides up to DP 4 by using GC of the alditol–trifluoroacetate derivatives. Examples of analyses of oligosaccharides in foods are given in Table 4.2.

Beadle (1969) was able to show, that, by application of a sufficiently stable column (he used a packed column with 3% JXR on 80–100-mesh Chromosorb W), the oligosaccharides in starch hydrolyzates up to DP 7 could be determined by GC of their TMS derivatives. West and Moskowitz (1977) showed how TMS derivatives of DPs 1 to 4 in corn syrup could be rapidly prepared (with TMSIM, 15 min at room temperature) and conveniently separated by GC. GC of TMS derivatives has been used to quantify the 'flatulent' or raffinose family of oligosaccharides: raffinose, stachyose and verbascose, which have DPs of 3, 4 and 5, respectively. Molnar-Perl *et al.* (1985) were able to extract these and other related oligosaccharides from the soy bean using 80% ethanol and then analyse them after oxime formation and silylation. The raffinosaccharides do not have reducing groups and are thus not converted to oximes prior to silylation. Sosulski *et al.* (1982) also used silylation for the GC quantification of the same oligosaccharides in the dehulled seeds of 11 legumes (Fig. 4.8). Ford (1979) determined raffinose and stachyose by analyzing the TMS derivatives of their reaction products from invertase treatment.

GC can also be used to determine the number-average molecular weight of oligosaccharide mixtures such as maltodextrins. By using sodium borohydride to convert the reducing ends to alditols, hydrolysing and then using the aldononitrile acetate derivatization procedure, Morrison (1975) was able to show that the ratio of aldononitrile acetate derivatives to alditol acetates, as determined by GC, gave the average degree of polymerization of the mixture. The high sensitivity of the GC technique allowed averaged DP determinations as high as 145.

4.7.4 Polysaccharides

The content and type of polysaccharide in food materials is usually determined by controlled degradation to the monomeric components of the polymer which can then be estimated by GC following suitable derivatization. Examples of analyses of polysaccharides in foods and the GC conditions used are given in Table 4.3. Available procedures of degradation, derivatization and GC separation conditions have been previously reviewed for natural polysaccharides (Bradbury and Halliday, 1984).

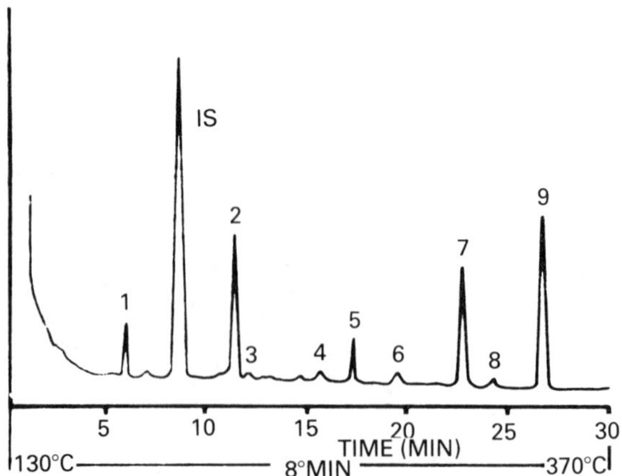

Fig. 4.8 — Analysis of oligosaccharides in soy bean flour as their TMS derivatives. Column: 1.8 m×2 mm i.d. glass column packed with 1.5% Dexsil 300 on Chromosorb W. Temperature-programmed: 130 to 350°C at 8°C/min. Peaks: IS internal standard; 2, sucrose; 3, galactopinitol; 4, galactinol; 5, raffinose; 7, stachyose; 9, verbascose. (From Sosulski, F. W. *et al.* (1982), *J. Food Sci.* **47**, 498–502, with permission.)

4.7.4.1 Sample preparation

Glueck and Thier (1983) outlined a convenient procedure for the isolation of thickeners and gums from a variety of foods. The food sample was first heated with dioxan at 60°C to solubilize fats and most of the sugars and organic acids. After washing with 70% ethanol, the residue was solubilized in water and then treated with amylase and amyloglucosidase enzymes to solubilize starch and maltodextrins. Excess protein was then precipitated from the aqueous solution with trichloroacetic acid. The polysaccharide was then precipitated from solution with acetone and washed with ethanol and then analysed. Preuss and Thier (1983) used the above method but substituted sulphosalicyclic acid to precipitate the protein for trichloroacetic acid, which does not extract milk glycoproteins and degrades agar and carrageenan. Lawrence and Iyenger (1985) used different procedures to extract gums from three different classes of foodstuffs: dairy products, non-starch-containing dressings and starch-containing sauces. They used Glueck and Thier's procedure for dairy products. For the dressings, they mixed with water and heated to 65–70°C, separated the emulsion with 50% trichloroacetic acid and precipitated the lower aqueous layer with 70% ethanol. The gum was redissolved and reprecipitated with alcohol. For the sauces, fat was extracted with light petroleum, starch degraded by iodine complexation and the gum precipitated as described above.

Several researchers have concerned themselves with determining the chemistry of 'dietary fibre', which is that fraction of foodstuffs that is not hydrolysed by the digestive enzymes of man. Amongst the most accepted methods for its determination are those based on chemical separation of the polysaccharides, which constitute the major part, if not all, of the dietary fibre fraction. These polysaccharides are estimated as their component monosaccharides using GC-based procedures. These methods all essentially involve extraction of finely ground material with solvents such as light petroleum and aqueous alcohol to remove fat and sugar respectively. Starch is removed enzymatically. Selvendran and Du Pont (1980) analysed the 'alcohol-insoluble residue' directly, while Englyst *et al.* (1982) and Theander and Westerlund (1986) identified uronic acid (predominantly pectin) and soluble and insoluble polysaccharide-containing fibre components. These fractions were all estimated by quantification of the derivatized hydrolysates by GC. Schweizer and Wuersch (1979) determined dietary fibre in food by comparing the neutral detergent extraction method of van Soest and Wine (1967) and a solvent extraction/enzyme treatment along the lines of those used above. They analysed the polysaccharide profiles of fractions prepared from 19 fibre preparations, vegetables and fruits.

Generally, polysaccharide profiles of foods or food ingredients are obtained by analysing the hydrolyzates of a sample that has been washed, heated or macerated in aqueous alcohol. Solvent extraction to remove fat or enzyme treatment to solubilize protein are sometimes used, depending on the nature of the sample. Various fractionation procedures are available for the isolation of individual polysaccharides from foods. Hemicelluloses can often be extracted with alkali and pectinaceous polymers under acidic conditions or by treatment with a chelating agent, such as ammonium oxalate. The cellulose content is often estimated as the glucose produced by acid hydrolysis of a residual fraction following extraction of the hemicellulose and pectinaceous fractions.

4.7.4.2 Polysaccharide degradation

For the purpose of GC analysis, polysaccharides or polysaccharide-containing substrates are hydrolysed to their component monosaccharides or converted with methanol to the corresponding methyl glycosides. These products are then derivatized and analysed by GC using procedures described above.

Non-cellulosic neutral polysaccharides are readily hydrolysed with trifluoroacetic acid (typically, 2 M, 120°C, 1.5 h) which can be conveniently removed by evaporation (Albersheim *et al.*, 1967). Other dilute acids used to hydrolyse polysaccharides include: 1 M sulphuric acid for 2.5 h at 100°C (Selvendran and Du Pont, 1980), or acidic ion exchange resin, 48 h at 100°C (Schaefer, 1983).

Preparations containing cellulose require stronger hydrolysis conditions and the Saeman procedure (Saeman *et al.*, 1954), or a variation of it, is usually used. This procedure utilizes cold, concentrated sulphuric acid (72% w/w) to solubilize the sample. The solution is then diluted (by about a factor of 10) with water, neutralized with barium carbonate, and the filtrate taken to dryness, derivatized and analysed by GC. Care must be taken in data interpretation because of errors introduced by the different rates of polysaccharide hydrolysis as well as the degradation of released monosaccharides due to acid degradation. Bittner *et al.* (1980) monitored the effect of both stages of the Saeman procedure on monosaccharide yields from hydrolysis. He suggested 30-min treatment at 22°C as optimum for the concentrated acid treatment. He also suggested the use of correction factors for samples giving xylose and glucose, as the yield of the former declined after 1 h whereas that of the latter first peaked at 3 h. Pectic polysaccharides are hydrolysed only slowly and Selvendran and Du Pont (1980) used a correction factor to estimate rhamnose yield which was only 60% released under their hydrolysis conditions. Hoebler *et al.* (1989) emphasized the importance of pre-grinding to small particle size to give efficient hydrolysis.

Because the variety of monosaccharides produced by hydrolysis is larger than the free monosaccharides usually present in food-related materials, a wider range of derivatives has been utilized for their separation. The different types of derivatives used are discussed in the above 'derivatives' section (4.3).

Pectin is an important constituent of fruits and its principal structural components are galacturonic acid and its methyl ester. These are difficult to quantify with the usual hydrolysis/GC techniques (see above) and they are typically analysed colorimetrically. Ford (1982) has developed an enzymatic hydrolysis/GC-based procedure for the analysis of pectin-containing substrates. After defatting and desugaring, the pectin was solubilised with 1% ammonium oxalate solution, and the pectinaceous extract hydrolysed with a pectinase preparation. The galacturonic acid produced was then converted to L-galactono-1.4-lactone by treatment with potassium borohydride and then silylated for analysis by GC. Alternatively, pectins can be determined by methanolysis/GC-based procedures.

Methanolysis, as opposed to hydrolysis, offers certain advantages for the degradation of food polysaccharides. The methyl glycosides formed are more stable than the reducing sugars produced by hydrolysis, and uronic acid residues present in the acidic polysaccharides are converted to methyl esters of the glycosides. These can then be analysed as their TMS ethers by GC. Although each monosaccharide is converted to at least two glycosides, application of high-resolution capillary GC allows separation of complex samples. A major advantage of this procedure is that a wide range of monosaccharides can be identified. These include galacturonic acid, glucuronic acid and 3,6-anhydrogalactose, all of which are components of gums and difficult to analyse by GC using other derivatization techniques. A disadvantage is the problem of quantification due to incomplete methanolysis, and Nozawa *et al.* (1969) have observed variation in yields on polysaccharide methanolysis according to hexose type. An improvement in quantification in the methanolysis of pectins has been recently shown if treatment with a commercial enzyme containing pectinase and hemicellulase activities is used (Quemener *et al.*, 1989). The methanolysis procedures utilized by the above workers use a mixture of methanol and hydrochloric acid. This reagent is not suitable for the methanolysis of cellulose, whereas a mixture of methanol and sulphuric acid can be used to estimate cellulose as well as non-cellulosic polysaccharides (Roberts *et al.*, 1987).

4.7.4.3 Analysis in foods

Polysaccharides are major components of foodstuffs where they occur either naturally, i.e. as the 'fibre' components (hemicellulose, cellulose or pectin), starch, and gums or as added components. The analysis in foods and the chemical nature of all these components has been investigated using GC-based techniques.

4.7.4.4 Gums and hydrocolloids

Because of its ability to monitor virtually all of the various constituent monomeric residues, methanolysis and GC of the TMS derivatives of the methyl glycosides formed has been shown to be an effective means of determining the polysaccharide gums and hydrocolloids used in food products (Schmolck and Mergenthaler, 1973; Preuss and Thier, 1983). Methanolysis has also been used to characterize the soluble polysaccharides in orange juice (Kauschus and Thier, 1985). Polysaccharide gums have also been determined in dairy foods (Glueck and Thier, 1983) and in dairy products, salad dressings and meat sauces (Lawrence and Iyenger, 1985) by analysing the neutral monosaccharides produced by trifluoroacetic acid hydrolysis as their aldononitrile acetate derivatives.

4.7.4.5 Polysaccharides in dietary fibre

The 'unavailable' carbohydrates. i.e. the structural polysaccharides that constitute the principal fraction of plant cell walls and are the major component of dietary fibre, have been determined for a number of foods and food ingredients by means of the hydrolysis/GC procedures outlined above. Polysaccharide profiles, expressed as the constituent monosaccharide residues, were obtained for a range of vegetables, legumes, cereals and fruits by Schweizer and Wuersch (1979) and for a variety of cereals, potatoes and vegetables (Englyst *et al.*, 1982; Theander and Westerlund, 1986). Englyst *et al.* (1983) characterized the polysaccharides in 37 cereal products and Bittner *et al.* (1982) did the same for a series of vegetables. Selvendran and coworkers have used hydrolysis/GC to characterize the polysaccharides in food ingredients, for example in cabbage (Stevens and Selvendran, 1980). Voragen *et al.* (1983) also used these procedures to determine the polysaccaride profiles of ethanol-insoluble residues of various fruits and vegetables. Fig. 4.9 shows a GC chromatogram of the hydrolysate of green coffee bean polysaccharides analysed as alditol acetates.

4.7.4.6 Other polysaccharide analyses

The softening of fruit during ripening is primarily due to degradation of the cell wall polysaccharides. Gross and Sams (1984) used acid hydrolysis and GC analysis of the aldononitrile derivatives to monitor the changes in the component neutral sugar residues in polysaccharides during the ripening of 17 different fruits.

Starch has been estimated by GC analysis of the glucose produced by acid hydrolysis of its constituent polyglucans, amylose and amylopectin (Englyst *et al.*, 1983). Banks *et al.* (1973) suggested that GC of the TMS ethers of isomaltose could be used as a means of estimating the number of branch points in amylopectin.

High-resolution capillary GC has proved invaluable in analysing the products of the classical methylation and hydrolysis method for the structural analysis of polysaccharides. GC data for derivatives of partially methylated monosaccharide derivatives have been tabulated (Churms, 1982).

4.8 REFERENCES

Adam, S. (1981). Separation of mono- and di-saccharide derivatives by high-resolution gas chromatography, *Z. Lebensm. Unters. Forsch.* **173**, 109–112.

Akhaven, I., Wrolstadt, R. E. and Richardson, D. G. (1980). Relative effectiveness of ion-exchange and lead acetate precipitation methods in isolating pear sugars and acids, *J. Chromatogr.*, **190**, 452–456.

Albersheim. P., Nevins, D. J., English, P. D. and Karr, A. (1967). A method for the analysis of sugars in plant cell-wall polysaccharides by gas-liquid chromatography, *Carbohyd. Res.* **5**, 340–345.

Banks, W., Greenwood, C. T. and Muir, D. D. (1973). The characterization of

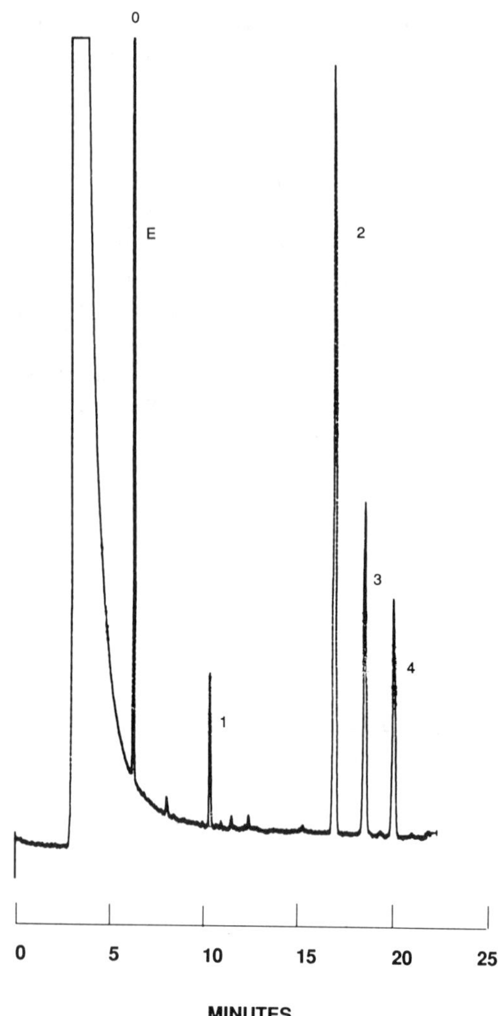

Fig. 4.9 — Analysis of polysaccharides in green coffee beans as the alditol acetate derivatives of their hydrolyzate. Column: SP 2330-coated fused-silica capillary column. Temperature-programmed: hold at 225°C for 2 min, to 245°C at 16°C/min, hold at 245°C. Peaks: 1, arabinose; 2, mannose; 3, galactose; 4, glucose.

starch and its components. Part 5. Observations on the quantitative acid hydrolysis of starch and glycogen, *Die Staerke* **25**, 405–408.

Beadle, J. B. (1969). Gas chromatographic determination of starch hydrolyzate distribution through maltoheptose, *J. Agric. Food Chem.* **17**, 904–906.

Birch, G. G. (1973). Gas–liquid chromatography of food carbohydrates with special reference to trimethylsilyl derivatives, *J. Food Technol.* **8**, 229–246.

Birkhed, A. D., Wange, B. and Edwardsson, S. (1980). Sockerarter och sockeralkoholer i livsmedel, *Var Foeda* **32**, 511–519.

Bittner, A. S., Harris, L. E. and Campbell, W. F. (1980). Rapid *N*-methylimidazole catalyzed acetylation of plant cell wall sugars, *J. Agric. Food Chem.* **28**, 1242–1245.

Bittner, A. S., Burritt, E. A., Moser, J., Street, J. C. (1982) Composition of dietary fibre : neutral and acidic sugar composition of the alcohol insoluble residue from human foods. *J. Food Sci.* **47,** 1469–1471.

Blakeney, A. B., Harris, P. J. Henry, R. J. and Stone, B. A. (1983). A simple and rapid preparation of alditol acetates for monosaccharide analysis, *Carbohyd. Res.* **113**, 291–299.

Blanc, M. B., Davis, G. E., Parchet, J.-M. and Viani, R. (1989). Chromatographic profile and carbohydrates in commercial soluble coffees, *J. Agric. Food Chem.* **37**, 926–930.

Bradbury, A. G. W. and Halliday, D. J. (1984). Structural characterization of polysaccharides using gas chromatography, in *Gums and Stabilizers for the Food Industry, Vol. 2*, Phillips, G. O., Wedlock, D. J. and Williams, P. A. (eds.) Pergamon, Oxford, pp. 45–53.

Brobst, K. M. and Lott, C. E. Jr. (1966). Determination of some components of corn syrup by gas–liquid chromatography of the trimethylsilyl derivatives, *Cereal Chem.* **43**, 35–43.

Brobst, K. M. and Scobell, H. D. (1981). Chromatographic analysis of sugars in cereals and cereal products, *Cereal Foods World*, **26**, 224–227.

Chepurnoi, I. P., Kunizher, S. M. and Katunkin, N. A. (1986). Gas chromatographic analysis of carbohydrates in liquid food products during glucose crystallization. *Izv. Vysch. Uchebn. Zaved. Pishch. Tekhnol.* **3**, 30–32. *Chem. Abs.* **105**:113 907f.

Churms, S. C. (1982). Gas chromatography tables, in *CRC Handbook of Chromatography, Carbohydrates, Volume I*, Zweig, G. and Sherma, J. (eds), CRC Press Inc., Boca Raton, Florida, pp. 3–68.

Corzo, N., Olano, A. and Martinez-Castro, I. (1986). Differentiating milks processed at different temperatures/times by GLC analysis of free disaccharide composition, *Rev. Agroquim. Technol. Aliment.* **26**, 565–570.

Daniels, H. D., Warner, C. R. and Fazio, T, (1982). Gas chromatographic determination of sorbitol, mannitol and xylitol in chewing gum and sorbitol in mints, *J. Assoc. Off. Anal. Chem.* **65**, 558–591.

Date, C., Tanaka, H., Yoshikawa, K., Ueda, Y., Baba, T., Hayashi, M., Tanaka, Y., Schimada, T., Yamamoto, H., Owada, K., Okazaki, K., Ishii, R. and Shoji, H. (1982). Quantitative determination of sugars in foods by gas–liquid chromatography of trimethylsilyl derivatives, *Jpn. J. Hyg.* **37**, 516–529.

Davies, H. V. (1988). Rapid determination of glucose, fructose and sucrose in potato fibres by capillary gas chromatography, *Potato Res.* **31**, 569–572.

Deifel, A. (1985). Gaschromatographische Bestimmung der Zucker im Honig, *Deut. Lebensm.-Rundsch.* **81**, 209–212.

Demaimay, M. (1978). Simplified method for analysis of silyl derivatives of sugar oximes by gas chromatography. Applications to foods, *Ann. Tech. Agric.* **27**, 455–467.

Donike, M. (1973). Acylierung mit bis(acylamiden). *N*-methyl-bis(trifluoroacetamid) and bis(trifluoroacetamid), zwei neue Reagenzien zur Trifluoroacetylierung, *J. Chromatogr.*, **78**, 273–279.

Drawert, F., Lessing, V. and Leupold, G. (1976). Gruppentrennung von organischen Saeuren, Kohlenhydraten und Aminosaeuren mit Ionenaustauschen, *Chromatographia* **9**, 373–379.

Ekland, G., Jofesson, B. and Roos, C. (1977). Gas–liquid chromatography of monosaccharides at the picogram level using glass capillary columns, trifluoracetyl derivatization and electron-capture detection, *J. Chromatogr.* **142**, 575–585.

Englmaier, P. (1985). Trifluoroacetylation of carbohydrates for glc using *N*-methylbis(trifluoroacetamide), *Carbohyd. Res.* **144**, 177–182.

Englyst, H., Wiggins, H. S. and Cummings, J. H. (1982). Determination of the non-starch polysaccharides in plant foods by gas–liquid chromatography of constituent sugars as alditol acetates, *Analyst* **107**, 307–318.

Englyst, H. N., Anderson, V. and Cummings, J. H. (1983). Starch and non-starch polysaccharides in some cereal foods, *J. Sci. Food Agric.* **34**, 1434–1440.

Folkes, D. J. (1985). Gas liquid chromatography in *Analysis of Food Carbohydrates*, Birch, G. G. (ed.), Elsevier, London, pp. 91–123.

Ford, C. W. (1979). Simultaneous quantitative determination of sucrose, raffinose, and stachyose by invertase hydrolysis and gas–liquid chromatography, *J. Sci. Food Agric.* **30**, 853–858.

Ford, C. W. (1982). A routine method of identification and quantitative determination by gas–liquid chromatography of galacturonic acid in pectic substances, *J. Sci. Food Agric.* **33**, 318–324.

Fox, A., Morgan, S. L. and Gilbert, J. (1989). Preparation of alditol acetates and their analysis by gas chromatography and mass spectrometry. In *Analysis of Carbohydrates by GLC and MS*, Bierman, C. J., and McGinnis, G. D. (eds), CRC Press Inc., Boca Raton, Florida, pp. 87–117.

Fuchs, G., Gawell, B.-M. and Lidhem, B. M. (1974). Quantitative determination of low-molecular weight carbohydrates in foods by gas–liquid chromatography, *Swed. J. Agric. Res.* **4**, 49–52.

Furneaux, R. H. (1983). Byproducts in the preparation of penta-*O*-acetyl-D-hexonitriles from D-galactose and D-glucose, *Carbohyd. Res.* **113**, 241–255.

Gianetto, A., Berruti, F. and Kempton, A. G. (1986). Lactose analysis in dilute solutions by gas chromatography using silylation, *Biotech. Bioeng.* **28**, 1119–1121.

Glueck, U. and Thier, H.-P. (1983). Isolierung natuerlicher Dickungsmitteln zur capillargaschromatographischen Bestimmung. *Z. Lebensm. Unters. Forsch.* **176**, 5–11.

Gross, K. C. and Sams, C. E. (1984). Changes in cell wall neutral sugar composition during fruit ripening: a species survey, *Phytochem.* **23**, 2457–2461.

Ha, Y. W. and Thomas, R. L. (1988). Simultaneous determination of neutral sugars and uronic acids in hydrocolloids, *J. Food Sci.* **53**, 574–577.

Harris, P. J., Blakeney, A. B., Henry, R. J. and Stone, B. A. (1988). Gas chromatographic determination of the monosaccharide composition of plant cell wall preparations. *J. Assoc. Off. Anal. Chem.* **71**, 272–275.

Harvey, C. D.; Jenness, R. and Morris, H. A. (1981). Gas chromatographic quantitation of sugars and non-volatile water-soluble organic acids in commercial Cheddar cheese, *J. Dairy Sci.* **64**, 1648–1654.

Henry, R. J. (1985). A comparison of the non-starch carbohydrates in cereal grains, *J. Sci. Food Agric.* **36**, 1243–1253.

Hoebler, C., Barry, J. L., David, A. and Delfort-laval, J. (1989). Rapid acid hydrolysis of plant cell wall polysaccharides and simplified quantitative determination of their neutral mono-saccharides by gas–liquid chromatography, *J. Agric. Food Chem.* **37**, 360–367.

Iverson, J. L. and Bueno, M. P. (1981). Evaluation of high pressure liquid chromatography and gas-liquid chromatography for quantitative determination of sugars in foods, *J. Assoc. Off. Anal. Chem.* **64**, 139–143.

Jaddou, H. and Al-Hakim, M. (1980). Gas–liquid chromatography of trimethylsilyl derivatives of sugars from Iraqi dates, *J. Agric. Food Chem.* **28**, 1208–1212.

Jamieson, A. M.(1976). A simple gas chromatographic procedure for the determination of sugars in brewing materials, *J. Am. Soc. Brew. Chem.* **34**, 44–48.

Kallio, H., Ahtonen, S., Raulo, J. and Linka, R. R. (1985). Identification of the sugars and acids in birch sap, *J. Food Sci.* **50**, 266–269.

Kamiyana, Y. and Sakai, Y. (1974). Quantitative determination of xylooligosaccharides by gas chromatography of the alditol trifluoroacetates, *Agric. Biol. Chem.* **38**, 2385–2390.

Kauschus, U. and Thier, H.-P. (1985). Zusammensetzung der geloesten Polysachariden von Fruechtsaeften, *Z. Lebensm. Unters. Forsch.* **181**, 395–399.

Kline, D. A., Fernandez-Flores, E. and Johnson, A. R. (1970). Quantitative determination of sugars in fruits by glc separation of TMS derivatives, *J. Assoc. Off. Anal. Chem.* **53**, 1198–1202.

Laine, R. A. and Sweeley, C. C. (1973). *O*-Methyl oximes of sugars. Analysis as *O*-trimethylsilyl derivatives by gas–liquid chromatography and mass spectrometry, *Carbohyd. Res.* **27**, 199–213.

Larson, P. A., Honold, G. R., Hobbs, W. E. (1974). Gas chromatographic separation of alpha-lactose and sucrose as the trimethylsilyl derivatives. *J. Chromatogr.* **90**, 345–349.

Lawrence, J. F. and Iyenger, J. R. (1985). Gas chromatographic determination of polysaccharide gums in foods after hydrolysis and derivatization, *J. Chromatogr.* **350**, 237–244.

Leblanc, D. J. and Ball, A. J. S. (1978). A fast one-step method for the silylation of sugars and sugar phosphates, *Anal. Biochem.* **84**, 574–578.

Lercker, G., Savioli, S., Vecchi, M. A., Sabatini, A. G., Nanetti, A. and Piana, L. (1986). Carbohydrate determination of royal jelly by high resolution gas chromatography (HRGC), *Food Chem.* **19**, 255–264.

Li, B. W. and Schuhmann, P. J. (1980). Gas–liquid chromatographic analysis of sugars in ready-to-eat breakfast cereals, *J. Food Sci.* **45**, 138–141.

Li, B. W. and Schuhmann, P. J. (1983). Sugar analysis of fruit juices: content and method, *J. Food. Sci.* **4**8, 633–653 (1983).

Li, B. W., Schuhmann, P. J. and Holden, J. M. (1983). Determination of sugars in yoghurt by gas–liquid chromatography, *J. Agric. Food Chem.* **31**, 985–989 (1983).

Long, A. R. and Chism, G. W. III (1987). A rapid direct extraction–derivatization method for determining sugars in fruit tissue, *J. Food Sci.* **52**, 150–154.

Low, N. H. and Sporns, P. (1988). Analysis and quantitation of minor di- and trisaccharides in honey using capillary G. C. *J. Food Sci.* **53,** 558–561.

Luke, M. A. (1971) Gas and thin layer chromatographic analysis of sugars in cocoa products, *J. Assoc. Off. Anal. Chem.*, **54**, 1432–1436.

Martin-Villa, M. C., Vidal-Valterde, C., Dabrio, M. V. and Rojas-Hidalgo, E. (1981). Chromatographic measurement of the carbohydrate content of some commonly used soft drinks, *Amer. J. Clin. Nutr.* **34**, 1432–1436.

Martinez-Castro, I. and Olano, A. (1978). Determination of lactulose in commercial milks, *Rev. Esp. Lech.* **110**, 213–217.

Mateo, R., Bosch, F., Pastor, A. and Jimenez, M. (1987). Capillary column gas chromatographic identification of sugars in honey as trimethylsilyl derivatives, *J. Chromatogr.* **410**, 319–328.

Mbugua, S. K., Ledford, R. A. and Steinkraus, K. H. (1983). Gas chromatographic determination of mono-, di-, trisaccharides in the UJI flour ingredients and during UJI fermentation. *Chem. Mikrobiol. Tech. Lebensm.* **8**, 40–45.

Mergenthaler, E. and Scherz, H. (1976). Analysis of polysaccharides used as food additives V. *Zeitschrift fur Lebensmittel-Untersuchung und Forschung* **162,** 159–162.

Molner-Perl, I. and Szakacs-Pinter, M. (1981). Gas–liquid chromatographic separation and determination of the components of maltitol syrups, *J. Chromatogr.* **216**, 219–228.

Molnar-Perl, I., Szakacs-Pinter, M., Kovago, A. and Petroczy, J. (1985). Extraction and quantification of the raffinosaccharides in soya bean, *Carbohyd. Res.* **138**, 83–89.

Morita, H. and Montgomery, W. G. (1978). Gas chromatography of silylated oxime derivatives of peat monosaccharides, *J. Chromatogr.* **155**, 195–197.

Morrison, I. M. (1975). Determination of the degree of polymerization of oligo- and polysaccharides by gas–liquid chromatography, *J. Chromatogr.* **108**, 361–364.

Mouillet, L., Luquet, F.-M. and Boudier, J.-F. (1977). Determination of sugars by gas chromatography. Application to dairy products, *Ann. Fals. L'Exp. Chim.* **70**, 145–155.

Mueller, B. and Goeke, G. (1972). Beitrag zur gaschromatographischen Bestimmung von Zuckern und Zuckeralkoholen in Lebensmitteln, *Deut. Lebensm. Rundsch.* **68**, 222–227.

Neeser, J.-R. and Schweizer, T. F. (1984). A quantitative determination of capillary gas chromatography of neutral and amino sugars (as *O*-methyloxime acetates) and a study on hydrolytic conditions for glycoproteins and polysaccharides in order to increase sugar recoveries, *Anal. Biochem.* **142**, 58–67.

Nozawa, Y. Hiraguri, Y. and Ito, Y. (1969). Studies on the acid stability of neutral monosaccharides by gas chromatography, with reference to the analysis of sugar components in the polysaccharides, *J. Chromatogr.* **45**, 244–249.

Nurok, D. and Reardon, T. J. (1977). The rapid gas-chromatographic determination of sucrose as its trimethylsilyl ether on an open tubular column, *Carbohyd. Res.* **56**, 165–167.

Oesterhelt, G., Vecci, M., Rucher, R. and Manz, U. (1980). Determination of sugar alcohols in chewing gum by gas chromatography, *Mitt. Geb. Lebensmittelunters. Hyg.* **71**, 419–426.

Olano, A. (1983). Presence of trehalose and sugar alcohols in sherry, *Am. J. Enol. Vitic.* **34**, 148–151.

Olano, A., Calvo, M. M., Reglero, G. (1986). Determination of free carbohydrates in milk using micropacked columns. *Chromatographia,* **21,** 538–540.

Pelletier, O. and Cadieux, S. (1982). Glass capillary or fused silica gas chromatography–mass spectrometry of several monosaccharides and related sugars: improved resolution, *J. Chromatogr.* **231**, 225–235.

Pierce, A. E. (1979). *Silylation of Organic Compounds*, Pierce Chemical Company, Rockford, Illinois.

Pierce (1989). Handbook and General Catalogue, Rockford, Illinois, pp. 152–164.

Prager, M. J. and Miskiewicz, M. A. (1979). GLC determination of individual sugars in confectionery products. *J. Assoc. Off. Anal. Chem.* **62,** 262–265.

Preuss, A. and Thier, H.-P. (1983). Isolierung natuerlicher Dickungsmittel aus Lebensmitteln zur capillargaschromatographischen Bestimmung, *Z. Lebensm. Unters. Forsch.* **176**, 5–11.

Preuss, A., Schulte, E. and Thier, H.-P. (1984). Nachweis von mono-, di-, and trisacchariden neben saccharose in Zwischen und Endprodukten der Zuckerindustrie, *Lebensm.-Wiss. Technol.* **1**, 163–166.

Quemener, B., Thibault, J.-F., Brunet, C. and Crepeau, M.-J. (1989). Improvements in the methanolysis of pectins by enzymic prehydrolysis, *Carbohyd. Res.* **190**, C7-C10.

Reineccius, G. A., Kavanagh, T. E. and Kenney, P. G. (1970). Identification and quantitation of free neutral carbohydrates in milk products by gas–liquid chromatography, *J. Dairy Sci.* **53**, 1018–1022.

Reyes, F. G. R., Wrolstad, R. E. and Cornwell, C. J. (1982). Comparison of enzymic, gas–liquid chromatographic, and high performance liquid chromatography methods for determining sugars and organic acids in strawberries at three stages of maturity, *J. Assoc. Off. Anal. Chem.* **65**, 126–131.

Roberts, E. J., Godshall, M. A., Clarke, M. A., Tsang, W. S. C. and Parrish, F. W. (1987). Methanolysis of polysaccharides: a new method, *Carbohyd. Res.* **168**, 103–109.

Rumpf, G. (1969). The silylation of substances occurring in natural products and detectable by gas chromatography, *J. Chromatogr.* **43**, 247–250.

Saeman, J. F., Moore, W. E., Mitchell, R. L. and Millett, M. A. (1954). Techniques for the determination of pulp constituents by quantitative paper chromatography, *Tappi*, **37**, 336–343.

Saura-Calixto, F., Canellas, J. and Garcia-Raso, A. (1984). Gas chromatographic analysis of sugars and sugar alcohols in the mesocarp, endocarp and kernel of almond fruit, *J. Agric. Food Chem.* **32**, 1018–1020.

Schaefer, H. (1983). Analysis of nonstarch polysaccharides in cocoa powder by a new micromethod, *J. Agric. Food Chem.* **31**, 1375–1376.

Schaeffler, K. J. and Morel du Boil, P. G. (1981). Quantitative gas chromatographic analysis of sucrose in the presence of sugar oximes using a buffered oximation reagent and glass capillary columns, *J. Chromatogr.* **207**, 221–229.

Schaeffler, K. J. and Morel du Boil, P. G. (1984). A review of gas chromatography in the South African sugar industry. Development and application of accurate methods for sugar analysis, *Sugar Technol. Rev.* **11**, 95–185.

Schmolck, W. and Mergenthaler, E. (1973). Beitraege zur Analytik von Polysacchariden, die als Lebensmittelnzusatzstoffe verwendet werden. II Gaschroma tographischer Nachweis nach Methanolyse und Trimethylsilylierung, *Z. Lebensm. Unters.-Forsch.* **152**, 263–273.

Schweizer, T. F. and Wuersch, P. (1979). Analysis of dietary fiber, *J. Sci. Food Agric.* **30**, 613–619.

Selvendran, R. R. and Du Pont, M. S. (1980). Simplified methods for the preparation and analysis of dietary fiber, *J. Sci. Food Agric.* **31**, 1173–1182.

Sennello, L. T. (1971). Gas chromatographic determination of fructose and glucose in syrups, *J. Chromatogr.* **56**, 121–125.

Sorensen, A. and Guldborg, M. (1984). Determination of sugars and sugar alcohols, *Publikation, Statens Levnedsmiddelinstitut*, **104**, 54 pp.

Sosulski, F. W., Elkowicz, L. and Reichert, R. D. (1982). Oligosaccharides in eleven legumes and their air-classified protein and starch fractions, *J. Food Sci.*, **47**, 498–502.

Stevens, B. J. H. and Selvendran, R. R. (1980). The isolation and analysis of cell wall material from the alcohol-insoluble residue of cabbage, *J. Sci. Food Agric.* **31**, 1257–1267.

Sullivan, J. E. and Schewe, L. R. (1977). Preparation and gas chromatography of highly volatile trifluoroacetylated carbohydrates using *N*-methyl-bis(trifluoroacetamide), *J. Chromatogr. Sci.*, **15** 196–197.

Sweeley, C. C., Bentley, R., Makita, M. and Wells, W. W. (1963). Gas–liquid chromatography of trimethylsilyl derivatives of sugars and related substances, *J. Am. Chem. Soc.* **85**, 2497–2507.

Tarrach, F. and Herrmann, K. (1987). Nachweis von Estern aus Fruchtsaeuren und Zuckern, in Fruchtsaftkonzentraten mittels GC–MS, *Z. Lebensm.-Unters. Forsch.* **184**, 381–384.

Theander, O. and Westerlund, E. A. (1986). Studies on dietary fiber. 3. Improved procedures for analysis of dietary fiber, *J. Agric. Food Chem.* **34**, 330–336.

Traitler, H., del Vedovo, S. and Schweizer, T. F. (1984). Gas chromatographic separation of sugars by on-column injection on glass capillary columns, *J. High Resolut. Chromatogr. Chromatogr. Commun.* **7**, 558–562.

Tsuda, T. and Nakanishi, H. (1983). Gas liquid chromatographic determination of sucrose fatty acid esters, *J. Assoc. Off. Anal. Chem.* **66**, 1050–1052.

Van Soest, P. J. and Wine, R. H. (1967) Use of detergents in the analysis of fibrous feeds. IV. Determination of plant cell wall constituents, *J. Assoc. Off. Anal. Chem.* **50**, 50–55.

Verwack, W., Foulon, M. and Vanbelle, M. (1978). Determination of mono-, di-, and trisaccharides in several foods by gas chromatography, *Revue Ferment. Ind. Aliment.* **33**, 48–58.

Voragen, F. G. J., Timmers, J. P. J., Linssen, J. P. H., Schols, H. A. and Pilnik, W. (1983), Methods of analysis for cell wall polysaccharides of fruit and vegetables, *Z. Lebensm. Unters. Forsch.* **177**, 251–256.

Weiss, A. H. and Tambawala, H. (1972). TMS derivatization in aqueous solutions, *J. Chromatogr. Sci.* **10**, 120–122.

West, I. R. and Moskowitz, G. J. (1977). Improved gas chromatography method for the quantification of saccharides in enzyme-converted corn syrups, *J. Agric. Food Chem.* **35**, 830–832.

Zuercher, K. and Hadorn, H. (1975). Optimierung der Versuchsbedingungen zur Silylierung der Zucker mit BSA und HMCS und der gaschromatographischen Trennung, *Deut. Lebensm. Rundssch.* **71**, 68–71.

Zuercher, K. and Hadorn, H. (1976). Vergleichende Zucker Bestimmungen mit gaschromatographischen, enzymatischen und reduktometrischen Methoden, *Deut. Lebensm. Rundsch.* **72**, 197–202.

5

Lipids

Michael H. Gordon

5.1 INTRODUCTION

Lipids constitute a diverse group of food components which are characterized by their high solubility in organic solvents, such as chloroform, and their relatively low solubility in water. They range from neutral lipids, e.g. triglycerides (or triacylglycerols according to the IUPAC–IUB nomenclature), which have a high solubility in low-polarity organic solvents, to polar lipids, e.g. glycolipids, which have a low solubility in these solvents. Gas chromatography has made an important contribution to the analysis of lipids with the analysis of fatty acid methyl esters (FAME) being one of the most common applications of GLC.

Many lipids must be derivatized before analysis because of their lack of volatility or low thermal stability, but some lipids, e.g. triglycerides, can be analysed without derivatization. The flame-ionization detector (FID) is used as the detector for most lipid analyses because lipids suitable for gas–liquid chromatography (GLC) analysis usually contain carbon, hydrogen and oxygen only.

Analysis of lipids in almost all foods except edible oils requires the initial extraction of the lipid fraction. In addition, in biological tissues containing active enzymes, inactivation of degradative enzymes, including lipase and lipoxygenase, may be required. Although free lipids may be extracted directly with organic solvents, lipids bound to non-lipid components often require a more severe treatment, e.g. acid hydrolysis, before extraction. Techniques for the extraction of lipids have been discussed in detail by Christie (1982), and Hitchcock and Hammond (1980).

5.2 TRIGLYCERIDE ANALYSIS

Triglycerides are the major components present in the lipid fraction of oils and fats, and analysis of triglycerides is of considerable importance in the food industry. The complexity of the triglyceride group varies, with some fats, including cocoa butter, being relatively simple in triglyceride composition but other fats being very complex, especially milk fat, which contains up to 64×10^6 individual triglycerides (Patton & Jensen, 1975). Complete separation of all the triglycerides in such complex mixtures is clearly not feasible, but GLC can still provide a partial analysis which is useful as a fingerprint of a fat. This has useful applications in the food industry in identifying fats or detecting adulteration or admixtures, for instance in the determination of cocoa butter equivalents in chocolate (Padley & Timms, 1980).

Packed-column GLC of triglycerides usually employs a non-polar stationary phase and achieves separation of triglycerides on the basis of carbon number (Fig. 5.1). The carbon number is the total number of carbon atoms in the acyl chains of the triglyceride.

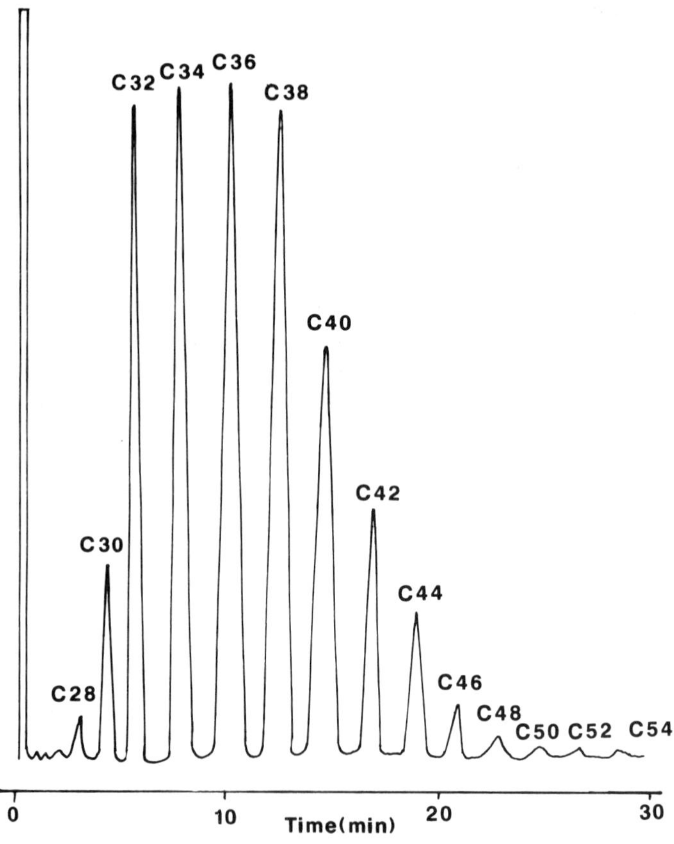

Fig. 5.1 — Triglycerides of coconut oil analysed on a 0.6 m × ¼″ (6 mm) external diameter glass column packed with 3% OV-1 on 100–120 Gas-Chrom Q. Temperature 250–355°C at 4° min^{-1}, hold at 355°C.

Short columns and high temperatures are required for the analysis because of the low volatility of triglycerides. A variety of conditions have been described in the literature, but these are usually quite similar to the IUPAC method 2.323 (IUPAC, 1987). This recommends a glass column 0.5 m–0.6 m in length, 2–4 mm i.d. filled with an acid-washed silanized support coated with up to 3% of a methyl polysiloxane, e.g. OV-1. Helium or nitrogen may be used as the carrier gas, with a flow rate of 50 ml min^{-1}. The oven is temperature-programmed from 220°C rising to 350°C at 4–5°C min^{-1}. Injector and detector temperatures are set at 350°C. The column should be conditioned prior to use by heating it at 350°C for at least 36 h with a carrier-gas flow rate of 5 ml min^{-1}. Quantification requires the injection of a standard mixture of triglycerides in order to calculate correction factors relative to trilaurin. Correction factors up to a value of 1.1 are acceptable.

Identification of the carbon number of unknown triglycerides may be achieved by interpolation of the plot of retention time against carbon number, which is a straight line for a temperature-programmed analysis with a single ramp rate and no holding times. With some stationary phases, dual-column operation is recommended because of the high degree of bleeding from the column (Padley & Timms, 1978), but the development of thermally stable phases, e.g. Dexsil 300, has reduced the requirement for this technique. Some gas chromatographs are also equipped with an automated bleed-compensation facility which overcomes any problems from a rising baseline. Conditioning of the column is essential for quantitative work, and Hammond (1981) has described an alternative procedure to that given in the IUPAC method.

If a highly polar stationary phase, e.g. Silar 10C, is used in packed-column GLC, partial separation of triglycerides varying in the degree of unsaturation can be achieved (Takagi & Itabashi, 1977). Selected stationary phases have been compared for use in the analysis of triglycerides by packed-column GLC (Aneja *et al.*, 1979). A polyphenyl ether sulphone (Poly-S-179) was preferred to Silar 10C for the partial separation of triglycerides according to the degree of unsaturation because of the short lifetime of columns prepared with the latter stationary phase.

Triglycerides† with a common carbon number which coelute in packed-column GLC with a non-polar stationary phase include four groups of compounds: (i) molecules with a common molecular weight with fatty acids which differ in chain length, e.g. MSP coelutes with PPP; (ii) triglycerides which differ in the positional distribution of fatty acids on the glycerol backbone, e.g. PSA coelutes with SPA; (iii) triglycerides which contain fatty acids varying in the degree of unsaturation, e.g. SOS coelutes with OLS; and (iv) triglycerides which include unsaturated fatty acids varying in the position or stereochemistry of the double bonds, e.g. SSO and SSE. The high resolution which is achievable with capillary columns has made it possible to separate some of these compounds.

Early applications of wall-coated open tubular (WCOT) columns in triglyceride analysis involved the use of very short columns (4–6 m) but Traitler & Prevot (1981a) used a 8-m column coated with SE30 to achieve a good separation of triglycerides with a carbon number of 54 but different degrees of unsaturation. The mixture SSS, SSO, SOO and OOO was separated into four peaks.

The resolution of triglycerides is improved by using longer capillary columns with

† Triglyceride structure is indicated by the following abbreviations for the acyl chains: B, butyryl; La, lauryl; M, myristyl; P, palmityl; S, stearyl; A, arachidyl; O, oleyl; L, linoleyl; E, elaidyl; Ln, linolenyl.

non-polar stationary phases up to 15–20 m in length (Grob *et al.*, 1980). The saturated C36 isomers BMS and LaLaLa could be separated on a 15 m × 0.30 mm capillary column coated with a film of 0.12 μm of OV-1. Little difference in selectivity between the non-polar methyl polysiloxane phases such as OV-1, SE-30, SE-52 and SE-54 was observed, although unsubstituted methyl polysiloxanes, e.g. OV-1, were preferred to substituted phases, e.g. SE-52, because of the lower elution temperature during the temperature-programmed analysis. Columns prepared with these stationay phases had reasonable lifetimes when used for triglyceride analysis involving temperature-programmed runs up to 370°C (Grob *et al.*, 1980; Monseigny *et al.*, 1979). The long retention times observed in triglyceride analysis with non-polar stationary phases encouraged the development of columns which could be used at higher temperatures.

The polyimide coating commonly used with fused-silica WCOT columns degrades at elevated temperatures above 370°C. This problem has been overcome by using aluminium-clad fused-silica WCOT columns, and this type of column has been used with a silanol-terminated polydimethylsiloxane stationary phase of improved thermal stability for triglyceride analysis at elevated temperatures (Lipsky & Duffy, 1986a). The column was found to be stable for isothermal analyses at 415°C, or for temperature-programmed analyses up to about 440°C. Recently a polyimide coating of improved thermal stability has been developed (Chrompack, 1988) and this appears to be a good alternative to aluminium cladding for high-temperature capillary columns.

Polarizable phenylmethylpolysiloxanes similar in character to OV-17 have been introduced as bonded stationary phases for the WCOT column analysis of triglycerides (Geeraert & Sandra, 1984, 1985). These columns give a better separation of triglycerides with a common carbon number than columns containing a non-polar stationary phase. SSO, SOO and OOO give three peaks with baseline separation in the analysis of butterfat triglycerides on a methyl 65% phenylpolysiloxane column (Fig. 5.2), whereas baseline separation is not achieved on a methyl polysiloxane column (Fig. 5.3). Whereas separations on non-polar stationary phases reflect mainly the number of unsaturated acyl groups in the triglyceride (Geeraert & de Schepper, 1982), polarizable columns achieve separation mainly on the basis of the number of double bonds and in some cases the distribution of chain lengths within the three acyl chains. Unsaturated triglycerides elute after saturated molecules in high-temperature analysis on a polarizable phase, whereas more saturated triglycerides elute later on a non-polar column. Retention times on a methylphenylpolysiloxane column increase in the sequence SSS < SOS < SOO < SLS < OOO < SLO < OLO < SLL < OLL < LLL < LLLn at high temperatures (Geeraert, 1987) but the elution sequence inverts at temperatures below 250°C, with more saturated triglycerides being more stongly retained at lower temperatures. It is therefore essential that the column should be maintained above this temperature when the triglyceride mixture is injected. Retention times on a methylphenylpolysiloxane column are relatively short. Hence a complete analysis of Brazil nut oil triglycerides can be achieved in 16 min (Fig. 5.4).

Grob (1979) found that a cold on-column injection technique was required for the accurate quantitative analysis of triglycerides. The standard deviation was 1–3% for

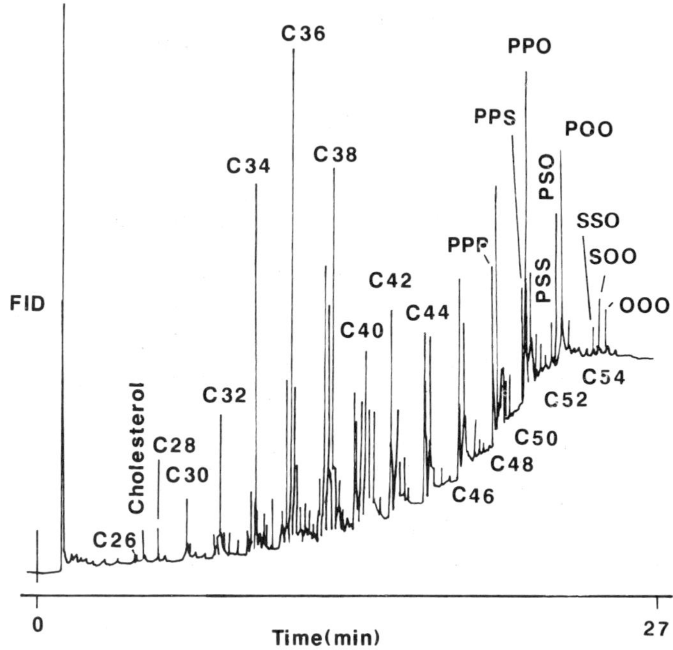

Fig. 5.2 — Triglycerides of milk fat. Column 25 m × 0.25 mm WCOT, 0.1-μm film, bonded methyl 65% phenyl polysiloxane; temperature 250–365°C at 5°C min^{-1}, hold at 365°C. Carrier gas: hydrogen. Reproduced from Lipsky & Duffy (1986) with permission from Aster Publishing Corporation.

cold on-column injection whereas it was about 10% for splitless injection and up to 35% for a split injection.

Cold on-column injection requires the front part of the column to be maintained at a temperature below the boiling point of the solvent. This is difficult to achieve with a conventional on-column capillary injector when the oven temperature is above 250°C. On-column injectors with extended cooling or a movable on-column injector that can be pulled out of the oven for introduction of the sample have been developed (Grob & Laubli, 1986a).

A programmed-temperature vaporizer (PTV) injector has also been developed. This involves injection of the sample into the cold injector prior to rapid heating of the injector, which introduces the sample onto the column. The PTV injector was compared with cold on-column injection for triglyceride analysis (Hinshaw & Seferovic, 1986), and it was found that the relative standard deviation of the response was slightly higher for the PTV injector than for cold on-column injection. The relative response factor for triolein was higher for PTV split injection than for cold on-column injection, but it was lowest for PTV splitless injection, suggesting some loss of triglyceride due to inlet-induced decomposition and mass discrimination with the latter technique (Table 5.1).

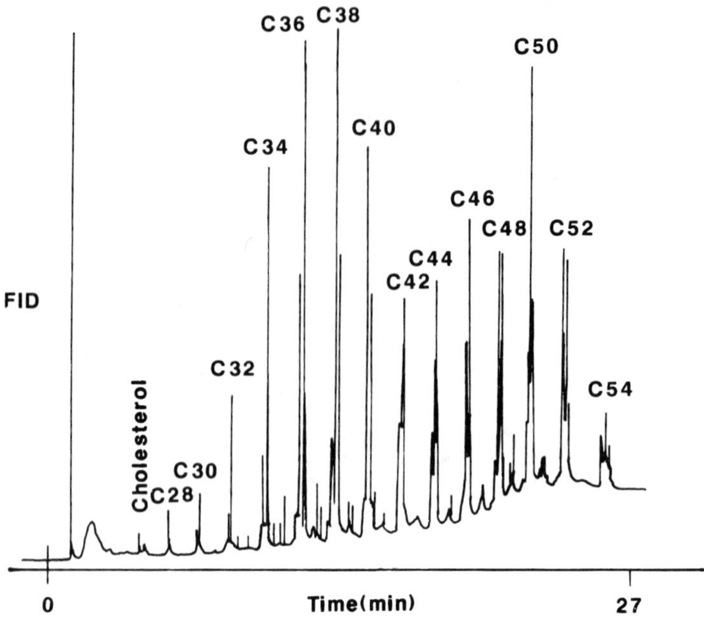

Fig. 5.3 — Triglycerides of milk fat. Column 25 m × 0.25 mm WCOT, 0.1-μm film, bonded methyl polysiloxane; temperature 250–365°C at 5°C min^{-1}, hold at 365°C; carrier gas: hydrogen. Reproduced from Lipsky & Duffy (1986) with permission from Aster Publishing Corporation.

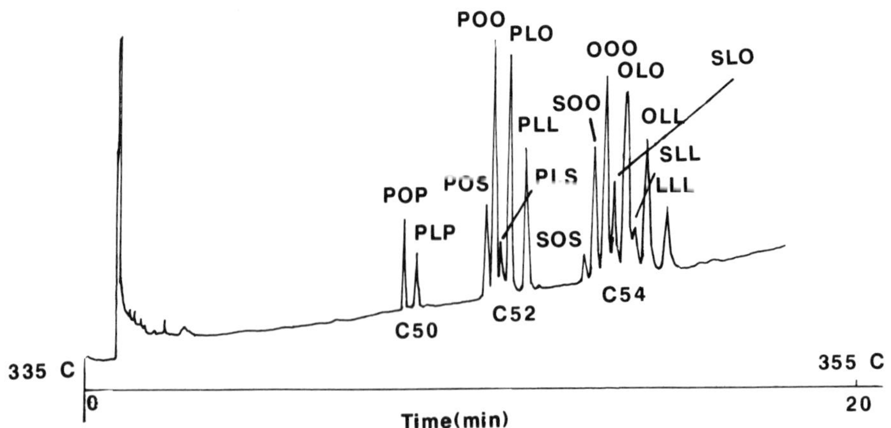

Fig. 5.4 — Analysis of Brazil nut oil on 25 m × 0.25 mm WCOT column coated with 0.12-μm phenylmethylsilicone. Temperature 335–355°C at 1°C min^{-1}. Carrier gas: hydrogen. Reproduced with permission from Geeraert & Sandra (1987).

Automated analysis of triglycerides may be achieved either with a short deactivated pre-column of wide-bore capillary tubing with extended cooling of the inlet (Termonia *et al.*, 1987) or with a PTV injector. Even with cold on-column injection of triglyceride mixtures, response factors should be used for accurate quantitative analysis. Capillary GC does not appear to be suitable for the quantitative analysis of highly unsaturated triglycerides because of the high correction factors which are required (Mares, 1988).

The separation of positional isomers of triglycerides, e.g. POP and PPO, is one of the more difficult separations in lipid analysis by GLC. At present HPLC appears more suitable for this separation, but Traitler and Rossier (1982) have reported separations of positional isomers using capillary columns coated with a dimethylpolysiloxane matrix containing inorganic salts, such as cupric chloride, cobalt chloride or silver nitrate.

5.3 MONO- AND DIGLYCERIDES

Analysis of mono- and diglycerides by gas chromatography can be performed on underivatized samples, but adsorption effects may lead to poor peak shapes. Acyl migration and disproportionation have also been reported at elevated analytical temperatures. The molecules are therefore usually derivatized before analysis. Acetates or TMS ethers are most commonly prepared, although other derivatives have also been used, including tert-butyldimethylsilyl ethers which are especially suitable for GC–MS because their mass spectra are more distinctive (Myher *et al.*, 1978). Engbersen & van Stijn (1976) found that silylation of mono- and diglycerides does not cause a significant degree of isomerization.

A mixture of hexamethyldisilazane, trimethylchlorosilane and pyridine (1 : 1 : 4) caused no isomerization, even after 3 days at room temperature. Partial isomerization of a 1-monoglyceride occurred in the presence of *N*-trimethylsilylimidazole in pyridine over a period of days. Thus it appears that partial glycerides are stable under normal silylation conditions.

A mono- or diglyceride sample may be silylated with a large excess of *N*-trimethylsilylimidazole in dry pyridine for 15 min at room temperature or with bis(trimethylsilyl)trifluoroacetamide in dry pyridine at 70°C for 30 min (Geeraert, 1987). The former reagent must be hydrolysed and removed by an aqueous wash, but the latter reagent is more volatile and may be injected onto the GC.

Short packed columns containing non-polar stationary phases are suitable for the separation of diglycerides according to carbon number (Christie, 1982). Some separation of components differing in the degree of unsaturation may occur with short packed columns containing non-polar stationary phases, but the separation of the molecules is improved with polar stationary phases. Mono- and diglycerides varying in the degree of unsaturation, and positional isomers of these molecules, may be separated on packed columns containing polar phases, such as Silar 5CP and Silar 10C (Myher & Kuksis, 1975; Myher *et al.*, 1978; Itabashi & Takagi, 1980; Fig. 5.5). 2-Monoglycerides elute before the 1-isomers, and 1,2-diglycerides elute before 1,3-diglycerides on all common stationary phases. The separation of the diglyceride positional isomers on the semipolar OV-17 phase was clearly shown to be better than on the non-polar SE-30 or OV-1 phases (Engbersen & van Stijn, 1976).

Table 5.1 — A comparison of injection techniques for triglyceride analysis (Hinshaw & Seferovic, 1986)

	Cold on-column		PTV Split		PTV Splitless	
	R	RSD	R	RSD	R	RSD
Tri-palmitolein	1.006	0.99%	1.038	1.86%	1.028	1.07%
Tri-olein	0.847	1.27%	0.893	2.70%	0.815	2.76%

R — Response factors relative to *n*-tetracosane.
RSD — Relative standard deviation of the response.

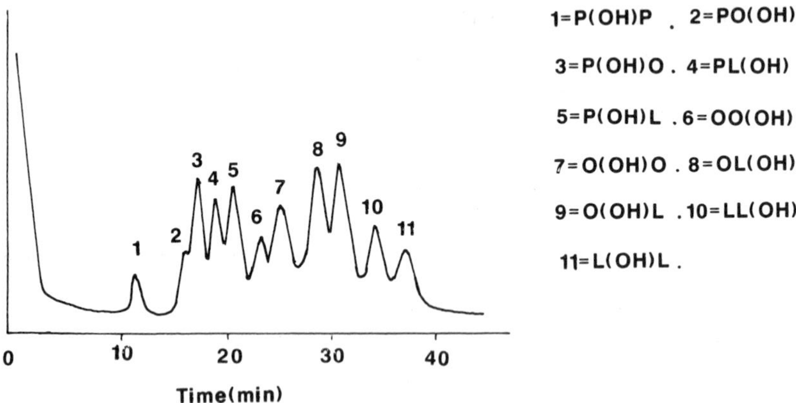

Fig. 5.5 — GC analysis of corn oil diglycerides on glass column (2 m × 3 mm i.d.) packed with 5% Silar 10C on 100–120 mesh Gas Chrom Q. Temperature 270°C. Reproduced with permission from Itabashi and Takagi (1980).

Diglycerides with a common carbon number but different acyl chain lengths may also be separated on packed columns. The acetate of *sn*-2-myristyl-3-caproylglycerol was separated from the *sn*-2-palmityl-3-butyrylglycerol derivative on a packed column containing EGGS-X (Kuksis *et al.*, 1973), but saturates and monoenes were not well resolved on this phase (Kuksis, 1972a, 1972b).

The superior resolution of capillary columns allows relatively simple mixtures of partial glycerides to be analysed more rapidly, and improves the separation of complex mixtures. The polar cyanopropylphenylpolysiloxane phases SP-2330 and Silar 5CP have been used for the analysis of monoglycerides and mixed acid diglycerides on the basis of molecular weight, positional substitution and degree of unsaturation. 1,3-Diglycerides are well separated from 1,2-diglycerides as TMS ethers on capillary columns coated with SP-2330, with unsaturated molecules being retained more strongly than saturated molecules. Unsaturated diglycerides usually elute before saturated molecules containing two additional carbon atoms, although polyunsaturated diglycerides may overlap with saturated diglycerides of a higher carbon number (Geeraert, 1987).

As in the case of triglycerides, unsaturated partial glycerides elute before saturated glycerides of the same carbon number on non-polar stationary phases, as in the analysis of methaneboronate derivatives of monoglycerides on an OV-1/Silanox capillary column (Gaskell & Brooks, 1977).

Isothermal conditions are commonly used in the analysis of monoglycerides and diglycerides, but temperature-programming is required when these components are present in mixtures with triglycerides. Fig. 5.6 demonstrates the separation of a

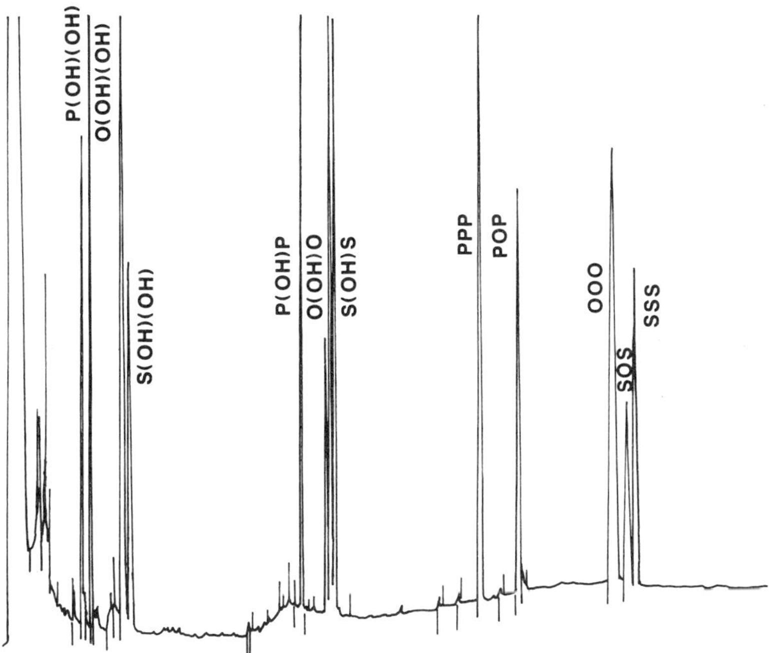

Fig. 5.6 — Analysis of a silylated mixture of mono-, di- and triglycerides. Column 13 m × 0.30 mm SE-30; temperature 190°C for 1 min, ramp to 230°C at 3°C min^{-1}, ramp to 300°C at 30°C min^{-1}, ramp to 330°C at 2°C min^{-1}, hold at 330°C. Carrier gas: hydrogen. Reproduced with permission from Traitler (1987).

mixture of glycerides after silylation on a capillary column coated with SE-30 (Traitler, 1987).

5.4 FATTY ACID METHYL ESTERS

5.4.1 Qualitative analysis

The complexity of the triglyceride composition of edible fats and other natural lipid mixtures has encouraged analysts to convert natural mixtures of lipids to FAMEs, since this simplifies the analytical problem. Methods of preparation of FAMEs have been reviewed (Sheppard & Iverson, 1975). *O*-Acyl lipids may be transesterified with a strong base, e.g. sodium methylate in methanol. Under these conditions, free fatty acids are not esterified, but if an acid catalyst, e.g. methanolic hydrogen chloride, is used, all fatty acids are converted to methyl esters. The standard IUPAC method 2.301 (IUPAC, 1987) involves the saponification of *O*-acyl lipids with sodium hydroxide followed by esterification with boron trifluoride in methanol, and a similar procedure is recommended by other international standards organizations. The boron trifluoride catalyst has a relatively short storage life, and artefacts may be observed in the chromatogram if the catalyst is not fresh. Preparation of FAMEs from fats containing short-chain fatty acids, e.g. butterfat, presents special problems.

Esters of short-chain fatty acids are saponified more rapidly than esters of longer chain fatty acids, and losses occur due to the increased solubility of short-chain esters in water or due to evaporation. A recommended procedure involves transesterification of the fat sample in iso-octane with 2 M methanolic potassium hydroxide for 6 min at room temperature, neutralization of the alkali and injection of the organic phase using methyl pentanoate as an internal standard (Bannon *et al.*, 1985). Transesterification with methanolic potassium hydroxide in a sealed tube has also been recommended for the preparation of methyl esters from milk fat (IUPAC method 2.301, section 5) (IUPAC, 1987).

There have been many studies on the analysis of FAME mixtures by packed-column GLC. Typically a 1.5–3.0 m column containing 3–20% (m/m) of liquid phase coated on an inert support is used. Both polar and non-polar stationary phases have been used but most analysts prefer polar phases because of the better separation of methyl esters varying in the degree of unsaturation. Unsaturated fatty acid methyl esters elute before saturated esters on non-polar stationary phases, but the elution sequence is reversed on polar phases. Haken (1975) has listed some of the preferred stationary phases.

Selection of a phase for FAME analysis is aided by consideration of the equivalent chain length (ECL) of the components of the mixture (Miwa *et al.*, 1960). The ECL value of a fatty acid is a number which indicates the carbon number of the straight chain saturated fatty acid with an identical retention time. ECL values can be calculated according to the following equation (Jamieson, 1975):

$$\mathrm{ECL} = 2\left[\frac{\log R_x - \log R_n}{\log \mathrm{R}_{n+2} - \log R_n}\right] + n$$

where R_x, R_n and R_{n+2} are the retention times of the unknown acid and saturated acids of chain length n and $n+2$.

A more common procedure is to determine the ECL values from the plot of the logarithm of the retention time of a homologous series of saturated straight-chain methyl esters against the number of carbon atoms in the acyl chain (Fig. 5.7).

Double bonds present in a fatty acid increase the ECL compared with that of a

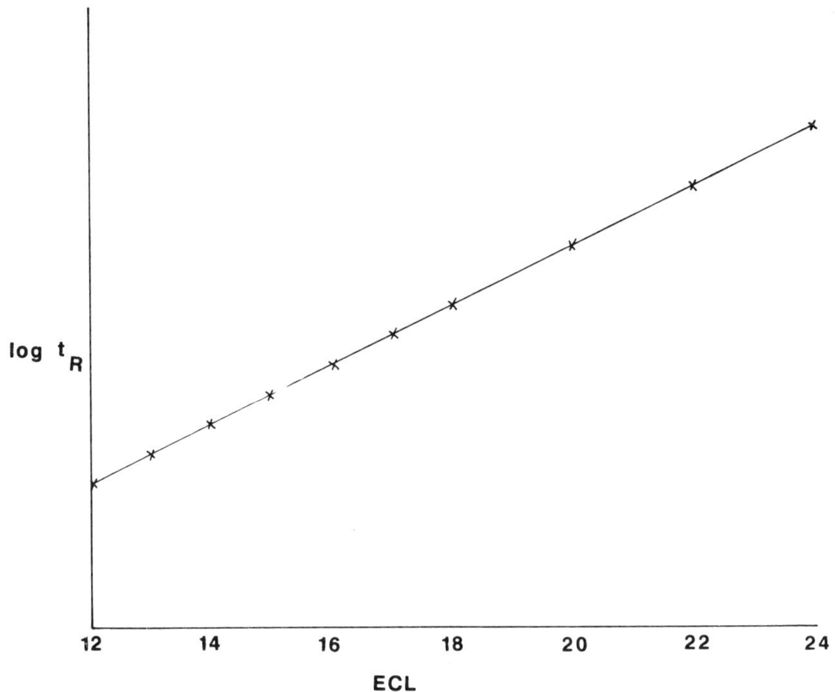

Fig. 5.7 — Plot of equivalent chain lengths versus log retention time for fatty acid methyl esters analysed on SP-2340.

saturated fatty acid on polar stationary phases. The extent of this increase in ECL is important if the sample contains polyunsaturated fatty acids as well as longer-chain saturated fatty acids. Table 5.2 compares the relative retention times of several fatty acid methyl esters on a number of stationary phases. The separation between 18:3 and 20:0 depends on the nature of the phase selected. In some cases 18:3 elutes before 20:0, whereas the elution sequence is reversed on other phases. The increase in ECL compared with the saturated esters of identical carbon number is dependent not only on the number of double bonds, but also on their position and their stereochemistry. Therefore although packed-column GC is often adequate for the analysis of natural vegetable oils which contain only 18:1*n*–9, 18:2*n*–6 and 18:3*n*–3 as unsaturated fatty acids, analysis of more complex mixtures including marine and animal fats, hydrogenated vegetable oils, or physiological samples of lipids usually requires capillary GC. Capillary GC is also usually more rapid than packed-column analysis. Proot and Sandra (1986) reduced the time of FAME analysis to 90 s using a 10 m × 100 μm i.d. WCOT column coated with a 0.1 μm 90% cyanopropylpolysiloxane phase. However, lengthy analytical times are common with high-resolution WCOT columns used in the analysis of mixtures containing long-chain fatty acids (Fig. 5.8). Wide-bore columns can reduce the analytical time to 13 min (Fig. 5.9) and these are very useful if the limited resolution is acceptable.

Table 5.2 — Variation in FAME retention times with type of liquid phase in packed-column operation (Haken, 1984)

Liquid phase and conditions	Relative retention times of methyl esters				
	18:0	18:1	18:2	18:3	20:0
10% phase, 200°C					
SP-2300 (50% phenyl, 50% cyanopropyl)	1.00	1.11	1.32	1.62	1.90
SP-2310 (25% phenyl, 75% cyanopropyl)	1.00	1.12	1.34	1.64	1.82
SP-2330 (10% phenyl, 90% cyanopropyl)	1.00	1.15	1.44	1.82	1.71
SP-2340 (100% cyanopropyl)	1.00	1.18	1.49	1.91	1.66
10% phase, 180°C					
SILAR-5CP (50% phenyl, 50% cyanopropyl)	1.00	1.13	1.35	1.68	1.93
SILAR-7CP (30% phenyl, 70% cyanopropyl)	1.00	1.17	1.46	1.88	1.80
SILAR-9CP (10% phenyl, 90% cyanopropyl)	1.00	1.19	1.52	1.99	1.74
SILAR-10C (100% cyanopropyl)	1.00	1.22	1.60	2.14	1.67
Separation at 175°C					
BDS (butanediol succinate)	1.00	1.11	1.36	1.75	1.90
EGA (ethyleneglycol adipate)	1.00	1.12	1.34	1.72	1.89
DEGS (diethyleneglycol succinate)	1.00	1.17	1.50	2.00	1.71
EGS (ethyleneglycol succinate)	1.00	1.20	1.56	2.14	1.07

Two of the most common phases for capillary GC analysis of fatty acid methyl esters are Carbowax 20M and SP-2340. Table 5.3 indicates the ECL of a number of FAMEs analysed on these stationary phases. The improved separation of unsaturated fatty acids, e.g. 18:2*n*–6 and 18:3*n*–3, on these polar phases compared to the non-polar phase SP-2100 can also be clearly seen in the Table. Separation of branched-chain fatty acids belonging to the iso or anteiso series is possible on WCOT capillary columns. These fatty acids are commonly found in low concentrations in animal fats.

Analysis of *cis*, *cis*-methylene-interrupted polyunsaturated fatty acids in fats is important because of the nutritional significance of these components. Athnasios *et al.* (1986) showed that a GLC method for the analysis of linoleic and linolenic acids using a 60 m SP-2340 capillary column represents a rapid, cost-effective and precise procedure. The results from this procedure agreed well with those from the AOAC standard enzymatic method.

Trans-unsaturated fatty acids occur naturally in small amounts in milk fat and in meat, but relatively large amounts are formed during the hydrogenation of edible fats. The double bond in unsaturated fatty acids migrates along the fatty acid chain during hydrogenation of a fat, so a complex mixture of *cis* and *trans* fatty acids may be present. It is usually not possible to achieve complete separation of all the isomers

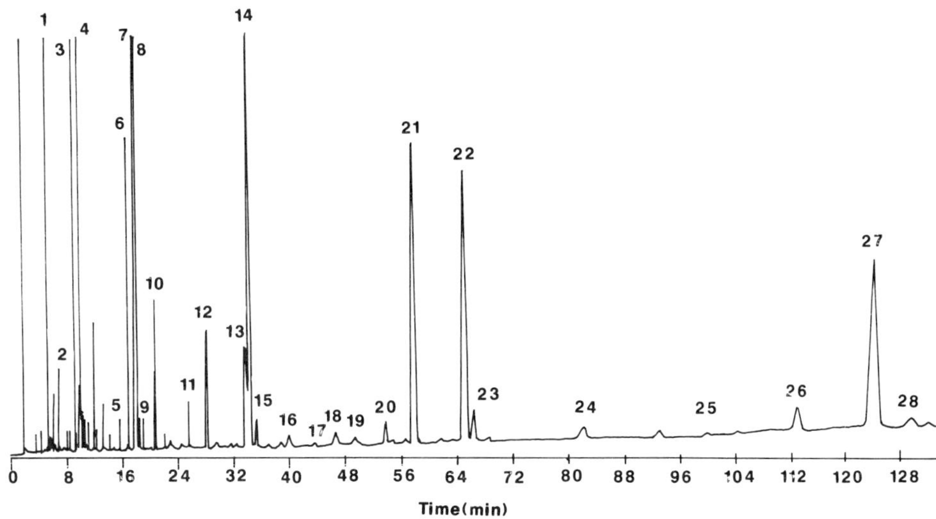

Fig. 5.8 — Analysis of cod liver oil fatty acid methyl esters. Column 30 m × 0.32 mm i.d. WCOT, 0.25 μm film, Supelcowax 10. Temperature 180°C, carrier gas helium. Reproduced with permission fron Supelco Inc, Bellafonte PA 16823. Figure from Supelco Chromatography Supplies International Catalogue 25 (1987). 1, 14:0; 2, 15:0; 3, 16:0; 4, 16:1*n*−7; 5, 16:4*n*−1; 6, 18:0; 7, 18:1*n*−9; 8, 18:1*n*−7; 9, 18:1*n*−5; 10, 18:2*n*−6; 11, 18:3*n*−3; 12, 18:4*n*−3; 13, 20:1*n*−11; 14, 20:1*n*−9; 15, 20:1*n*−7; 16, 20:2*n*−6; 17, 20:3*n*−6; 18, 20:4*n*−6; 19, 20:3*n*−3; 20, 20:4*n*−3; 21, 20:5*n*−3; 22, 22:1*n*−11; 23, 22:1*n*−9; 24, 21:5*n*−3; 25, 22:5*n*−6; 26, 22:5*n*−3; 27, 22:6*n*−3 & 24:0; 28, 24:1*n*−9.

produced during partial hydrogenation of an edible oil. *Cis*-monoenes may be separated as a group from *trans*-monoenes on a packed column containing OV-275, and this has been developed into a standard method for total *trans* fatty acids in margarine (Gildenberg & Firestone, 1985).

Maximum separation of *cis*- and *trans*-isomers is achieved using long cyanopolysiloxane WCOT capillary columns. Thc polar phase SP-2560 gives better separation than SP-2330, SP-2340 or OV-275. Several positional isomers of *cis*- and *trans*-unsaturated fatty acids can be separated with a 100 m column coated with SP-2560, as seen in Fig. 5.10. The separation is strongly temperature-dependent on this column, as is evident from the chromatograms at 170 and 175°C.

Preparative thin-layer chromatography (TLC) on silver nitrate-treated plates can be used to separate *cis*- and *trans*-unsaturated FAMEs prior to capillary GLC on a 60 m SP-2340 WCOT column and this procedure provides an excellent correlation with infra-red spectroscopy for the determination of the total *trans*-unsaturated fatty acids in foods (Lanza & Slover, 1981). This study showed that direct capillary GLC on a 100 m SP-2340 WCOT column without prior TLC separation was not satisfactory because two *trans* isomers coeluted with 18:1*n*−9*c* and 18:1*n*–7*c*. The determination of *trans*, *trans*-isomers of octadecadienoic acids has also been described (Mutter & Homan, 1987). The procedure involves silver nitrate high-performance liquid chromatography (HPLC) separation of FAMEs into several fractions prior to

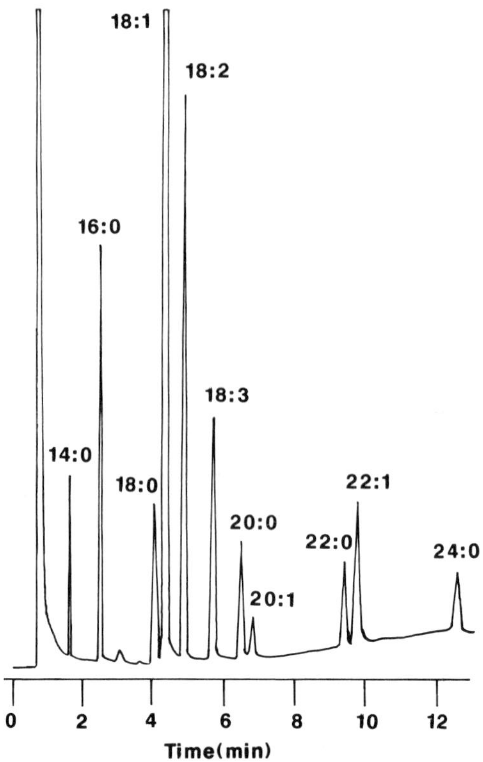

Fig. 5.9 — Analysis of rapeseed oil fatty acid methyl esters. Column 30 m × 0.75 mm i.d., 1.0 μm film. Supelcowax 10. Temperature 180°C, carrier gas helium. Reproduced with permission from Supelco Inc., Bellefonte PA 16823. Figure from Supelco Chromatography Supplies International Catalogue 25 (1987).

GLC analysis on a 50 m × 0.22 mm CP–Sil 8 (100% cyanopropylpolysiloxane) WCOT column. Analysis of conjugated *cis* and *trans* isomers of octadecatrienoic acids has been reported (Takagi & Itabashi, 1981). These fatty acids occur in a few minor seed oils, including tung oil

Various other minor fatty acids occur in edible oils and some of these have been analysed by GLC. Cyclopropene fatty acids occur in a few oils including cotton seed and Kapok seed oils. These molecules may polymerize or partially decompose at the elevated temperatures required for GLC analysis. Therefore, the fatty acids are usually converted into ether and ketone derivatives with silver nitrate or hydrogenated prior to analysis (Conway *et al.*, 1985). However, some cyclopropene FAMEs can be analysed directly by GLC without decomposition if on-column injection is used with a packed or capillary glass column (Fisher & Schuller, 1981; Bianchini *et al.*, 1981; Ralaimanarivo *et al.*, 1982).

The ECL values of the methyl esters of cyclopropane fatty acids have been determined on polar and non-polar stationary phases (Christie *et al.*, 1968). Separations of ester mixtures containing cyclopropane fatty acids have also been reported

Table 5.3 — ECL values of FAMEs on three stationary phases

Fatty acid	Carbowax 20 M[a]	SP-2340[b]	SP-2100[c]
14:0	14.00	14.00	14.00
16:0	16.00	16.00	16.00
18:0	18.00	18.00	18.00
18:1*n*−11	18.12	—	—
18:1*n*−9	18.18	18.68	17.70
18:1*n*−7	18.27	18.76	17.76
18:1*n*−5	18.39	—	17.87
18:2*n*−6	18.62	19.61	17.61
18:3*n*−6	18.92	20.28	17.43
18:3*n*−3	19.23	20.64	17.66
18:4*n*−3	19.53	—	17.47
20:0	20.00	20.00	20.00
20:1*n*−9	20.18	20.60	19.71
20:1*n*−7	20.26	—	19.77
20:2*n*−6	20.62	21.50	19.62
20:3*n*−6	20.88	22.14	19.40
20:4*n*−6	21.07	22.59	19.21
20:3*n*−3	21.24	22.52	—
20:4*n*−3	21.50	—	19.45
20:5*n*−3	21.68	23.63	19.25
22:0	22.00	22.00	22.00

[a] Ackman & Sipos (1981).
[b] Calculated from Lanza & Slover (1981).
[c] Gillan (1983).

on capillary columns (Lercker, 1983). Cyclopropane fatty acids elute just before the corresponding straight-chain acid on an OV 1 capillary column. Hydroxy fatty acids, including ricinoleic and dihydroxystearic acids, which occur in castor oil, may be analysed as methyl esters on a 15 m WCOT capillary column coated with Carbowax 20M (Lercker, 1983). The hydroxy group is commonly derivatized before analysis. Kawamura and Gagorian (1988) converted the methyl esters of hydroxy fatty acids to TMS ethers before separating isomers on a DB-5 capillary column.

Cyclic fatty acid monomers are formed during the heating of vegetable oils at elevated temperatures. The size of the ring and the nature of substituents can be determined by GC–MS after the total hydrogenation with a platinum oxide catalyst (Meltzer *et al.*, 1981; Sebedio, 1985). The position of the ethylenic bonds can be determined by GC–MS of the TMS derivatives (Saito & Kaneda, 1976). The determination of the degree of unsaturation of the cyclic fatty acid monomers can be performed by capillary GC–MS, while the stereochemistry of the double bonds can be assessed by GC–FTIR (Fourier transform infrared spectroscopy) (Sebedio *et al.*, 1987).

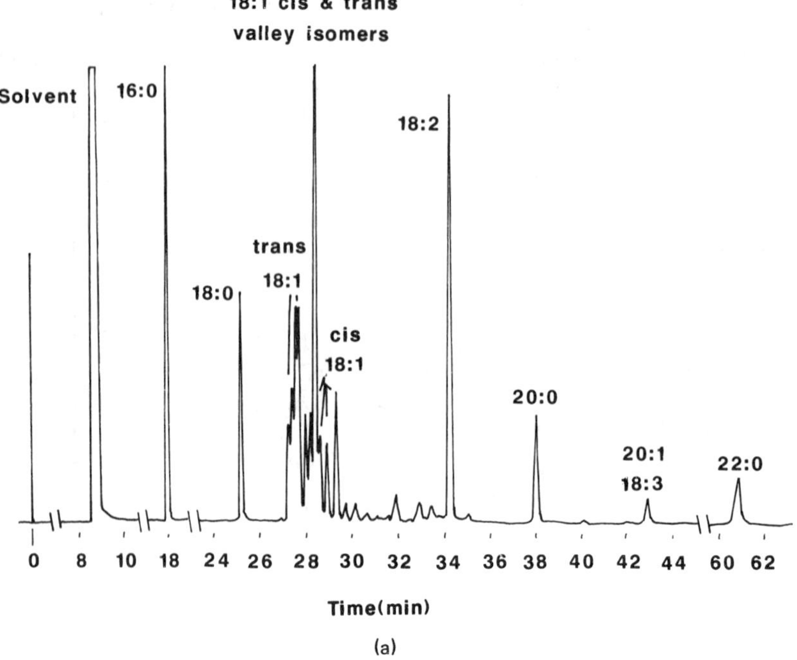

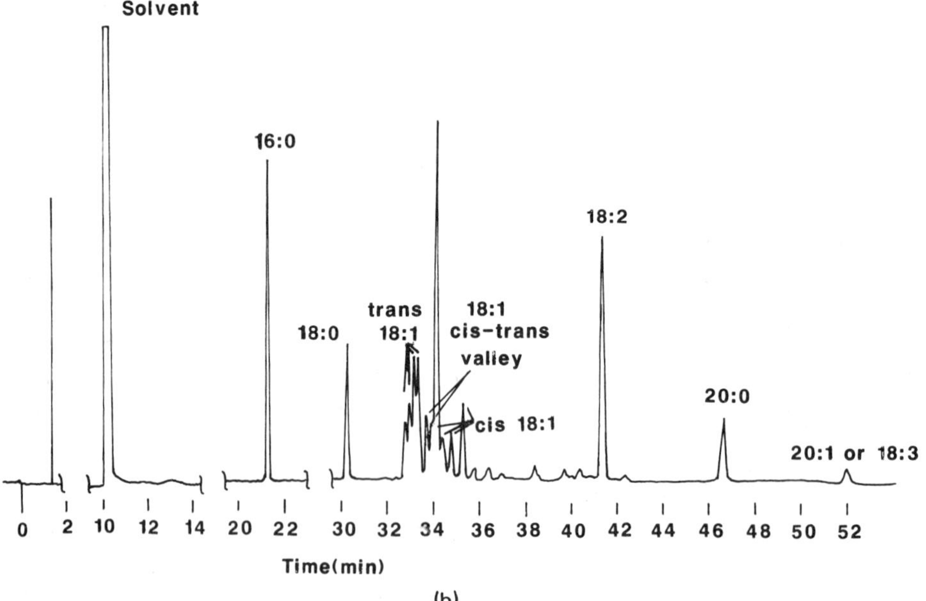

Fig. 5.10 — Analysis of positional isomers of *cis* and *trans* fatty acid methyl esters. Column 100 m × 0.25 mm i.d. FSOT, 0.20-μm film. SP-2560. Temperature: (a) 175°C (b) 170°C. Carrier gas: helium. Reproduced with permission from Supelco Inc, Bellefonte PA 16823. Figure from Supelco Inc GC Bulletin 822.

5.4.2 Quantitative analysis of fatty acid methyl esters

The FID responds to ions generated by the combustion of the C–H components of the molecule. It does not respond to the carboxylic acid carbon atom. Therefore, correction factors have to be applied to FAME mixtures to allow for the greater relative contribution of the carboxylic acid group to the molecular mass of shorter chain fatty acids. Theoretical relative response factors have been calculated (Ackman & Sipos, 1964) and subsequent studies have demonstrated that these factors are correct if care is taken in the preparation and analysis of methyl ester mixtures (Craske & Bannon, 1987).

5.5 FREE FATTY ACIDS

Free fatty acids are usually present at low levels in edible fats. Although they may make an important contribution to flavour in some products such as cheese, they are perceived as off-flavours in oils and fats if present at excessive levels. Therefore the analysis of free fatty acids is an important factor in assessing the quality of oils and fats. Free fatty acids are commonly analysed by separation from the glycerides and conversion to methyl esters with a reagent such as boron trifluoride in methanol. Diazomethane treatment or silylation are alternative derivatization procedures that allow free fatty acids to be analysed in the presence of glycerides (Goh & Timms, 1985; Lercker, 1983). Severe tailing and irreversible adsorption may occur with packed-column analysis of underivatized free fatty acids, although this may be overcome by using an acidic stationary phase, pretreating the column with orthophosphoric acid (Christie, 1982) or saturating the carrier gas with formic acid (Zijlstra *et al.*, 1977). Free fatty acids may be analysed without derivatization on mixed OV-1-FFAP-coated glass capillaries (Porschmann *et al.*, 1984), or on a bonded FFAP capillary column (MacNamara & Burke, 1985). Good peak shapes were observed with these columns.

Volatile free fatty acids in mixed organic aqueous samples are difficult to analyse because the solvent may tail badly and the peaks may elute close to the solvent. Kim *et al.* (1987) added triethylamine to aqueous samples to convert free fatty acids to salts before evaporating the water and derivatizing the free fatty acids to *tert*-butyldimethylsilyl ethers, which were well resolved from the solvent when analysed on a 15 m × 0.32 mm DB5 capillary column. Traitler *et al.* (1988) found that the use of solvents comprising acetone–water (8:2) or acetonitrile–water (8:2) gave good peak shapes and a rapid return to the baseline after the solvent when volatile free fatty acids from biological specimens were analysed without derivatization on a 25 m × 0.32 mm i.d. Carbowax 20M capillary column. Good reproducibility was achieved in quantitative analysis in the latter study when isocaproic acid was used as an internal standard.

5.6 STEROLS, METHYLSTEROLS, DIMETHYLSTEROLS AND TRITERPENES

Sterols occur as relatively simple mixtures in edible oils in small concentrations (0.1–5.6%). They may readily be analysed by packed-column or capillary GLC after separation from other lipid classes. The IUPAC method 2.403 (IUPAC, 1987) involves saponification of the fat, extraction with diethyl ether and TLC using chloroform to separate the sterol fraction. The sterols are then analysed by packed-column GLC using a support coated with OV-17 or an alternative silicone rubber (SE-30 or JXR) at a level of 2–5%.

Sterols may be analysed without derivatization, but the peaks of TMS ethers or acetates show less tailing than those of the free sterols. Packed OV-1 or SE-30 columns do not separate ß-sitosterol from Δ^5-avenasterol but this can be achieved on packed OV-17 columns. However, no separation of Δ^5-sterols from the corresponding saturated stanol, which may occur in hydrogenated fats, can be achieved with OV-17 even on capillary columns. WCOT capillary columns coated with OV-1 can separate cholestanol and cholesterol as well as the common sterols found in edible oils (Kornfeldt & Croon, 1981; Fig. 5.11).

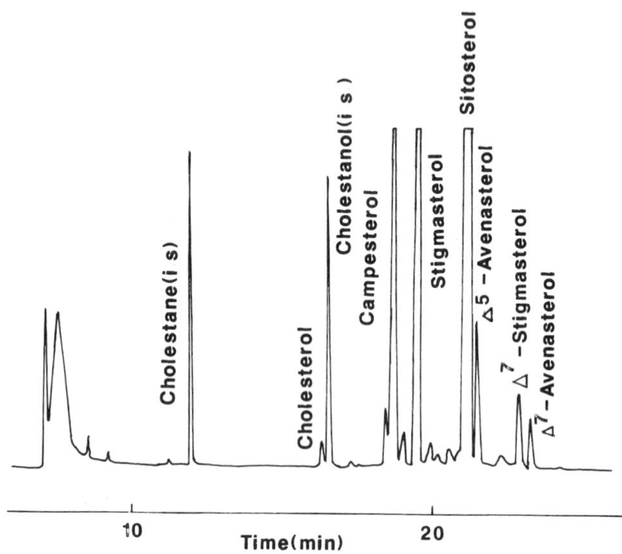

Fig. 5.11 — Analysis of TMS ethers of sterols from soybean oil on 50-m WCOT glass capillary column coated with OV-1. Temperature: 270°C. Reproduced with permission from Kornfeldt & Croon (1981).

The determination of ß-sitosterol in butter requires a capillary column to separate the sterol from the triterpene alcohol lanosterol, which coelutes on a packed column containing SP-2401 (Smith *et al.*, 1987). The effects of variations in sterol structure on retention properties on a variety of packed and capillary columns have been discussed by Patterson (1971), Thompson *et al.* (1980) and Itoh (1982). Although many workers prepare derivatives of sterols for capillary GC analysis, Geeraert (1985) found that the mixtures of free sterols could be analysed accurately on a 4 m × 0.3 mm HTS-OV-1 WCOT column. Good peak shape and coefficients of variation of less than 1% were observed using on-column injection.

4-Methylsterols, 4,4-dimethylsterols (triterpene alcohols) and triterpene hydrocarbons accompany sterols at relatively low levels in edible oils. These components can be separated from the sterols by TLC and analysed on WCOT capillary columns coated with OV-1 (Kornfeldt & Croon, 1981) or OV-17 (Itoh *et al.*, 1981). Packed columns containing OV-17 coated supports have also been used for methylsterol analysis (Itoh *et al.*, 1973). Retention times of 4-methylsterols and triterpenes have been reported for OV-1 and OV-17 phases (Homberg & Bielefeld, 1982).

Disteryl ethers, which are formed during the bleaching of fats and oils, may also be analysed by GC on a 20-m capillary column coated with a non-polar stationary phase (Schulte & Weber, 1987).

A method for simultaneously determining tocopherols and sterols has been described (Slover *et al.*, 1983). The method required saponification of the fat followed by extraction with cyclohexane and formation of the TMS ethers which were separated on a 50 m × 0.25 mm glass capillary column coated with Dexsil 400. Sterols may also be separated in intact lipid mixtures after silylation. Sterols elute before diglycerides on a methylpolysiloxane WCOT column (Fig. 5.13; Geeraert, 1985).

5.7 STERYL ESTERS

Sterols occur both free and esterified to fatty acids in edible oils and fats. While steryl esters are commonly saponified and analysed as sterols and FAMEs the analysis of intact steryl esters by GC has developed as techniques for analysing triglycerides have been introduced, since the problems of low volatility are similar for the two lipid classes. Analysis of steryl esters on short packed columns containing non-polar stationary phases leads to separation of molecules according to the number of carbon atoms in the molecule, and no separation of molecules varying in the degree of unsaturation can be achieved (Ross *et al.*, 1984). Improved separation of steryl esters can be achieved on capillary columns coated with non-polar and polar stationary phases.

Short capillary columns coated with non-polar stationary phases give a poor separation of some steryl esters in which the fatty acid chain varies in unsaturation, e.g. cholesteryl oleate and cholesteryl linoleate, but molecules in which the sterol component varies in unsaturation, e.g. stigmasteryl stearate and sitosteryl stearate, can be separated on a 12-m column coated with BP-1 (Evershed *et al.*, 1987). Other pairs of steryl esters that vary in the degree of unsaturation, e.g. avenasteryl linoleate and avenasteryl linolenate, are well separated on a 15-m non-polar capillary column (Lusby *et al.*, 1984). Cholesteryl stearate could be separated from cholesteryl oleate on a 25 m WCOT column coated with OV-1, but the latter ester was not separated from cholesteryl linoleate (Smith, 1982).

The advantages of using more polar stationary phases in the GC analysis of steryl esters were clearly demonstrated by Smith (1983). A 10-m SP-2330 column allowed the separation of all cholesteryl esters, including those of the sterol esterified with stearic (18:0), oleic (18:1), linoleic (18:2) and linolenic (18:3) acids. Unsaturated steryl esters were partly hydrogenated on this column if hydrogen was used as the carrier gas and therefore it was necessary to use helium instead (Smith, 1982). Polarizable stationary phases suitable for analysis of triglycerides are also appropriate for the analysis of steryl esters. Analysis of steryl esters on a 25-m 50% phenyl

50% methylpolysiloxane WCOT column allows good separation of a large number of steryl esters on the basis of both carbon number and degree of unsaturation (Fig. 5.12). There is no evidence for hydrogenation of unsaturated steryl esters when hydrogen is used as the carrier gas on this column.

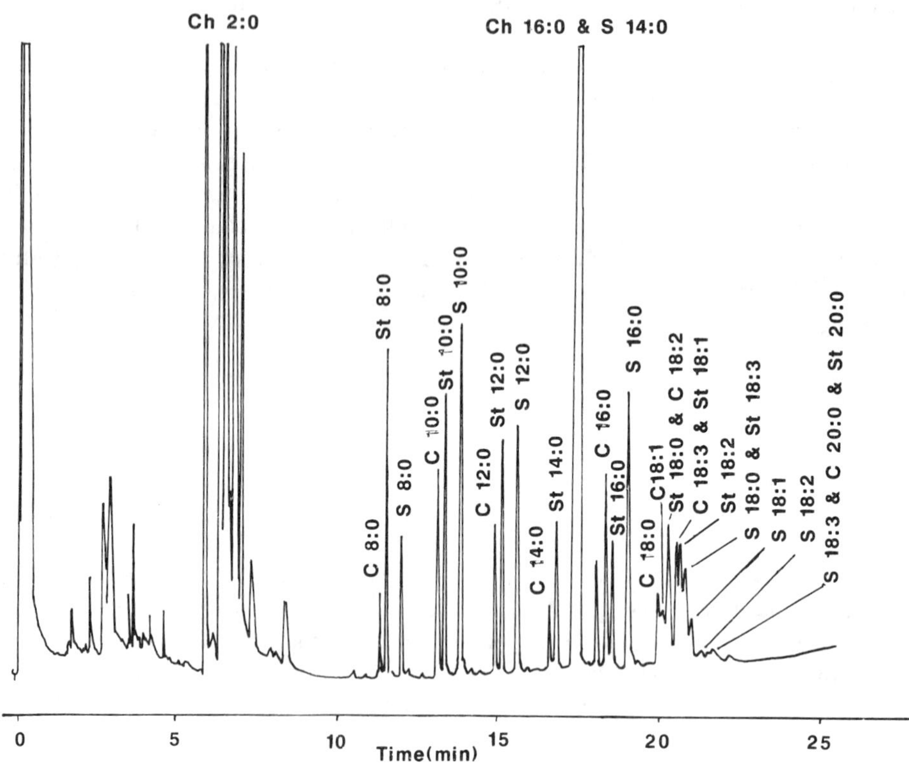

Fig. 5.12 — Analysis of mixture of esters of campesterol, stigmasterol and ß-sitosterol on a 25 m × 0.25 mm TAP (phenylmethylpolysiloxane) FSOT column. Temperature 120°C (0.2 min), ramp to 250°C at 30°C min^{-1}, ramp to 355°C at 5°C min^{-1}. C, campesterol; S, sitosterol; St, stigmasterol; Ch, cholesterol. Carrier gas: hydrogen.

Quantification of cholesteryl esters by capillary GLC has been investigated by Smith (1983) on both a non-polar 15-m DB-1 and a 10-m polar SP-2330 column. Despite on-column injection, the response factors fell sharply for cholesteryl esters eluting after cholesteryl stearate on both columns. Polyunsaturated steryl esters with long elution times show particularly low response factors on a 25 m 50%-phenyl 50%-methylpolysiloxane WCOT column (Griffiths, 1988).

5.8 WAXES

The major components of plant cuticular waxes are hydrocarbons, acids, aldehydes, alcohols and esters. The surface wax of wild mustard and rapeseed can be analysed by capillary GLC using a 10-m OV-101 WCOT capillary column (Andrew *et al.*, 1987). This has been developed into a method of distinguishing between the two seeds.

Wax esters varying in chain length and degree of unsaturation can be separated on a 25-m methylpolysiloxane capillary column. (Rezanka & Podojil, 1986). Isomers of saturated wax esters differing in the position of the ester group may be separated on a 15-m capillary column coated with the polar stationary phase SP-2340 if the ester group is shifted by four carbon atoms along the chain (Itabashi & Takagi, 1984). Good separations of wax esters according to the degree of unsaturation were also achieved in this work. Coupling of an HPLC to a capillary GC allows the rapid and accurate analysis of wax esters in edible oils (Grob & Laubli, 1986c).

5.9 INTACT LIPID MIXTURES

On-column injection of intact lipid mixtures onto thermally stable capillary columns allows the analysis of oils without molecular weight discrimination or degradation of thermally ıabile samples. In the analysis of coffee oil on a 25 m × 0.25 mm i.d. HTS-OV-1 capillary column, lipids eluted in the order free fatty acids, diterpenes, sterols, diterpene esters and diglycerides, steryl esters and triglycerides (Geeraert, 1985; Fig. 5.13).

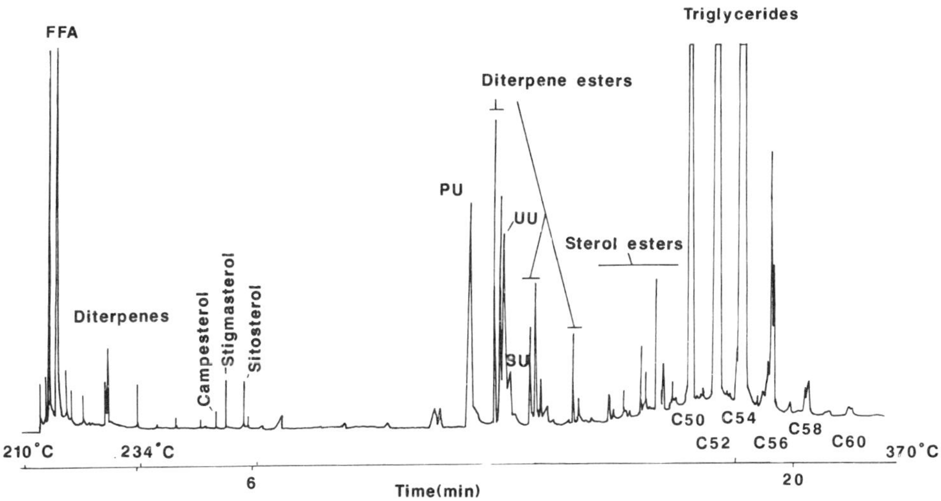

Fig. 5.13 — Analysis of coffee oil from raw beans. Column 25 m × 0.25 mm i.d. FSOT, HTS-OV-1. Temperature: 210°C, 0.1 min, 8°C min^{-1} to 234°C, 4°C min^{-1} to 246°C, 10°C min^{-1} to 370°C, 4 min isothermal. Reproduced with permission from Geeraert (1985).

Cholesteryl esters overlapped with triglycerides when a 10-m capillary column coated with SE-52 was used for the analysis of total serum lipids (Lercker, 1983). Overlap of cholesteryl esters with triglycerides has also been reported with polarizable capillary columns (Mares, 1988). Although underivatized mono- and diglycerides can be separated from triglycerides by capillary GLC (Traitler & Prevot, 1981), monoglycerides suffer from poor resolution and adsorption, with the consequence that high correction factors are required. Silylation of intact lipid mixtures should therefore be performed before GC analysis. *N,O*-Bistrimethylsilyltrifluoroacetamide has been recommended for this purpose (Traitler, 1987). At present the combination of HPLC and capillary GLC appears to be a more powerful approach for the analysis of complex lipid mixtures, but improvements in column selectivity in the future may increase the separation of lipids by GLC alone.

5.10 PHOSPHOLIPIDS

The FAMEs of phospholipids are commonly analysed after acid- or base-catalysed derivatization. Intact glycerophospholipids cannot be analysed by GC because of their lack of volatility and thermal instability. Enzymic hydrolysis allows partial glycerides to be prepared, and the use of phospholipase C, which liberates a 1,2-diglyceride and phosphate ester is particularly useful. Phospholipase A liberates 2-monoglycerides from glycerophospholipids and this provides a technique for distinguishing substituents at *sn*-1 and *sn*-2 positions. Ceramides may be converted to permethyl, *O*-TMS or methyl boronate derivatives for analysis by GC or GC–MS (Gaskell, 1985). Ceramide methylboronates were separated primarily according to carbon number on a 40-m OV-1 capillary column (Gaskell *et al.*, 1976).

Pyrolysis of phospholipids provides characteristic fragments which allow phosphatidyl ethanolamine, phosphatidyl choline, phosphatidyl inositol and phosphatidyl serine to be detected in biological samples by GC analysis. This technique allowed Caboni *et al.* (1984) to quantify the phospholipid classes in cow's milk.

5.11 OXIDIZED AND POLYMERIZED LIPIDS

Oxidation of lipids occurs during storage and processing of foods. Volatile oxidation products are produced at moderate temperatures and may be manifested as off-flavours in food products. Polymers form in significant concentrations at elevated temperatures, e.g. 180°–190°C which are reached during the frying of foods. Particularly important with regards to food safety are oxidized sterols.

Formation of cholesterol oxides may occur in egg products, cheese and meat products and these have been subjected to considerable investigation because of evidence that they are angiotoxic, cytotoxic, atherogenic, mutagenic and carcinogenic.

5.11.1 Volatile oxidation products

Direct injection of oxidized oil onto a packed GSC column with an inlet at 260°C causes breakdown of the hydroperoxides and allows the determination of pentane which can be used to monitor oil oxidation (Evans *et al.*, 1969). Dupuy *et al.* (1973) adsorbed oil samples on glass wool in a glass tube, which was then inserted into the inlet of a gas chromatogram. Volatiles were flushed from the glass tube onto a porous polymer prior to GC analysis. This technique was later modified (Min, 1981) by injection of the oil sample into a specially constructed U-tube to isolate the volatile components. The tube was later connected to the GC for separation and quantification of the volatile components. Direct injection of oil onto glass wool in a glass liner from which the volatiles were swept onto a WCOT capillary column allowed good separation of complex mixtures of volatiles which were identified by a mass spectrometer (Dupuy *et al.*, 1985). Dynamic headspace analysis has also been used to trap volatiles from oxidized oils on a porous polymer prior to analysis (Jackson *et al.*, 1986).

Direct sampling of the static headspace of edible oils has been used by several workers to introduce volatiles into a GC. Snyder *et al.* (1985) used this technique to separate and identify complex mixtures of volatiles arising from lipid oxidation. Static headspace sampling suffers from the disadvantage that it is less sensitive than a technique in which all volatiles are introduced onto the GC column. However, contamination of the column by involatile components is less likely to occur, and the technique is readily automated.

5.11.2 Sterol oxides

Complex mixtures of sterol oxides are formed during the oxidation of sterols. They are only present at low concentrations in food and are commonly concentrated by saponification of oil prior to analysis of the unsaponifiable sterol oxides. However, decomposition of some oxidation products including 7-ketocholesterol occurs during saponification under normal conditions. Concentration of the oxidation products by HPLC has recently been employed prior to GC analysis (Missler *et al.*, 1985; Maerker & Unruh, 1986). An alternative approach has been employed in which the fat is saponified at room temperature. 7-Ketocholesterol and cholesterol α-epoxide were recovered with greater than 95% recovery after this procedure (Park & Addis, 1986).

The cholesterol oxidation products have been analysed either by direct GC injection without derivatization (Tsai *et al.*, 1980; Gumulka *et al.*, 1982) or, more usually for complex mixtures of sterol oxidation products, as the silyl ethers (Missler *et al.*, 1985; Korahani *et al.*, 1981; Park and Addis, 1986; Nourooz-Zadeh & Appelqvist, 1987). Silylation changed the order of elution of the sterol oxidation products and allowed the separation of 25-hydroxycholesterol from 7-hydroxycholesterol on a 25 m × 0.2 mm i.d. 5% phenylpolysiloxane bonded-phase WCOT column (Maerker & Unruh, 1986). Thirteen oxidation products were separated as silyl ethers on this column. Several oxidation products coeluted when analysed as silyl ethers on a 25 m × 0.25 mm i.d. WCOT column coated with OV-101 (Korahani *et al.*, 1981).

5.11.3 Polymers

Polymers formed during the thermal degradation of oils are usually analysed by gel permeation chromatography. However, packed-column GLC analysis on non-polar columns can be used to monitor the formation of dimeric fatty acids produced by heated fats (Perkins & Wantland, 1973). Analysis of the methyl esters of dimeric fatty acids can be achieved in 6 min, and the speed and simplicity of the analysis make this a promising procedure for monitoring the deterioration of frying oils (Paradis & Nawar, 1981).

5.11.4 Ozonolysis products

Reductive ozonolysis represents a useful method of identifying complex mixtures of unsaturated fatty acids varying in positions of the double bonds. The mixtures of aldehydes, aldehyde esters and dialdehydes can be analysed on a packed column (Johnston *et al.*, 1978) but separation is improved on a 50 m × 0.2 mm FFAP fused-silica column (Dutton *et al.*, 1988). Capillary GC has also been applied to ozonized triglycerides using a 5 m HTS-OV-1 column (Geeraert, 1985; Geeraert and de Schepper, 1982; Verzele *et al.*, 1983).

5.12 CONCLUSION

Clearly the analysis of lipids by GLC has developed rapidly in recent years. Development of stationary phases with increased thermal stability and improved injection techniques for involatile solutes have extended the range of samples that can be studied by this technique. Further developments in polar bonded phase columns are likely to improve GC separations of lipids in the future. Multidimensional GC analysis, and the coupling of HPLC and capillary GC columns also appear to be developments that may have considerable significance for lipid analysis.

REFERENCES

Ackman, R. G. and Sipos, J. C. (1964). Application of specific response factors in the GC analysis of methyl esters of fatty acids with flame ionisation detectors. *J. Am. Oil Chem. Soc.*, **41**, 377–378.

Ackman, R. G. and Sipös, J. C. (1981). Standard wall-coated open-tubular columns for fatty acid analyses: is flexible fused silica the answer? *Prog. Lipid Res.*, **20**, 773–775.

Andrew, M., Hamilton, R. J. and Rossell, J. B. (1987). The chemical differentiation between *Sinapis arvensis* and *Brassica napus* seeds by surface wax analysis. *Fett. Wissenschaft Technologie*, **89**(1), 7–15.

Aneja, R., Bhati, A., Hamilton, R. J., Padley, F. B., Steven, D. A. (1979). Evaluation of selected stationary phases suitable for the GLC analysis of triglycerides, *J. Chromatogr.*, **173**, 392–397.

AOAC (1980). *Official methods of analysis*, 13th edition, section 28.071, Arlington, VA.

Athnasios, A. K., Healy, E. J., Gross, A. F. and Templeman, G. J. (1986). Determination of *cis,cis*-methylene interrupted polyunsaturated fatty acids in fats and oils by capillary gas chromatography. *J. Assoc. Off. Anal. Chem.*, **69**, 65–67

Bianchini, J. P., Ralaimanarivo, A. and Gaydou, E. M. (1981). Determination of cyclopropenoic and cyclopropanoic fatty acids in cottonseed and kapok seed oils by gas–liquid chromatography. *Anal. Chem.*, **53**, 2194–2201.

Bannon, C. D., Craske, J. D. and Hilliker, A. E. (1985). Analysis of FAME with high accuracy and reliability. IV. Fats with fatty acids containing 4 or more carbon atoms. *J. Am. Oil Chem. Soc.*, **62**, 1501–1507.

Caboni, M. F., Lercker, G. and Ghe, A. M. (1984). Analysis of phospholipids in cow's milk by high-temperature injection GC and HPLC. *J. Chromatogr.*, **315**, 223–231.

Christie, W. W. (1982). *Lipid Analysis*, 2nd edition, Pergamon Press, Oxford.

Christie, W. W., Gunstone, F. D., Ismail, I. A. and Wade, L. (1968). Synthesis and chromatographic and spectroscopic properties of cyclopropane esters derived from all methyl octadecenoates (2–17). *Chem. Phys. Lipids*, **2**, 196–202.

Chrompack News. (1988). Polyimide HT: a new high temperature column coating. **15**(2), 8.

Conway, J., Ratnayake, W. M. N. and Ackman, R. G. (1985). Hydrazine reduction in the GLC analysis of the methyl esters of cyclopropenoic fatty acids. *J. Am. Oil Chem. Soc.*, **62**(9), 1340–1343.

Craske, J. D. and Bannon, C. D. (1987). GLC analysis of the fatty acid composition of fats and oils: a total system for high accuracy. *J. Am. Oil Chem. Soc.*, **64**, 1413–1417.

Dupuy, H. P., Fore, S. P. and Goldblatt, L. A. (1973). Direct gas chromatographic examination of volatiles in salad oils and shortenings. *J. Am. Oil Chem. Soc.*, **50**, 340–342.

Dupuy, H. P., Flick, G. J. Jr, Bailey, M. E., St Angelo, A. J., Legendre, M. G. and Sumrell, G. (1985). Direct sampling capillary GC of volatiles in vegetable oils. *J. Am. Oil Chem. Soc.*, **62**, 1690–1693.

Dutton, H. J., Johnson, S. B., Pusch, F. J., Lie Ken Jie, M. S. F., Gunstone, F. D. and Holman, R. T. (1988). Composition of mixed octadecadienoates via ozonolysis, chromatography and computer solution of linear equations. *Lipids*, **23**(5), 481–489.

Engbersen, J. A. W. and van Stijn, F. (1976). Gas chromatographic analysis of synthetic glycidol esters, mono-, di-, and triglycerides. *Chem. Phys. Lipids*, **16**, 133–141.

Evans, C. D., List, G. R., Hoffmann, R. L. and Moser, H. A. (1969). Edible oil quality as measured by thermal release of pentane. *J. Am. Oil Chem. Soc.*, **46**, 501–504.

Evershed, R. P., Male, V. L. and Goad, L. J. (1987). Strategy for the analysis of steryl esters from plant and animal tissues. *J. Chromatogr.*, **400**, 187–205.

Fisher, G. S. and Schuller, W. H. (1981). Gas chromatographic analysis of cyclopro-prenoid acids in cottonseed oils. *J. Am. Oil Chem. Soc.*, **58**, 943–946.

Gaskell, S. J. (1985) in *Glass Capillary Chromatography in Clinical Medicine and Pharmacology* Jaeger, H. (ed.) Marcel Dekker, New York, pp 333.

Gaskell, S. J. and Brooks, C. J. W. (1977). Gas–liquid chromatography–mass spectrometry of phospholipid mixtures after enzymic hydrolysis. *J. Chromatogr.*, **142**, 469–480.

Gaskell, S. J., Edmonds, C. G. and Brooks, C. J. W. (1976). Applications of boronate derivatives in the study of ceramides by gas–liquid chromatography –mass spectrometry. *J. Chromatogr.*, **126**, 591–599.

Geeraert, E. (1985) in *Sample Introduction in Capillary GC*, Volume 1 Sandra, P. (ed.) Dr Alfred Huethig Verlag, Heidelberg, pp 133–158.

Geeraert, E. (1987) Chapter 2 in *Chromatography of lipids in biomedical research and clinical diagnosis* Kuksis, A. (ed.) Elsevier, Amsterdam.

Geeraert, E. and Sandra, P. (1984). On the potential of CGC in triglyceride analysis. *J. High Resolut. Chromatogr. Chromatogr. Commun.*, **7**, 431–432.

Geeraert, E. and Sandra, P. (1985). Capillary GC of triglycerides in fats and oils using a high temperature phenylmethylsilicone stationary phase. Part 1. *J. High Resolut. Chromatogr. Chromatogr. Commun.*, **8**, 415–422.

Geeraert, E. and Sandra, P. (1987). Capillary GC of triglycerides in fats and oils using a high temperature phenyl methylsilicone stationary phase. *J. Am. Oil Chem. Soc.*, **64**(1), 100–105.

Geeraert, E. and de Schepper, D. (1982). Structure elucidation of triglycerides by chromatographic techniques. Part 1. Capillary GC of ozonised triglycerides. *J. High Resolut. Chromatogr. Chromatogr. Commun.*, **5**, 80–84.

Geeraert, E., Sandra, P. and de Schepper, D. (1983(a)). On-column injection in the capillary gas chromatographic analysis of fats and oils. *J. Chromatogr.*, **279**, 287–295.

Geeraert, E., de Schepper, D. and Sandra, P. (1983(b)). Design and operation of a simple movable on-column injector. *J. High Resolut. Chromatogr. Chromatogr. Commun.*, **6**, 386–387.

Gildenberg, L. and Firestone, D. (1985). GC determination of *trans*-unsaturation in margarine: collaborative study. *J. Assoc. Offic. Anal. Chem.*, **68**, 46–51.

Gillan, F. T. (1983). Analysis of complex FAME mixtures on non-polar capillary gas chromatography columns. *J. Chromatogr. Sci.*, **21**, 293–297.

Goh, E. M. and Timms, R. E. (1985). Determination of mono- and diglycerides in palm oil, olein and stearin. *J. Am. Oil Chem. Soc.*, **62**(4), 730–734.

Griffiths, R. E. (1988). Personal communication.

Grob, K. Jr. (1979). Evaluation of injection techniques for triglycerides in capillary gas chromatography. *J. Chromatogr.*, **178**, 387–392.

Grob, K. Jr. and Laubli, T. (1986(a)). High oven temperature on-column injection in capillary gas chromatography. I. Sample introduction. *J. Chromatogr.*, **357**, 345–355.

Grob, K. Jr. and Laubli, T. (1986(b)). High oven temperature on-column injection in capillary gas chromatography. II. Avoidance of peak diffusion. *J. Chromatogr.*, **357**, 357–369.

Grob, K. Jr. and Laubli, T. (1986(c)). Determination of wax esters in olive oil by coupled HPLC–HRGC. *J. High Resolut. Chromatogr. Chromatogr. Commun.*, **9**, 593–594.

Grob, K. Jr, Neukom, H. P. Battaglia, R. (1980). Triglyceride analysis with glass capillary GC. *J. Am. Oil Chem. Soc.*, **57**, 282–286.

Gumulka, J., St. Pyrek, J. and Smith, L. L. (1982). Interception of discrete oxygen species in aqueous media by cholesterol: formation of cholesterol epoxides and secosterols. *Lipids*, **17**, 197–203.

Haken, J. K. (1975). Stationary phases and the applicability of classifications and preferred phase schemes to fatty esters. *J. Chromatogr. Sci.*, **13**, 430–439.

Haken, J. K. (1984). Developments in polysiloxane stationary phases in gas chromatography. *J. Chromatogr.*, **300**, 1–77.

Hammond, E. W. (1981). Quantitative chromatographic analysis of lipids in foods. *J. Chromatogr.*, **203**, 397–403.

Heckers, H., Dittmar, K., Melcher, F. W. and Kalinowski, H. O. (1977). Silar 10C,

Silar 9CP, SP 2340 and OV-275 in the GLC of FAME on packed columns. Chromatographic characteristics and molecular structures. *J. Chromatogr.*, **135**, 93–107.

Hinshaw, J. V. Jr. and Seferovic, W. (1986). Programmed temperature split–splitless injection of triglycerides; comparison to cold on-column injection. *J. High Resolut. Chromatogr. Chromatogr. Commun.*, **9**, 731–736.

Hitchcock, C. and Hammond, E. W. (1980). *Developments in food analysis techniques* 2, King, R. D. (ed.) Elsevier Applied Science Publishers, London, pp 185–224.

Homberg, E. and Bielefeld, B. (1982). Sterine und Methylsterine in Kakaobutter und Kakaobutter-Ersatzfetten. *Deutsche Lebensmittel-Rundschau*, **78**(3), 73–77.

IUPAC (International Union of Pure and Applied Chemistry) (1987). *Standard methods for the analysis of oils, fats, and derivatives*. 7th Revised and enlarged edition. Blackwell Scientific Publications, Oxford.

Itabashi, Y and Takagi, T. (1980). Gas chromatographic separations of di- and monoacylglycerols based on the degree of unsaturation and positional placement of acyl groups. *Lipids*, **15**, 205–215.

Itabashi, Y and Takagi, T. (1984). GC resolution on polar open-tubular columns of saturated and unsaturated wax ester isomers differing in combinations of acyl and alcoholic groups. *J. Chromatogr.*, **299**, 351–363.

Itoh, T. (1982). Structure retention relationship of sterols and triterpene alcohols in gas chromatography on a glass capillary column. *J. Chromatogr.*, **234**(1), 65–76.

Itoh, T., Tamura, T. and Matsumoto, T. (1973). Methylsterol compositions of 19 vegetable oils. *J. Am. Oil Chem. Soc.*, **50**, 300–303.

Itoh, T., Yoshida, K., Yatsu, T., Tamura, T. and Matsumoto, T. (1981). Triterpene alcohols and sterols of spanish olive oil. *J. Am. Oil Chem. Soc.*, **58**, 545–550.

Iverson, J. L. and Sheppard, A. J. (1977). Butyl ester preparation for gas–liquid chromatographic determination of fatty acids in butter. *J. Assoc. Offic. Analyt. Chem.* **60**, 284–288.

Jackson, H. W., Mirmira, G. K. and Wagner, W. M. (1986). Oxidative oil stability using volatiles methodology. *J. Am. Oil Chem. Soc.*, **63**, 117–118.

Jamieson, G. R. (1975). GLC identification technique for long-chain unsaturated fatty acids. *J. Chromatogr. Sci.*, **13**, 491–497.

Johnston, A. E., Dutton, H. J., Scholfield, C. R. and Butterfield, R. O. (1978). Double bond analysis of dienoic fatty acids in mixtures. *J. Am. Oil Chem. Soc.*, **55**, 486–490.

Kawamura, K. and Gagorian, R. P. (1988). Identification of isomeric hydroxy fatty acids in aerosol samples by capillary GC–MS. *J. Chromatogr.*, **438**, 309–317.

Kim, K., Zlatkis, A., Horning, E. C. and Middleditch, B. S. (1987). GC of volatile fatty acids as *tert*-butyldimethylsilyl ethers. *J. High Resolut. Chromatogr. Chromatogr. Commun.*, **10**, 522–523.

Korahani, V., Bascoul, J. and Crastes de Paulet, A. (1981). Capillary column GLC analysis of cholesterol derivatives. *J. Chromatogr.*, **211**, 392–397.

Kornfeldt, A. and Croon, L.-B. (1981). 4-Demethyl,4-monomethyl and 4,4-dimethylsterols in some vegetable oils. *Lipids*, **16**(5), 306–314.

Kuksis, A. (1972a). Gas–liquid chromatographic fractionation of neutral diacylglycerols on organosilicone polyester liquid phases. *Can. J. Biochem.*, **49**, 1245–1250.

Kuksis, A. (1972a). Gas–liquid chromatographic fractionation of natural diacylglycerols on stabilised polyester liquid phases. *J. Chromatogr. Sci*, **10**, 53–56.

Kuksis, A., Marai, L., Myher, J. J. (1972). Triglyceride structure of milk fats. *J. Am. Oil Chem. Soc.*, **50**, 193–201.

Lanza, E. and Slover, H. T. (1981). The use of SP2340 glass capillary columns for the estimation of the trans fatty acid content of foods. *Lipids*, **16**(4), 260–267.

Lercker, G. (1983). Short capillary columns in the analysis of lipids. *J. Chromatogr.*, **279**, 543–548.

Lipsky, S. R. and Duffy, M. L. (1986a). High temperature GC: the development of new aluminium clad flexible fused silica glass capillary columns coated with thermostable non-polar phases. Part 1. *J. High Resolut. Chromatogr. Chromatogr. Commun.*, **9**, 376–382.

Lipsky, S. R. and Duffy, M. L. (1986b). New advances in capillary gas chromatography. *LC-GC*, **4**(9), 898–906.

Lusby, W. R., Thompson, M. J., Kochansky, J. (1984). Analysis of sterol esters by capillary gas chromatography–electron impact and chemical ionisation–mass spectrometry. *Lipids*, **19**, 888–901.

Macnamara, K. and Burke, N. (1985). Determination of FFA and other major congeners in distilled spirits by direct analysis on a bonded FFAP capillary column. *J. High Resolut. Chromatogr. Chromatogr. Commun.*, **8**, 853–855.

Maerker, G. and Unruh, J. Jr. (1986). Cholesterol oxides I. Isolation and determination of some cholesterol oxidation products. *J. Am. Oil Chem. Soc.*, **63**, 767–771.

Mares, P. (1988). High temperature capillary gas liquid chromatography of triacylglycerols and other intact lipids. *Prog. Lipid. Res.*, **27**, 107–133.

Meltzer, J. B., Frankel, E. N., Bessler, T. R., Perkins, E. G. (1981). Analysis of thermally abused soybean oils for cyclic monomers. *J. Am. Oil Chem. Soc.*, **58**, 779–784.

Min, D. (1981). Correlation of sensory evaluation and instrumental analysis of edible oils. *J. Food Sci.*, **46**, 1453–1456.

Missler, S. R., Wasilchuk, B. A. and Meritt, C. Jr. (1985). Separation and identification of cholesterol oxidation products in a dried egg preparation. *J. Food Sci.*, **50**, 595–598, 646.

Miwa, T. K., Mikolajczak, K. L., Earle, F. R. and Wolff, I. A. (1960). Gas chromatographic characterisation of fatty acids. *Analyt. Chem.*, **32**, 1739–1742.

Monseigny, A., Vigneron, P. V., Levacq, M. and Zwoboda, F. (1979). Analysé chromatographique quantitative des triglycerides sur colonne capillaire de verre. *Rev. Franc. Corps Gras*, **26**, 109–120.

Mutter, M. and Homan, H. R. (1987). Estimation of amount of *trans,trans*-isomers of octadecadienoic acids in hydrogenated oils and fats. *J. High Resolut. Chromatogr. Chromatogr. Commun.*, **10**, 672–673.

Myher, J. J. and Kuksis, A. (1975). Improved resolution of natural diacylglycerols by GLC on polar siloxanes. *J. Chromatogr. Sci.*, **13**, 138–145.

Myher, J. J., Kuksis, A., Marai, L. and Yeung, S. K. F. (1978). Microdetermination of oligosaccharides and polyunsaturated diacylglycerols by gas-chromatography–mass spectrometry of their *tert*-butyl dimethylsilyl ethers. *Analyt. Chem.*, **50**, 557–561.

Nourooz-Zadeh, J. and Appelqvist, L.-A. (1987). Cholesterol oxides in Swedish foods and food ingredients: fresh eggs and dehydrated egg products. *J. Food Sci.*, **52**, 57–62, 67.

Padley, F. B. and Timms, R. E. (1978). Analysis of confectionary fats II. Gas liquid chromatography of triglycerides. *Lebensm. Wiss. und Technol.*, **11**, 319–322.

Padley, F. B. and Timms, R. E. (1980). The determination of cocoa butter equivalents in chocolate. *J. Am. Oil Chem. Soc.*, **57**, 286–293.

Park, S. W. and Addis, P. B. (1986). Identification and quantitative estimation of oxidised cholesterol derivatives in heated tallow. *J. Agric. Food Chem.*, **34**, 653–659.

Paradis, A. J. and Nawar, W. W. (1981). A GC method for the assessment of used frying oils: comparison with other methods. *J. Am. Oil Chem. Soc.*, **58**, 635–638.

Patterson, G. W. (1971). Relation between structure and retention time of sterols in gas chromatography *Analyt. Chem.*, **43**, 1165–1170.

Patton, S. and Jensen, R. G. (1975). Lipid metabolism and membrane functions of the mammary gland. *Prog. Chem. Fats and other Lipids*, **14**, 163–277.

Perkins, E. G. and Wantland, L. R. (1973). Characterisation of nonvolatile compounds formed during thermal oxidation of 1-linoleyl-2,3-distearin III. Evidence for the presence of dimeric fatty acids. *J. Am. Oil Chem. Soc.*, **50**, 459–461.

Porschmann, J., Welsch, T., Engewald, W. and Gy. Vigh (1984). Characterisation of OV-1-FFAP mixture coated glass capillary columns used for the separation of FFA. *J. High Resolut. Chromatogr. Chromatogr. Commun.*, **7**, 509–514.

Proot, M. and Sandra, P. (1986). High speed capillary GC on 10 m × 100 μm i.d. FSOT columns. *J. High Resolut. Chromatogr. Chromatogr. Commun.*, **9**, 618–623.

Ralaimanarivo, A., Gaydou, E. M., Bianchini, J. P. (1982). Fatty acid composition of seed oils from six *Adansonia* species with particular reference to cyclopropane and cyclopropene acids. *Lipids*, **17**, 1–10.

Rezanka, T. and Podojil, M. (1986). Identification of wax esters of the fresh water green alga *Chlorella kessleri* by GC–MS. *J. Chromatogr.*, **362**, 399–406.

Ross, P. E., Kouroumalis, E., Clarke, A., Hopwood, D. and Bouchier, I. A. D. (1984). Cholesteryl esters in human gallbladder bile and mucosa. *Clin. Chim. Acta*, **144**, 145–154.

Saito, M. and Kaneda, T. (1976). Studies on the relation between the nutritive value and the structure of polymerised oils. *X. Yukagaku*, **25**, 79–86.

Schulte, E. and Weber, N. (1987). Analysis of disteryl ethers, *Lipids*, **22**, 1049–1052.

Sebedio, J. L. (1985). Application of methoxy-bromomercuric-adduct fractionation to the study of cyclic fatty acid monomers from a heated linseed oil. *Fette Seifen Anstrichmittel*, **87**, 267–273.

Sebedio, J. L., Le Quere, J. L., Semon, E., Morin, O., Prevost, J. and Grandgirard, A. (1987). Heat treatment of vegetable oils II. GC–MS and GC–FTIR spectra of some isolated cyclic fatty acid monomers. *J. Am. Oil Chem. Soc.*, **64**, 1324–1333.

Sheppard, A. J. and Iverson, J. L. (1975). Esterification of fatty acids for gas–liquid chromatographic analysis. *J. Chromatogr. Sci*, **13**, 448–452.

Slover, H. T. and Lanza, E. (1979). Quantitative analysis of food fatty acids by capillary GC. *J. Am. Oil Chem. Soc.*, **56**, 933–943.

Slover, H. T., Thompson, R, H. Jr. and Merola, G. V. (1983). Determination of tocopherols and sterols by capillary GC. *J. Am. Oil Chem. Soc.*, **60**, 1524–1528.

Smith, N. B. (1982). Hydrogenation of cholesteryl esters and of cholesteryl and epicholesteryl silyl ethers by capillary GC. *Lipids*, **17**, 464–468.

Smith, N. B. (1983). Gas–liquid chromatography of cholesteryl esters on non-polar and polar capillary columns following on-column injection. *J. Chromatogr.*, **254**, 195–202.

Smith, R. L., Sullivan, D. M. and Richter, E. F. (1987). Determination of phytosterols in butter samples by using capillary GC. *J. Assoc. Offic. Anal. Chem.*, **70**, 912–915.

Snyder, J. M., Frankel, E. N. and Selke, E. J. (1985). Capillary gas chromatographic analyses of headspace volatiles from vegetable oils. *J. Am. Oil Chem. Soc.*, **62**, 1675–1679.

Takagi, T. and Itabashi, Y. (1977). Gas chromatographic resolution of polyunsaturated wax esters based on their degree of unsaturation on Silar 10C. *J. Chromatogr. Sci.*, **15**, 121–124.

Takagi, T. and Itabashi, Y. (1981). Occurrence of mixtures of geometrical isomers of conjugated octadecatrienoic acids in some seed oils: analysis by open-tubular gas–liquid chromatography and high performance liquid chromatography. *Lipids*, **16**, 546–551.

Termonia, M., Munari, F. and Sandra, P. (1987). High oven temperature cold on-column injection for the automated CGC analysis of high molecular weight compounds such as triglycerides. *J. High Resolut. Chromatogr. Chromatogr. Commun.*, **10**, 263–268.

Thompson, M. J., Patterson, G. W., Dutky, S. R., Svoboda, J. A. and Kaplanis, J. N. (1980). Techniques for the isolation and identification of steroids in insects and algae. *Lipids*, **15**, 719–733.

Traitler, H. (1987). Recent advances in capillary gas chromatography applied to lipid analysis. *Prog. Lipid Res.*, **26**, 257–280.

Traitler, H. and Prevot, A. (1981a). GC separation of triglycerides according to their degree of unsaturation on glass capillary columns. *J. High Resolut. Chromatogr. Chromatogr. Commun.*, **4**, 109–114.

Traitler, H. and Prevot, A. (1981b). Thermostable persilylated polar glass capillary columns by the ammonia-etching method. *J. High Resolut. Chromatogr. Chromatogr. Commun.*, **4**, 433–436.

Traitler, H. and Rossier, M. (1982). Influence of some inorganic salts on the polarity of apolar polymer phases in capillary gas chromatography. *J. High Resolut. Chromatogr. Chromatogr. Commun.*, **5**, 189–191.

Traitler, H., Fleming, S. E. and Koellreuter, B. (1988). Analysis of volatile fatty acids in biological specimens by capillary GC. *J. High Resolut. Chromatogr. Chromatogr. Commun.*, **11**, 124–128.

Tsai, L.-S., Ijichi, K., Hudson, C. A. and Meehan, J. J. (1980). A method for the quantitative estimation of cholesterol α-oxide in eggs. *Lipids*, **15**, 124–128.

Verzele, M., David, F., Van Roelenbosch, M., Diricks, G. and Sandra, P. (1983). *In situ* gummification of methylphenylsilicones in fused-silica capillary columns. *J. Chromatogr.*, **279**, 99–102.

Zijlstra, J. B., Beukema, J., Wolthers, B. G., Byrne, B. M., Green, A. and Dankert, J. (1977). Pretreatment methods prior to GC analysis of volatile fatty acids from faecal samples. *Clin. Chim. Acta*, **78**, 243–250.

6

Amino acids and other nitrogen-containing compounds

Michael H. Gordon

6.1 INTRODUCTION

Amino acid analysis is a very important area in food science because nine amino acids are essential components of the diet, namely histidine, isoleucine, leucine, lysine, methionine, phenylalanine, threonine, tryptophan and valine. In addition, the amino acid composition of proteins is an important aspect of the characterization of proteins, and free amino acids are of significance in the flavour of some foods e.g. cheese (Wood *et al.*, 1985).

Amino acids are commonly analysed in an amino acid analyser which relies on ion-exchange chromatography with post-column reaction with ninhydrin, and in recent years HPLC (high-performance liquid chromatography) has become a common alternative method of analysis.

However, the high cost of an amino acid analyser, and the ready availability of GLC (gas–liquid chromatography) facilities have encouraged some groups to investigate GLC as a technique for analysing amino acids. The main problem to be overcome in the analysis of amino acids by GLC is the low volatility and thermal instability of these compounds. Derivatization of the carboxylic acid and amino groups is required for all amino acids, and for some compounds additional functional groups must be derivatized. Thus serine, threonine and tyrosine contain hydroxyl groups, while cysteine, cystine and methionine contain active sulphur groups. Tryptophan contains an indole ring, and asparagine and glutamine contain carboxamide groups. The increase in volatility and stability achieved on derivatization is such that di- to hexapeptides may also be analysed by capillary GLC (Koenig, 1985).

6.2 DERIVATIZATION

Various volatile derivatives of amino acids suitable for GLC were prepared following the first report of the analysis of aldehydes derived from amino acids in 1956 (Hunter *et al.*, 1956). However, Bayer showed that *N*-trifluoroacetyl amino acid methyl esters (I) were very suitable for GC analysis (Bayer, 1958).

$$CF_3\overset{\overset{\displaystyle O}{\|}}{C}\text{–}NH\underset{\underset{\displaystyle R}{|}}{C}H\text{–}\overset{\overset{\displaystyle O}{\|}}{C}\text{–}OCH_3 \tag{I}$$

Subsequent developments have established *N*-acyl amino acid alkyl esters as the main class of derivatives for GC (Husek & Macek, 1975).

The different functional groups in amino acids vary in their reactivity with derivatizing reagents, and in their sensitivity to experimental conditions. Trimethylsilylation of amino acids has been investigated as a simple derivatization procedure (Gehrke & Leimer, 1980), since amino, carboxyl, hydroxyl, carbonyl and thiol groups can all be derivatized, but some amino acids, including glycine, glutamic acid, lysine, arginine, histidine and tryptophan, form more tha one derivative owing to excessive trimethylsilylation. Active sites on the GC column must be completely inactivated by silylation if partial decomposition of the silylated amino acids is not to occur on the column.

The most common derivatization procedure requires two reaction steps. Initially the carboxyl groups are esterified by heating with an alcohol in the presence of an acidic catalyst (6.1). Then amino acid esters are converted to the *N*(*O*,*S*)-acyl amino acid esters by reaction with an acid anhydride (6.2).

$$H_3\overset{+}{N}\underset{\underset{\displaystyle R}{|}}{C}HCOOH + R^1OH \xrightarrow{H^+} H_3\overset{+}{N}\underset{\underset{\displaystyle R}{|}}{C}HCOOR^1 + H_2O \tag{6.1}$$

$$H_3\overset{+}{N}\underset{\underset{\displaystyle R}{|}}{C}HCOOR^1 + (R''CO)_2O \rightarrow R''CONH\underset{\underset{\displaystyle R}{|}}{C}HCOOR^1 + R''COOH + H^+ \tag{6.2}$$

The alcohol used in the esterification is normally isopropanol, *n*-propanol, isobutanol or *n*-butanol, since methyl esters are rather volatile and may be lost during evaporation of the acylation reagent, while amino acids are sparingly soluble in higher alcohols. Hydrochloric acid is normally used as an acid catalyst for the esterification. Trifluoroacetyl, pentafluoropropionyl or heptafluorobutyryl anhydrides are normally used as the acylating reagents, since these produce stable volatile products. Alkaline glass surfaces may promote decomposition of *N*-acyl amino acid esters, and steps must be taken to avoid this.

A typical procedure for the derivatization of amino acids to *N*-heptafluorobutyryl isobutyl esters was described by Pearce (1977). A sample containing 25–250 nmol of each amino acid is freeze-dried, and dichloromethane (100 μl) is added and evaporated to azeotrope off any residual water. Esterification is then carried out with isobutanol and hydrochloric acid (3 M, 100 μl) at 110° for 30 minutes under nitrogen. The alcohol is evaporated and the sample is dried azeotropically with methylene chloride. Ethyl acetate (25 μl) containing butylated hydroxytoluene is added, followed by heptafluorobutyric anhydride (10 μl). The mixture is then heated at 110° for 10 min under nitrogen and cooled.

Tryptophan can form multiple derivatives resulting in more than one peak in the chromatogram but Moodie (1981) has shown that this can be overcome if the heptafluorobutyric anhydride is present in excess.

If histidine determination is required, the addition of ethoxyformic anhydride (5 μl) is recommended as a final step in Pearce's procedure. The mixture is then heated under nitrogen with ethyl acetate at 110°C for 5 min. The sample is evaporated almost to dryness and redissolved in ethyl acetate for analysis.

6.3 GC COLUMNS FOR ANALYSIS OF NATURAL AMINO ACIDS

N(*O*,*S*)-Heptafluorobutyryl amino acid esters can be separated on packed columns with methyl silicone stationary phases such as SE30 or OV-1 (Mackenzie & Tenaschuk, 1979; Siezen & Mague, 1977). However the development of wall-coated open tubular (WCOT) capillary columns has improved the separation of amino acid derivatives. Usually a methyl silicone stationary phase is used. In one example, a 50 m × 0.22 mm internal diameter (i.d.) WCOT fused-silica capillary column coated with SE-30 was used to analyse the *n*-heptafluorobutyryl isobutyl derivatives of free amino acids in Cheddar cheese (Wood *et al.*, 1985). Good separation of all amino acids was achieved in 30 min (Fig. 6.1).

6.4 QUANTITATIVE ANALYSIS OF AMINO ACIDS

Sample preparation prior to amino acid analysis involves several steps at which losses may occur, and therefore an internal standard should be added at the start of the procedure. Free amino acids must be separated from other food components or proteins must be hydrolysed. The amino acids must then be derivatized quantitatively, but since amino acids vary in their reactivity with derivatizing reagents, further errors may be introduced at this stage. Amino acids that do not occur naturally are commonly used as internal standards, e.g. norleucine, cycloleucine, pipecolic acid and tranexamic acid. However, if a mass spectrometer is used as the GC (gas chromatography) detector, amino acids labelled with stable isotopes of hydrogen, carbon, nitrogen or oxygen may be used as internal standards (Zagalak *et al.*, 1977). Alternatively if a chiral GC column is used, an enantiomer of a naturally occurring amino acid may be used as internal standard (Frank *et al.*, 1980).

Since losses of isotopically labelled amino acids or enantiomers are identical to those of the naturally occurring amino acid, these compounds are particularly suitable as internal standards.

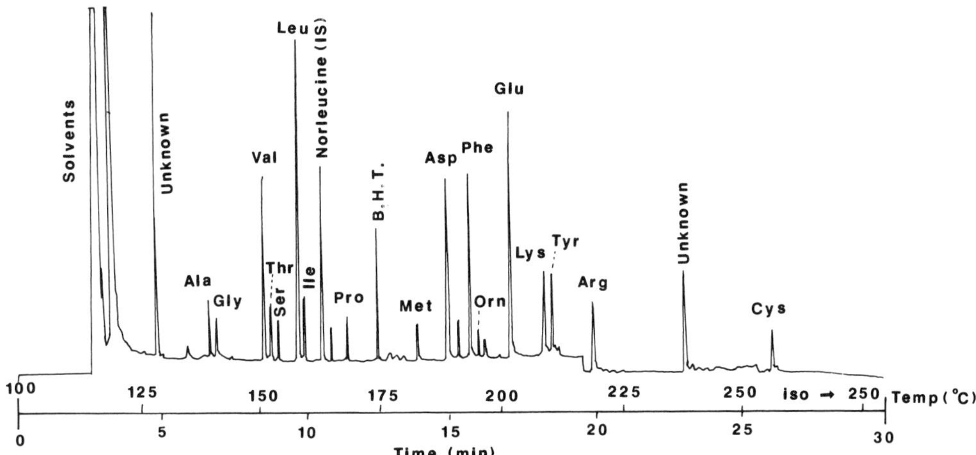

Fig. 6.1 — Analysis of the *n*-heptafluorobutyryl isobutyl derivatives of the free amino acids in 8-month-old Cheddar cheese (with permission from Wood *et al.*, 1985). (50 m × 0.22 mm i.d. WCOT column coated with SE30, temperature programmed from 100°C to 250°C at 6°C/min, final holding time 3 min).

6.5 ANALYSIS OF PRODUCTS FORMED FROM NATURAL AMINO ACIDS DURING FOOD PROCESSING

Only 20 amino acids are commonly present in natural food proteins, and these amino acids are exclusively L-enantiomers because of the stereospecificity of protein biosynthesis. However, several amino acid reactions may occur during food processing, namely cross-linking, degradation and racemization of amino acids. Volatile amino acid degradation products may be analysed by techniques such as those described in Chapter 3. However, analysis of cross-linked and racemic mixtures of amino acids has required the development of new techniques which are discussed in the following sections.

6.5.1 Cross-linked amino acids

Thermal treatment of proteins at alkaline pH or severe heat treatment near neutrality lead to the development of intra- or intermolecular cross-links between amino acid residues with the formation of cross-linked amino acids such as lysinoalanine, lanthionine, and ornithinoalanine in the protein hydrolysate. Cross-links under mild conditions mainly result from the condensation of residues of lycine, cysteine or ornithine with a residue of dehydroalanine, although other cross-links may form under severe conditions. Cross-linked amino acids may be analysed by GLC, with lysinoalanine formation having been studied in detail by this technique (Hasegawa & Okamoto, 1980; Buser & Erbersdopler, 1984; Hasegawa *et al.*, 1987). Fig. 6.2 shows the chromatogram of the *N*-trifluoroacetyl-*O*-butylester derivatives of a mixture of amino acids containing lysinoalanine analysed on a packed column with a mixed stationary phase comprising OV-17 and OV-210 (Hasegawa *et al.*, 1987).

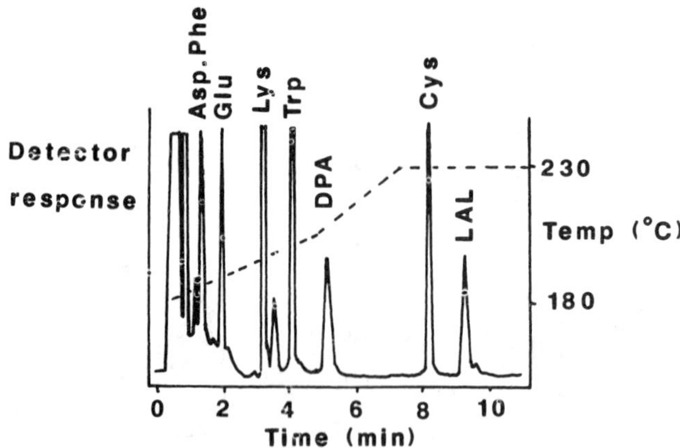

Fig. 6.2 — Analysis of the *N*-trifluoroacetyl-*O*-butylester derivatives of a mixture of amino acids containing lysinoalanine (with permission from Hasegawa *et al.*, 1987). (1.5 m × 3 mm packed column containing 2% OV-17 and 1% OV-210 on Supelcoport (100–200 mesh), temperature programmed from 180°C to 205°C at 5°C/min and from 205°C to 230°C at 10°C/min).

6.5.2 Racemic mixtures of amino acids

Although amino acids in natural proteins occur as L-enantiomers, partial racemization may occur in alkali-treated food proteins. Racemization can impair the nutritional value of food proteins by reducing the amount of essential amino acids and by reducing digestibility of the proteins. In addition certain D-enantiomers are toxic. Concern has been expressed about the racemization of amino acids during the solubilization of soy proteins with alkali (Bankhead *et al.*, 1978). Racemization of amino acids in casein, lactalbumin, soy protein and wheat gluten has been observed after heating the proteins with 0.1 M sodium hydroxide at 65° for 3 h. Aspartic acid and glutamic acid and their amides; phenylalanine and alanine were mainly affected (Masters & Friedman, 1979).

Polydimethylsiloxanes with optically active side-chains have been developed as stationary phases for the analysis of the enantiomers of amino acids (Frank *et al.*, 1977). Chirasil-Val, which is a polysiloxone containing *N*-propionyl-L-valine *tert*-butylamide, is a particularly versatile stationary phase for the analysis of amino acid enantiomers. Although GLC analysis of amino acid enantiomers has mainly been studied in clinical chemistry, the excellent separations achievable (see Fig. 6.3) indicate that this is potentially a very useful technique for food analysis.

6.6 ALKALOIDS

Alkaloids are plant metabolites of amino acids. They are significant in foods because of their toxicity, or specific physiological effects. Some alkaloids, e.g. quinine, also contribute bitterness to foods, and the betalaines are significant for their colour. The main method of analysing alkaloids in food is HPLC, but GC analysis has been applied to several alkaloids, including caffeine (Abdel-Moety, 1988; Weerasinghe *et al.*, 1982), theobromine (Ishida *et al.*, 1986), piperine (Sasakawa & Kato, 1985)

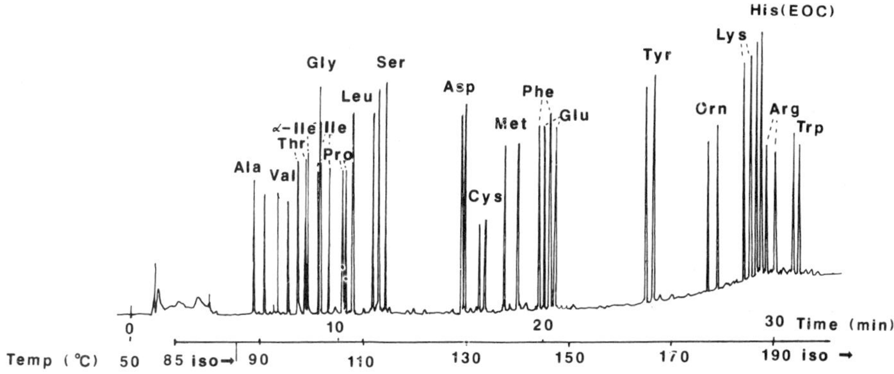

Fig. 6.3 — GC separation of the enantiomers of 19 amino acids as *N*(*O*,*S*)-pentafluoropropionyl amino acid isopropyl esters on a glass capillary, 20 m × 0.28 mm, coated with Chirasil–Val. Temperature program indicated in the figure (with permission from Frank *et al.*, 1980).

and quinine (Klatsmanyi & Zala, 1982). Sample preparation commonly involves extraction into organic solvents from alkaline aqueous solutions. The analysis may be performed on packed columns containing non-polar methyl silicone polymers as stationary phases. The detector is commonly a FID (flame ionization detector), although the sensitivity of detection is improved with an alkali flame ionization detector (AFID) (Vitzthum *et al.*, 1974).

6.7 GLUCOSINOLATES

Glucosinolates (II) are thioglucosides that occur in plants of the family Cruciferae, with most plants of significance as food being in the genus *Brassica*. Glucosinolates are of interest because products formed after hydrolysis contribute a pungent taste to horseradish and mustard, and are also important in the flavour of many vegetables, including brussel sprouts, cabbage, broccoli and turnips. In addition, rearrangement of the hydrolysis products may produce harmful products, including thiocyanates, which are goitrogenic; nitriles, which are toxic; and isothiocyanates, which may be carcinogenic. Cattle consuming feed containing glucosinolate breakdown products exhibit slow growth. The development of oilseed rape as cattle feed required plant breeding to reduce the glucosinolate content, and therefore good analytical methods were required to determine glucosinolate levels.

$$\beta\text{-D-glucose} - S - \underset{\displaystyle R}{\overset{}{C}} = NO - SO_3^- \qquad \text{(II)}$$

Glucosinolates have been commonly analysed by HPLC and GLC (McGregor *et al.*, 1983). Early GC investigations involved myrosinase-catalysed hydrolysis of the glucosinolates and analysis of the isothiocyanates and oxazolidinethiones. Identification of products derived from glucosinolates can be achieved by comparison of retention times with those of standards using two columns varying in polarity (van Etten *et al.*, 1976), or by GC–MS (mass spectrometry). However, this method does not allow aromatic glucosinolates, including sinalbin and glucobrassicin (3-indolylmethyl glucosinolate), to be determined because of the instability of the isothiocyanates formed from these compounds. The method is also unsuitable for glucosinolates with hydrophilic side-chains (Olsen & Sorensen, 1979). Trimethylsilyl derivatives can be prepared from glucosinolates but sulphate is released during the derivatization, and therefore Thies (1979) introduced a procedure in which glucosinolates are separated by ion-exchange chromatography and sulphate is then removed prior to derivatization. Desulphoglucosinolates can be silylated with various reagents including a mixture of *N*-methyl-*N*-trimethylsilylheptafluorobutyramide (MSHFBA) and trimethylchlorosilane (TMCS) (Heaney & Fenwick, 1980). Good separation of silylated desulphoglucosinolates can be achieved using packed columns containing low-polarity stationary phases such as OV-1 or OV-7, as seen in Fig. 6.4. 3-Indolylmethyl and 4-hydroxy-3-indolylmethyl glucosinolates could not be

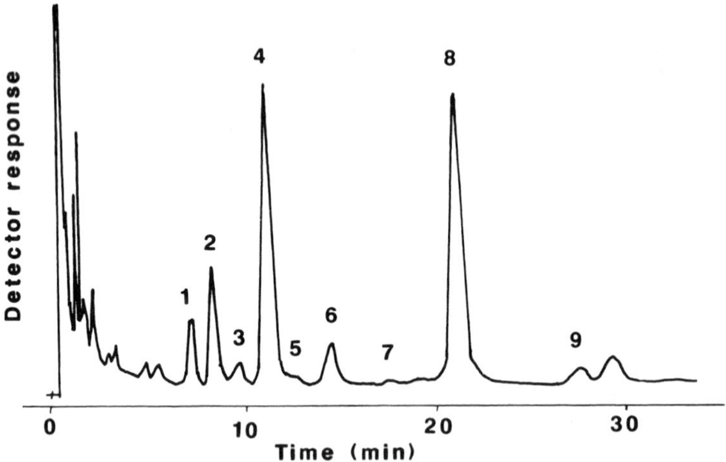

Fig. 6.4 — GLC chromatogram of the trimethylsilyl derivatives of glucosinolates isolated from *Brassica napus* L.CV. Tower (with permission from Olsen & Sorensen, 1980); (1.5 m × 4 mm i.d. glass column packed with 3% OV-1 on Chromosorb WHP 80/100 mesh, programmed from 200°C at 5°C/min to the final temperature of 280°C). 1, Sinigrin; 2, gluconapin; 3, glucobrassicanapin; 4, progoitrin; 5, napoleiferin; 6, glucotropaeolin; 7, gluconasturtiin; 8, sinalbin; 9, glucobrassicin.

separated on an OV-7 packed column, but separation was achieved with a SE-54 capillary column (Truscott *et al.*, 1982). 3-Methylsulfinylpropyl glucosinolate forms several peaks on silylation (Heaney and Fenwick, 1980). A rapid method for the determination of glucosinolates in canola meal was developed by Sosulski and Dabrowski (1984). The method involves extraction and preparation of the desulphoglucosinolates followed by derivatization with MSHFBA and TMCS in pyridine. Analysis of the derivatized desulphoglucosinolates could be achieved within 10 min using a 25-m OV-1 WCOT capillary column.

6.8 NITRATES AND NITRITES

Nitrates and nitrites occur ubiquitously in nature and are added to some foods as preservatives. Since nitrates may be reduced to nitrites, which may react with amines or amides to form carcinogenic *N*-nitroso compounds under conditions similar to those in the human stomach, accurate monitoring of these ions is required. Liquid chromatography is commonly used, but several GC techniques have been developed for nitrite or nitrate in various matrices. The ions must be derivatized to achieve suitable volatility before GC analysis.

Nitrates are commonly analysed as nitrobenzene. Nitrate levels of 100–200 μg/kg can be detected using a thermal energy analyser as GC detector (Ross and Hotchkiss, 1985). The thermal energy analyser is a detector in which the sample is pyrolysed at 300°C, where a nitrosyl radical is released. The nitrosyl radical is then oxidized with ozone to give electronically excited nitrogen dioxide. The excited nitrogen dioxide decays back to its ground state with the emission of characteristic radiation. The intensity of the emission is proportional to the nitrosyl radical concentation (Fine *et al.*, 1975).

Nitrite can be oxidized to nitrate and also analysed as nitrobenzene (Wu & Saschenbrecker, 1977). However, a number of alternative derivatization techniques have been developed. Nitrite can be converted into the trimethylsilyl derivative of 1*H*-benzotriazole by a series of chemical reactions (Tanaka *et al.*, 1980). Nitrite can also be analysed as tetrazolophthalazine (Tanaka *et al.*, 1981), which is formed quantitatively by ring closure of hydralazine with nitrite. Alternative procedures include derivatization with 3,4-dichlorobromobenzene (Funazo *et al.*, 1979) or *p*-bromochlorobenzene (Funazo *et al.*, 1980) by the Sandmeyer reaction.

An elegant one-step derivatization procedure involves the reaction of pentafluorobenzyl bromide with nitrite in alkali at 50°C for 90 min (Wu *et al.*, 1983). This derivative can be prepared in 80% yield and it allows an electron capture detection (ECD) to be used. Good agreement was found in the analysis of nitrite in meat and river water when the GC–ECD and spectrophotometric methods were compared (Wu *et al.*, 1984). Non-polar stationary phases such as SE-30 are commonly used in packed columns for the chromatographic separation of the nitrite derivatives. A detection limit of 0.31 p.p.m. was observed using an FID to detect nitrite as silylated benzotriazole (Tanaka *et al.*, 1980), but this was reduced to 0.0046 p.p.m. by using an ECD to detect the pentafluorobenzyl derivative (Wu *et al.*, 1984). Detection limits of 0.04 p.p.m. and 0.05 p.p.m. were quoted for the detection of nitrite and nitrate as nitrobenzene with an ECD (Wu & Saschenbrecker, 1977).

6.9 N-NITROSAMINES

N-Nitrosamines are carcinogenic compounds that have been detected on occasions in various foods including fish, meat, cheese and beer. They are formed by the reaction of nitrite with amines, especially secondary amines (6.3).

$$R_2NH + NO_2 \rightarrow \underset{N\text{-nitrosamine}}{R_2N{-}NO} + H_2O \tag{6.3}$$

N-nitrosamines are carcinogenic at low levels, and therefore a sensitive analytical procedure is required. Many detectors have been used in the GC analysis of *N*-

nitrosamines including chemiluminescence (Gough *et al.*, 1977), thermal energy analysis (Song and Hu, 1988; Gavinelli *et al.*, 1986; Osterdahl, 1983), MS (Rucka *et al.*, 1985), alkali-flame ionization (Gough & Sugden, 1973; Riedmann, 1974; Fazio *et al.*, 1971) and electrolytic conductivity detectors (Rhoades & Johnson, 1970; Palframan *et al.*, 1973; Anderson & Hall, 1980). The thermal energy analyser is commonly used in the analysis of *N*-nitrosamines because of its specificity and high sensitivity. Detection limits for *N*-nitrosamines are 0.1 p.p.b. with a thermal energy analyser (Osterdahl, 1983; Song & Hu, 1988) compared with 1 p.p.b. by GC–MS (Rucka *et al.*, 1985). Good separation of *N*-nitrosamines can be achieved with a Carbowax 20M packed (Gough *et al.*, 1977) or an equivalent capillary column (Gavinelli *et al.*, 1986) (Fig. 6.5).

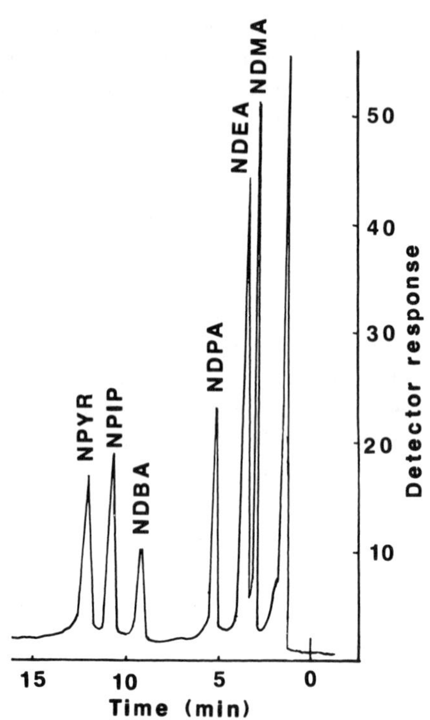

Fig. 6.5 — Analysis of standard *N*-nitrosamines by GLC with a chemiluminescent detector (with permission from Gough *et al.*, 1977), involving conversion to nitric oxide and detecting the infra-red emission resulting from reaction with ozone; (4 m × 1.8 mm i.d. stainless steel column packed with 5% Carbowax 20 M on Diatomite C; 150°C isothermal).

CONCLUSIONS

The discussion in this chapter illustrates some of the nitrogen compounds that can be analysed by GC. Although the requirement for derivatization of some compounds extends analysis time, the widespread availability of GC equipment and the high sensitivity and specificity of some detectors are factors that persuade many laboratories to use this technique for these compounds even if HPLC methods are also available.

REFERENCES

Abdel-Moety, E. M. (1988). First-derivative spectrophotometric and GLC determination of caffeine in foods and pharmaceuticals. *Zeitschrift fur Lebensmittel-Untersuchung und Forschung*, **186**(5) 412–416.

Anderson, R. J. and Hall, R. C. (1980). Hall bipolar pulse, differential electrolytic conductivity detector for GC: design and applications. *Am. Lab.*, **12**, 108–124.

Bankhead, R. R., Weingartner, K. E., Kuntz, D. A. and Erdman, J. W. (1978). Effects of sodium bicarbonate blanch on the retention of micronutrients in soy beverage. *J. Food Sci.*, **43**, 345–348.

Bayer, E. (1958). In *Gas Chromatography 1958*, D. H. Desty (ed), Butterworths Scientific, London, pp. 333.

Buser, W. and Erbersdopler, H. F. (1984). Determination of lysinoalanine as the heptafluorobutyryl isobutyl ester derivative by GLC. *J. Chromatogr.*, **303**, 234–237.

Fazio, T., Damico, J., Howard, J. W., White, R. H. and Watts, J. (1971). Gas chromatographic determination and mass spectrometric confirmation of *N*-nitrosodimethylamine in smoke processed marine fish. *J. Agric. Food Chem.*, **19**, 250–253.

Fine, D. H., Lieb, D. and Rufeh, F. (1975). Principle of operation of the thermal energy analyzer for the trace analysis of volatile and non-volatile *N*-nitroso compounds. *J. Chromatogr.*, **107**, 351–357.

Frank, H., Nicholson, G. J. and Bayer, E. (1977). Rapid gas chromatographic separation of amino acid enantiomers with a novel chiral stationary phase. *J. Chromatogr. Sci*, **15**, 174–176.

Frank, H., Rettenmeier, A., Weicker, H., Nicholson, G. J. and Bayer, E. (1980). A new gas chromatographic method for determination of amino acid levels in human serum. *Clin. Chim. Acta*, **105**, 201–211.

Funazo, K., Tanaka, M. and Shono, T. (1979). Analysis of nitrite ion in water by electron capture gas chromatography. *Chem. Lett.*, 309–310.

Funazo, K., Tanaka, M. and Shono, T. (1980). Determination of nitrite at parts-per-billion levels by derivatization and electron capture gas chromatography. *Anal. Chem.*, **52**, 1222–1224.

Gavinelli, M., Airoldi, L. and Fanelli, R. (1986). A new method for quantitative analysis of volatile nitrosamines in food by simultaneous distillation–extraction. *J. High Resolut. Chromatogr. Chromatogr. Commun.*, **9**(4), 257–259.

Gehrke, C. W. and Leimer, K. (1970). Trimethylsilylation of amino acids. Effect of solvents on derivatization using bis-(trimethylsilyl) trifluoroacetamide. *J. Chromatogr.*, **53**, 201–208.

Gough, T. A. and Sugden, K. (1973). A study of the stability of a nitrogen-selective thermionic detector. *J. Chromatogr.*, **86**, 65–71.

Gough, T. A., Webb, K. S. and Eaton, R. F. (1977). Simple chemiluminescent detector for the screening of foodstuffs for the presence of volatile nitrosamines. *J. Chromatogr.*, **137**, 293–303.

Hasegawa, K. and Okamoto, N. (1980). Studies on the gas chromatographic analysis of lysinoalanine in alkali-treated food proteins. *Agric. Biol. Chem.*, **44**, 649–655.

Hasegawa, K., Mukai, K., Gothoh, M., Honjo, S. and Matoba, T. (1987). Determination of the lysinoalanine content in commercial foods by gas chromatography –selected ion monitoring. *Agric. Biol. Chem.*, **51**(11), 2889–2894.

Heaney, R. K. & Fenwick, G. R. (1980). The analysis of glucosinolates in *Brassica* species using gas chromatography. Direct determination of the thiocyanate ion precursors, glucobrassicin and neoglucobrassicin. *J. Sci. Food Agric.*, **31**, 593–599.

Hunter, I. R., Dimick, K. P. and Corse, J. W. (1956). Determination of amino acids by ninhydrin oxidation and gas chromatography. Separation of leucine and isoleucine. *Chem. Ind.*, **16**, 294–295.

Husek, P. and Macek, K. (1975). Gas chromatography of amino acids. *J. Chromatogr.*, **113**, 139–230.

Ishida, H., Sekine, H., Kimura, S. and Sekiya, S. (1986). Gas chromatographic determination of theobromine in foods. *J. Food Hygienic Soc. of Japan*, **27**(1), 75–80. (Food Science and Technology Abstracts (1987), 6K.10).

Klatsmanyi, J. and Zala, P. (1982). Gas chromatographic determination of the quinine content of tonic beverages. *Elelmiszervizsgalati–Kozlemenyek*, **28**(1/2), 49–53. (Food Science and Technology Abstracts (1984) 8H1680).

Koenig, W. A. (1985). Analysis of peptide antibiotics by glass capillary GC–MS. In *Glass capillary chromatography in clinical medicine and pharmacology*, Jaeger, H. (ed.), Marcel Dekker Inc., New York, p. 551.

McGregor, D. I., Mullin, W. J. and Fenwick, G. R. (1983). Analytical methodology for determining glucosinolate composition and content. *J. Assoc. Off. Anal. Chem.*, **66**(4), 826–849.

Mackenzie, S. L. and Tenaschuk, D. (1979). Quantitative formation of *N*(*O*,*S*)-heptafluorobutyryl isobutyl amino acids for GC analysis. *J. Chromatogr.*, **171**, 195–209.

Masters, P. M. and Friedman, M. (1979). Racemization of amino acids in alkali-treated food proteins. *J. Agric. Food Chem.*, **27**, 507–511.

Moodie, I. A. (1981). Gas–liquid chromatography of amino acids. The heptafluorobutyryl–isobutyl ester derivative of tryptophan. *J. Chromatogr.*, **208**, 60–66

Olsen, O. and Sorensen, H. (1979). Isolation of glucosinolates and the identification of *O*(α-L-rhamnopyranosyloxy) benzylglucosinolate from *Reseda odorata*. *Phytochem.* **18**, 1547–1552.

Olsen, O. and Sorensen, H. (1980). Sinalbin and other glucosinolates in seeds of double low rape species and *Brassica napus* cv Bronowski. *J. Agric. Food Chem.*, **28**, 43–48.

Osterdahl, B. G. (1983). Volatile *N*-nitrosamines in beer and other alcoholic beverages. *Var Foda*, **35**(5), 221–230.

Palframan, J. F., MacNab, J. and Crosby, N. T. (1973). An evaluation of the alkali flame ionization detector and the Coulson electrolytic detector in the analysis of *N*-nitrosamines in food. *J. Chromatogr.*, **76**, 307–319.

Pearce, R. J. (1977). Amino acid analysis by GLC of *N*-heptafluorobutyryl isobutyl esters. *J. Chromatogr.*, **136**, 113–126.

Rhoades, J. W. and Johnson, D. E. (1970). Gas chromatography and selective detection of *N*-nitrosamines. *J. Chromatogr. Sci.*, **8**, 616–617.

Rieďmann, M. (1974). Gaschromatographisches Screening von Nitrosaminen in Lebensmitteln mit den Stickstoff-Flamnenionisations detektor. *J. Chromatogr.*, **88**, 376–381.

Ross, H. D. and Hotchkiss, J. H. (1985). Determination of nitrate in dried foods by GC–thermal energy analyzer. *J. Assoc. Offic. Anal. Chem.*, **68**(1), 41–43.

Rucka, I., Kocan, A. and Madaric, A. (1985). Determination of *N*-nitrosodimethylamine in brewers malt and beer by gas chromatography–mass spectrometry. *Bulletin Potravinarskeho Vyskumu*, **24**(2/3), 105–111. (Food Science and Technology Abstracts (1986), 3H139.)

Sasakawa, K. and Kato, T. (1985). Determination of piperine in seasoned meat by gas chromatography. Reports of the Central Customs Laboratory, **25**, 19–24. (Food Science and Technology Abstracts (1987), 8S13.)

Siezen, P. J. and Mague, T. H. (1977). GLC of the *N*-heptafluorobutyryl isobutyl esters of fifty biologically interesting amino acids. *J. Chromatogr.*, **130**, 151–160.

Song, P. J. and Hu, J. F. (1988). *N*-Nitrosamines in Chinese foods. *Food and Chem. Toxicol.* **26**(3), 205–208.

Sosulski, F. W. and Dabrowski, K. J. (1984). Determination of glucosinolates in canola meal and protein products by desulphation and capillary gas–liquid chromatography. *J. Agric. Food Chem.*, **32**(5), 1172–1175.

Tanaka, A., Nose, N. and Watanabe, A. (1980). Gas chromatographic determination of nitrite in foods as trimethylsilyl derivative of 1*H*-benzotriazole. *J. Chromatogr.*, **194**, 21–31.

Tanaka, A., Nose, N., Yamada, F., Saito, S. and Watanabe, A. (1981). Determination of nitrite in human, cow and market milks by GLC and electron-capture detection. *J. Chromatogr.*, **206**, 531–540.

Thies, W. (1979). Detection and utilization of a glucosinolate sulfohydrolase in the edible snail, *Helix* pomatia. *Die Naturwissenschaften* **66**, 364–365.

Truscott, R. J. W., Minchinton, I. R., Burke, D. G. & Sang, J. P. (1982). A novel methoxyindole glucosinolate. *Biochem. Biophys. Res. Commun.*, **107**, 1368–1375.

van Etten, C. H., Daxenbichler, M. E., Williams, P. H. and Kwolek, W. F. (1976). Glucosinolate and derived products in cruciferous vegetables. Analysis of the edible part from twenty-two varieties of cabbage. *J. Agric. Food Chem.*, **24**(3), 452–455.

Vitzthum, O., Barthels, M. and Kwasny, H. (1974). Rapid gas-chromatographic determination of caffeine in caffeine-containing and decaffeinated coffees with the nitrogen-sensitive detector. *Z. Lebensm. Unters. Forsch.* **154**, 135–141.

Weerasinghe, D. K., Fernando, R. S. and Chandradasa, P. B. (1982). A convenient rapid estimation of caffeine in tea. *Tea Quarterly*, **51**(4), 175–179.

Wood, A. F., Aston, J. W. and Douglas, G. K. (1985). The determination of free amino acids in cheese by capillary column gas–liquid chromatography. *Aust. J. Dairy Technol.* **40**, 166–169.

Wu, W. S. and Saschenbrecker, P. W. (1977). Nitration of benzene as method for determining nitrites and nitrates in meat and meat products. *J. Assoc. Offic. Anal. Chem.* **60**, 1137–1141.

Wu, H.-L., Chen, S.-H., Lin, S.-J., Hwang, W.-R., Funazo, K., Tanaka, M. and Shono, T. (1983). Gas chromatographic determination of inorganic anions as pentafluorobenzyl derivatives. *J. Chromatogr.*, **269**, 183–190.

Wu, H., Chen, S., Funazo, K., Tanaka, M. and Shono, T. (1984). Electron-capture gas chromatographic determination of nitrite as the pentafluorobenzyl derivative. *J. Chromatogr.*, **291**, 409–415.

Zagalak, M. J., Curtis, H.-Ch., Leimbacher, W., Redweik, U. (1977). Quantitation of deuterated and non-deuterated phenylalanine and tyrosine in human plasma using the selective ion monitoring method with combined GC–MS. *J. Chromatogr.*, **142**, 523–531.

7

Vitamins

Jiří Davídek and **Jan Velíšek**

7.1 INTRODUCTION

The literature on gas–liquid chromatography (GLC) of vitamins which has accumulated over the last two to three decades has shown that the technique has become a powerful analytical tool for the qualitative as well as quantitative analysis of vitamins in many situations where it offers certain advantages over other analytical techniques (Carrol and Herting, 1964; de Ritter, 1967; Sheppard *et al.*, 1972; Dickes and Nicholas, 1976; Christie and Wiggins, 1978; de Leenheer *et al.*, 1985; Velíšek and Dorídek, 1986; Davídek and Velísěk, 1986; Spiegel and Teply, 1984).

Most of the fat-soluble and water-soluble vitamins may be analysed by GLC, in the form of suitable volatile derivatives, suitable degradation products or suitable volatile derivatives of these degradation products. Appropriate GLC evaluation procedures have been developed and performed on pure vitamins and various relatively simple samples, such as multivitamins and other pharmaceutical preparations containing vitamins and, more rarely, on foods; currently used methods still include mainly biological, microbiological, chemical and physicochemical assay methods.

That GLC has achieved a prominent place, especially in the determination of fat-soluble vitamins in foods, is mainly due to a major research effort in developing convenient and suitable methods for the evaluation of vitamin E. For this group of compounds, GLC methods have developed into practical and fairly routine techniques, and they are now widely accepted and employed in many food-quality research laboratories. The application of GLC to compounds assigned as vitamin D, provitamins D, their analogues, metabolites and degradation products has also been developed for practical use. The GLC separation and determination of vitamin K is also feasible, but the GLC method, in general, has not yet been fully developed and accepted. The GLC separation of vitamin A, its analogues and isomers as well as its provitamins has not developed to any appreciable practical extent. Almost exclusively, especially in the case of vitamin A and its provitamin forms, other chromatographic methods are to be preferred, e.g. high-performance liquid chromatography (HPLC) (de Leenheer *et al.*, 1985; Van Viekirk, 1982; Brubacher *et al.*, 1986).

For the evaluation of the water-soluble vitamins, vitamins of the B complex, vitamin C and other biologically active compounds which are no longer classified as true vitamins, such as orotic acid (vitamin B_{13}), pangamic acid (vitamin B_{15}), 4-aminobenzoic acid, inositol (*myo*- or *meso*-inositol), rutin (vitamin P) and *S*-methylmethionine (vitamin U), the most suitable assay methods for food include bioassay and microbiological assay methods, spectrophotometric (including fluorometric), HPLC and other miscellaneous methods (Christie and Wiggins, 1978; de Leenheer *et al.*, 1985; Brubacher *et al.*, 1986). The utilization of GLC for the evaluation of water-soluble vitamins in foods has been developed to a practical extent in only a few research and quality-control laboratories. The technique has been successfully used on a relatively limited scale and has been mainly confined to thiamin, nicotinic acid and nicotinamide, pantothenic acid, pyridoxine and ascorbic and dehydroascorbic acids, where it may offer some advantages over other (classical) analytical techniques; e.g. selectivity and speed in the determination of thiamin in comparison with its determination with the thiochrome method; speed and much higher accuracy of the determination of nicotinic acid, pantothenic acid and pyridoxine in comparison with commonly used microbiological methods; the specificity of the determination of ascorbic acid in the presence of other carboxylic acids, reductones and sulphydryl compounds; the possibility of specific determination of *myo*-inositol in mixtures with its biologically inactive isomers, etc.

7.2 THE FAT-SOLUBLE VITAMINS

Solvent extraction and other convenient methods, e.g. the AOAC methods, (*Official Methods*, 1985) are used for the isolation of lipids from the sample, isolation of unsaponifiable matter from oils and fats, etc.

7.2.1 Vitamin A and retinoids

Because of their high thermolability and low volatility, retinol as well as all provitamins A and other carotenoids are not, in general, directly amenable to GLC. Only a limited number of compounds have been successfully analysed by this technique. GLC analysis of vitamin A has been overall the least satisfactory of all the vitamins of the fat-soluble group

Due to the instability of the conjugated polyene chains at elevated temperatures (even at reduced pressure) these compounds tend to undergo isomerization, polymerization and degradation. For instance, the determination of retinol (which is unstable at temperatures exceeding 150°C) requires the preliminary conversion to the corresponding esters or ethers, or esters and ethers of their perhydro derivatives. Retinyl acetate and free retinol, however, give rise to anhydro derivatives, while hydrogenation of retinol or its acetate partly leads to the formation of the corresponding hydrocarbons, and trimethylsilyl ethers tend to isomerize and decompose. Provitamins A, as well as other carotenes, can be chromatographed after hydrogenation to the corresponding perhydro derivatives. Non-polar and slightly polar silicones are used as stationary phases.

Derivatization, as well as preliminary hydrogenation, offers an alternative approach, which has been employed in the GC–MS analysis of retinoids and other carotenoids. The GLC method itself or GC–MS is used mainly for the determination of the structure of carotenoids, as a method for determining the number of carbon atoms present and also for the identification of their functional groups.

In routine food analysis, however, the method has not received much attention as modern liquid chromatography (LC) separation procedures seem to be much more promising (Sheppard *et al.*, 1972; Christie and Wiggins, 1978; de Leenheer *et al.*, 1985; Van Niekerk, 1982; Brubacher *et al.*, 1986; Taylor and Ikewa, 1971, 1980; Taylor and Davies, 1975; Vecchi *et al.*, 1973; Wiggins, 1976; Fenton *et al.*, 1973).

7.2.2 Vitamin D

Ergocalciferol (vitamin D_2) and cholecalciferol (vitamin D_3) have been the most frequently determined members of the D group. At present GLC and HPLC seem to be the most useful existing techniques for the determination of these compounds in foods (Heftmann, 1976; De Vries and Borsje, 1982). Gas chromatography with an FID has the sensitivity to cope with most fortified foods and those unfortified foods that contain large quantities of vitamin D (such as certain fish, dried eggs, etc.). The ECD has the sensitivity to determine vitamin D even in unfortified foods, such as butter.

Ergocalciferol as well as cholecalciferol can be chromatographed either as such or preferably (the chromatographic behaviour is improved) in the form of suitable derivatives, e.g. ethers (methyl- and trimethylsilyl ethers have often been employed) and esters (acetates, propionates, etc.), using an FID as a detector. Methyl ethers can also be used for quantitative mass-fragmentographic measurements. During GLC the free vitamins as well as corresponding provitamins (ergosterol, i.e. provitamin D_2, and 7-dehydrocholesterol, i.e. provitamin D_3) undergo irreversible thermal rearrangement (isomerization) to cyclic products, the so-called pyro- and isopyrovitamins D, at temperatures higher than approximately 150°C (Scheme 1). The ratio of these two isomers, which can be easily separated by GLC, is constant under given experimental conditions and both of the peak areas can be used for quantitative estimation of vitamin D; however, the sensitivity of the determination is then lower, for two products are formed. Quantitative measurements are, however, possible, either by determining only one peak area or by summing up the areas of both peaks as a measurement of the vitamin D content.

In comparison with spectrophotometric methods the sensitivity of this assay method is lower and has also other disadvantages, i.e. partial adsorption of the analysed compounds, simultaneous determination of provitamins D and the thermal isomerization of vitamins and provitamins already discussed. Adsorption can be avoided by either silanization of all parts of the injector and of the column or by derivatization of the analysed vitamins prior to the injection (esters or ethers). The derivatization, however, does not prevent the thermal rearrangement discussed earlier. To make the method more sensitive, trifluoroacetates or even heptafluorobutyrates have been chromatographed, employing an ECD, which enables as little as a few nanograms of the vitamins to be determined.

The only way of preventing the thermal isomerization of vitamins during GLC analysis, and the preferred approach, is a complete conversion of vitamins D to isovitamins D and further to quite stable isotachysterols (all-*trans* vitamins D) prior

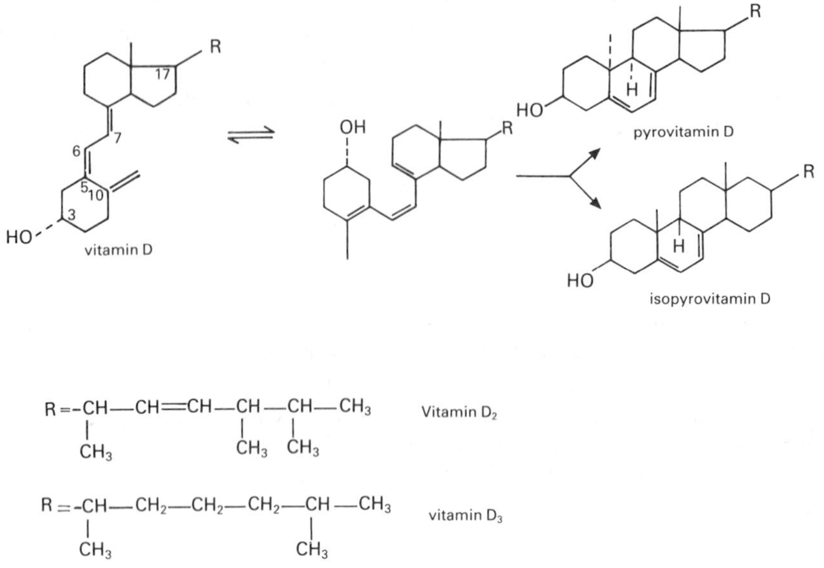

vitamin D

pyrovitamin D

isopyrovitamin D

R = -CH(CH$_3$)—CH=CH—CH(CH$_3$)—CH(CH$_3$)—CH$_3$ Vitamin D$_2$

R = -CH(CH$_3$)—CH$_2$—CH$_2$—CH$_2$—CH(CH$_3$)—CH$_3$ vitamin D$_3$

Scheme 1

to injection into the gas chromatographic column. This may be achieved by the action of strong Lewis acids, such as acetyl chloride, trifluoroacetic acid, antimony chloride, etc. (Scheme 2). The isotachysterols do not form cyclic products upon

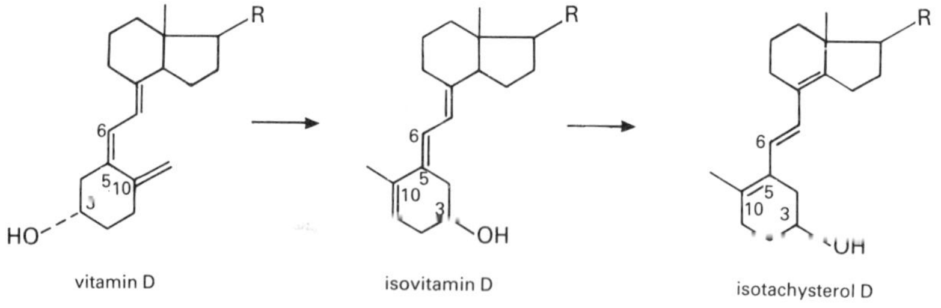

vitamin D

isovitamin D

isotachysterol D

Scheme 2

heating; therefore the resulting chromatogram shows only one peak for each vitamin. The volatility as well as the sensitivity of the determination can again be significantly increased by esterification of the isotachysterols formed with heptafluorobutyric anhydride, thus enabling as little as a few nanograms (or even less) of the vitamins to be determined with an ECD.

Vitamins D, their thermal rearrangement products (pyro- and isopyrovitamins D), as well as esters of isotachysterols can be separated employing either non-polar or polar stationary phases (silicones, polyesters, etc.). Dihydrotachysterol, as it is well separated from the two main forms of vitamin D chromatographed as such, from their trimethylsilyl ethers or from the derivatives of the corresponding isotachysterols, can be used as an internal standard. As it is very rare to find both forms of vitamin D in one sample of food, it is possible that vitamin D_2 may be employed as an internal standard in the determination of vitamin D_3 and vice versa.

Methods for the determination of vitamin D in fatty fish (Wiggins, 1976; De Vries and Borsje, 1982; Hommes and van der Mijll Dekker, 1973; Bell and Christie, 1973, 1974; Kobayashi *et al.*, 1976; Christie, 1975), eggs (Wiggins, 1976), milk (De Vries and Borsje, 1982; Bell and Christie, 1974; Panalaks, 1970; Janecke and Brendel, 1971), butter (Wilson *et al.*, 1969), margarine (Wiggins, 1976), fortified infant formulas (Touw *et al.*, 1972), mushrooms (Kobayashi, 1980; Takeuchi *et al.*, 1985), etc., have been described. In biological samples 25-hydroxycholesterol (a metabolite of vitamin D_3) can also be determined (Sklan, 1980)). Lumisterol, tachysterol and toxisterols resulting as irradiation products from ergosterol in model solutions may be separated from each other and estimated as such without any derivatization (Mermet-Bouvier, 1972).

Due to the fact that vitamin D is a very heat- and light-sensitive compound, all the analytical operations must be carried out at lowered temperatures, in the absence of light, air and acids. Larger proportions of retinol, tocopherols and sterols, possibly accompanying vitamin D, must be removed prior to the GLC analysis. Their total removal, as in the case of spectrophotometric methods, is not necessary. The individual procedures differ from each other, but they all include alkaline hydrolysis of the sample, isolation of the unsaponifiable matter and its further purification. The hydrolysis is usually carried out with aqueous solutions of potassium hydroxide, often under an inert gas and/or in the presence of antioxidants. The unsaponifiable material is then extracted into a suitable, previously purified solvent, such as diethyl ether, light petroleum, benzene, etc. Sterols may be removed by precipitation with digitonin, by chromatography on sorbents impregnated with digitonin, preferably by TLC, or by other techniques.

Procedure (according to Kobayashi (1980), Sheppard and Hubbard (1971), de Leenheer and Cruyl (1980)

To 1 g of fat or oil in a distillation flask 50 ml of ethanol (95%, v/v), 20 ml of 20% (m/m) pyrogallol ethanolic solution and 8 ml of 90% (m/m) potassium hydroxide are added (the amount of samples other than pure fats and oils is 10 g to which 100 ml of ethanol, 40 ml of pyrogallol solution and 16 ml of potassium hydroxide solution are added).

The sample is saponified under reflux on a steam bath for 30 min and cooled; 100 ml of benzene are added and the whole mixed and transferred to a separating funnel; then 40 ml of potassium hydroxide solution (6%, m/m) are added and the whole mixed and allowed to stand until two layers separate. The lower layer is then discarded, the upper one washed with 40 ml of potassium hydroxide solution

(3%, m/m), several times with water (until neutral), filtered and made up to volume in a volumetric flask. An aliquot corresponding to about 0.2 mg of vitamin D is evaporated to dryness with a stream of nitrogen, and the residue dissolved in 3 ml of hexane (if the sample contained more than 1 g of lipids, an additional 3 ml hexane are added per 1 g of lipids).

Separation of sterols

Celite 545 (10 g) dried at 110°C for 6 h is well mixed with 5 ml digitonin solution containing 300 mg digitonin, and 3 g of the resulting material are transferred to a 200×14.5 mm glass column half filled with hexane. Onto the column (the sorbent is covered with 2–3 ml of hexane) 3 ml of hexane solution of the unsaponifiable matter are transferred, eluted twice with 2 ml of hexane under a slight pressure of nitrogen then three times with benzene, and the combined eluates are evaporated with a stream of nitrogen.

Isomerization to isotachysterols

The residue is dissolved in 15 ml of 1,2-dichloroethane, 0.3 ml of acetylchloride is added, the mixture is bubbled with nitrogen, the flask is closed, its contents mixed for 30 min and the solvent evaporated to dryness using a rotary evaporator at 40°C. The residue is dissolved in 1 ml of absolute ethanol containing 100 μg of dihydrotachysterol and the solvent again evaporated.

Esterification of isotachysterols

The residue is dissolved in 200 μl of benzotrifluoride, and 5 μl of heptafluorobutyric anhydride is added and the reagents mixed. After 30 min of standing, the solution is evaporated with a stream of nitrogen, the residue dissolved in hexane and analysed by GLC.

Gas–liquid chromatography

Detector: ECD; column: glass, 1800×2 mm; stationary phase: 1% FFAP; solid support: Gas Chrom Q (0.125–0.149 mm); temperatures: column 200°C, injector 230°C (Fig. 7.1).

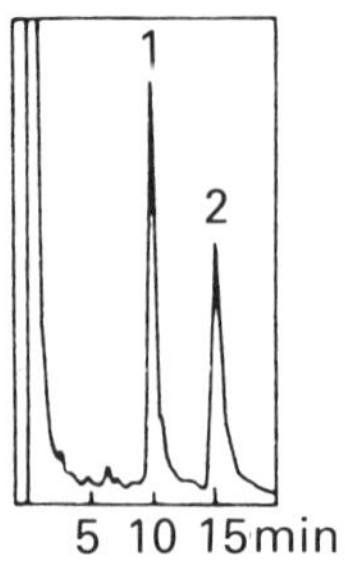

Fig. 7.1 — Determination of vitamin D in fish oil (according to de Leenheer and Cruyl (1980)). 1, dihydrotachysterol D_2 (internal standard); 2, isotachysterol D_3 (heptafluorobutyrates).

Notes

Stock solutions of isotachysterols are prepared by dissolving 100 mg of ergocalciferol and/or cholecalciferol in 15 ml of 1,2-dichloroethane and isomerized to the corresponding isotachysterols according to the procedure described above. The isotachysterols are dissolved in 10 ml of absolute ethanol and an aliquot of this solution is esterified.

All the stock solutions are bubbled with nitrogen and stored in a refrigerator at 5°C in flasks covered with aluminium foil.

Absolute ethanol is purified as follows: 2 g of potassium hydroxide and 1 g of potassium permanganate are added per 1000 ml of ethanol and boiled under reflux for about 30 min, distilled, dried over anhydrous calcium sulphate and distilled a second time with protection against the ingress of moisture.

The pyrogallol ethanolic solution may be replaced by a solution consisting of 1.5 g of *L*-ascorbic acid in 20 ml of ethanol (prepared immediately before use).

The isomerization procedure employing antimony chloride is similar to that employing acetylchloride (de Leenheer and Cruyl).

7.2.3 Vitamin E

The determination of vitamin E, i.e. of the eight naturally occurring active compounds, α-tocopherol (5,7,8-trimethyltocol), β-tocopherol (5,8-dimethyltocol), γ-tocopherol (7,8-dimethyltocol), δ-tocopherol (8-methyltocol) and the four corresponding tocotrienols, has always posed a serious analytical problem. But since the individual tocopherols and tocotrienols differ in their vitamin potency and antioxidant properties a precise knowledge of the individual forms is often necessary.

Nowadays many successful and generally accepted methods are based on GLC since they are rapid and achieve the separation and subsequent quantification of the eight individual compounds if this knowledge is required. HPLC has also been used in this area of vitamin analysis (de Leenheer *et al.*, 1985; Van Niekerk, 1982; Brubacher *et al.*, 1986).

Tocopherols and tocotrienols can be chromatographed as free compounds, as the corresponding trimethylsilyl ethers or as other derivatives, such as esters (acetates, propionates, butyrates, etc., with FID as a detector, or trifluoroacetates and other perfluoroesters with the aid of an ECD). β-and γ-tocopherols are not separated from each other when analysed as such or as any suitable derivatives, except trimethylsilyl ethers on packed columns. Their separation is, however, not required in many cases. If necessary, they may be separated to some extent as the corresponding trimethylsilyl ethers or as the corresponding tocopherylquinones on packed columns (Nair *et al.*, 1966), and very easily as trimethylsilyl ethers, employing capillary columns (Mordret and Laurent, 1978).

During recent years GLC methods for tocopherols have been applied to a wide range of biological materials, including foodstuffs, such as wheat, flours and other cereals and, especially, fats and oils (Mordret and Laurent, 1978; Glover *et al.*, 1967, 1969, 1983; Sheppard *et al.*, 1971; Slover, 1971; Christie *et al.*, 1973; Riera, 1975; Dampert and Beringer, 1976; Meijboom and Jongenotter, 1979; Ishiguro, 1983; Katsui, 1981; Rao and Perkins, 1972; Mariani and Fedeli, 1983; Nelson *et al.*, 1970). Their determination in feedstuffs has been studied as well (Ranfft, 1973; Kawamoto and Mochida, 1980; Mancas, 1984; Uebersax, 1970; IUPAC, 1981). Together with an alternative spectrophotometric method (Emmerie–Engel method), the GLC procedure has been recommended as a IUPAC method (IUPAC, 1981) and has found general application for the determination of the various components of the tocopherol fraction in animal and vegetable fats and oils, namely tocopherols and tocotrienols chromatographed as their trimethylsilyl derivatives. The GLC method with packed columns, where non-polar methylpolysiloxanes or phenylmethylpolysiloxanes are employed as stationary phases, cannot be used to separate β- and γ-tocopherols, which are determined together, nor the corresponding tocotrienols, which are eluted with the corresponding tocopherols. The GLC method with a capillary column coated with the same stationary phases as given above distinguishes the individual compounds.

The procedure, as in the case of vitamin D, involves an alkaline hydrolysis of the sample; extraction of the total unsaponifiable matter into an appropriate solvent under conditions which prevent the oxidation of vitamin E; fractionation of the total unsaponifiable material by TLC (some other techniques such as column chromatography, sublimation, etc., may also be used); recovery of the tocopherols from the silica gel removed from the developed plate; and extraction with an appropriate solvent. All solvents must be free from peroxides, and therefore require special purification. All the operations must be carried out in all-glass apparatus, away from daylight.

GLC is also a convenient technique for the resolution and quantitative analysis of diastereoisomers of α-tocopherol produced by synthesis using natural phytol, synthetic phytol or isophytol, or produced by hydrogenation of natural α-tocotrienol (Slover and Thompson, 1981).

Procedure (according to Meijboom and Jongenotter (1979), IUPAC (1986))

About 1 g of the oil or fat is weighed into a 50-ml round-bottom flask, 4 ml of pyrogallol solution in absolute ethanol free from aldehydes are added (5%, m/m) and heated under a reflux condenser. When the boiling starts, 1 ml of potassium hydroxide solution (16 g KOH per 10 ml) is added and the mixture boiled for 3 min, cooled and diluted with 25 ml of water.

The contents of the flask are transferred to a separating funnel; the flask is rinsed with 40 ml of diethyl ether, which is also used for the first extraction (vigorous shaking must be avoided). Two additional extractions, each with 25 ml of diethyl ether are made; the combined extracts are washed with 20 ml portions of water until neutral. The organic layer is evaporated using a rotary evaporator and the residual water is removed by the addition of an absolute ethanol–benzene mixture (1:4, v/v) and by the evaporation of the solvents using a nitrogen stream. The residue is dissolved in 1 ml of heptane and purified by TLC.

Thin-layer chromatography

An aliquot of about 50 μl of the solution of unsaponifiables in heptane is applied to a thin-layer plate coated with silica gel G together with 1 μl of the reference solution of tocopherols, which contains 20 μg of α-tocopherol and 20 μg of δ-tocopherol. The plate is developed away from daylight in benzene–ethyl acetate (96:4, v/v) or hexane–diethyl ether mixture (70:30, v/v) until the solvent front reaches 150 mm from the start. The tocopherols and tocotrienols separated from other unsaponifiables are visualized with 2′,7′-dichlorofluorescein in ethanol (0.2%, m/v) and scraped off the plate as one band. An appropriate amount (about 1 mg) of an internal standard, for example hexadecyl stearate, octacosane, squalane, cholesteryl isovalerate or some other high-boiling compound, is added to the scraped material. After extraction with several 10-ml portions of diethyl ether or chloroform–methanol 7:3, v/v) the extract is filtered, the solvents evaporated with nitrogen and the residue derivatized by silylation.

Derivatization

The residue is let stand with occasional shaking with 1 ml of pyridine–hexamethyldisilazane (HMDS)-trimethylchlorosilane (TMCS) mixture (10:9:6, v/v/v) for 15 min at room temperature and then analysed by GLC.

Gas–liquid chromatography

Detector: FID; column: glass capillary, 15 m×0.25 mm; stationary phase: OV-17; temperatures: column 240°C, injector 270°C (Fig. 7.2).

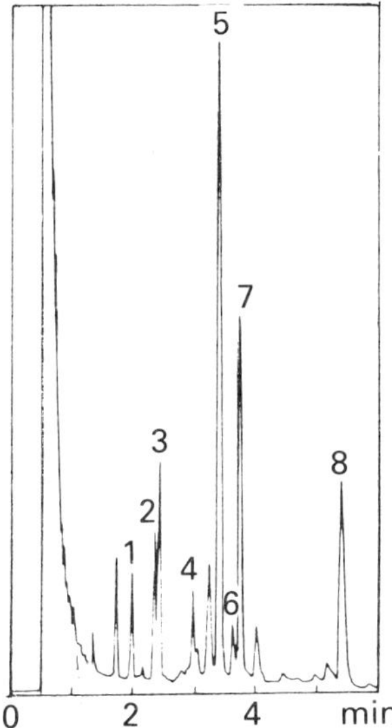

Fig. 7.2 — Determination of vitamin E in palm oil (according to Lercker and Caboni (1985)). 1, δ-tocopherol; 2, β-tocopherol; 3, γ-tocopherol; 4, δ-tocotrienol; 5, α-tocopherol; 6, β-tocotrienol; 7, γ-tocotrienol; 8,α-tocotrienol (trimethylsilyl ethers).

Notes

Ethanol is purified according to the procedure described earlier (p. 195). Diethyl ether purification proceeds as follows: to 1000 ml of diethyl ether in a separating funnel 100 g of ferrous sulphate (heptahydrate) and 200 ml of 10% (m/m) sulphuric acid are added and the mixture shaken for 15 min. The upper layer is decanted to a flask containing 30 g of potassium hydroxide and 20 g of potassium permanganate, allowed to stand for 1 h, filtered and distilled.

The reference tocopherols on the developed plate can be visualized by spraying the plate with a solution prepared from equal parts of 0.2% (m/m) ferric chloride and 0.2% (m/m) 2,2′-bipyridyl in absolute ethanol. Silica gel G can be replaced by silica gel HF_{254}, in which case the detection with 2′,7′-dichlorofluorescein is omitted.

Stock solutions of tocopherols in absolute ethanol contain 1 mg of compound per 1 ml.

7.2.4 Vitamin K

An easy and convenient method for the separation and determination of naphthoquinones, such as phylloquinone, vitamin K_1 (2-methyl-3-phytyl-1,4-naphthoquinone); menaquinone, vitamin K_2 (2-methyl-3-difarnesyl-1,4-naphthoquinone) and its homologues; menadione, vitamin K_3 (2-methyl-1,4-naphthoquinone) and its derivatives, that have been the most frequently analysed members of the vitamin K group is also provided by GLC. The comprehensive review on the fat-soluble vitamins by Sheppard *et al.*, (1972) discussed a number of reports on the analysis of vitamin K and showed that these compounds were analysed in a few research laboratories with a limited success, but the method has not been developed anywhere on a practical basis. It seems that the main reason for scant attention being given to GLC assay of naturally occurring vitamin K_1 in foods is due to the fact that vitamin K_2 from intestinal microflora appears to provide adequate quantities of vitamin K for human nutritional requirements.

In general, vitamin K can be chromatographed as such or in the form of ethers (methyl- or trimethylsilyl) of the corresponding hydroquinones on non-polar silicones. The diastereoisomers of vitamin K_1 in the form of the corresponding methyl ethers have been determined by GLC using a glass capillary column coated with Silar 10C (Vecchi *et al.*, 1981). In the form of trimethylsilyl ethers, *cis*- and *trans*-isomers of dihydrophylloquinone and some homologues of dihydromenaquinone may also be separated (Vetter *et al.*, 1967). The disadvantage of this separation is the ready hydrolysis of the derivatives to hydroquinones and oxidation of the latter substances to the original quinones (Dialameh and Olson, 1969).

The only relatively simple and generally feasible GLC method has been developed for the assay of vitamin K_1 in green leafy vegetables (Seifert, 1979) but, nevertheless, microbiological and HPLC methods are still to be preferred.

GLC has also been employed in the studies of photodegradation of phylloquinone (Nakata and Tsuchida, 1980; Nakata *et al.*, 1976) and menaquinone (Mee *et al.*, 1975), and separation of phylloquinone and its 2,3-epoxide (Becktold *et al.*, 1984). Methods for the evaluation of phylloquinone (Dialameh and Olson, 1969; Sheppard, 1971) in liver tissue and for the determination of a complex of menadione with sodium bisulphite in feedstuffs (Winkler, 1973) have been described as well.

Procedure (according to Seifert (1979), Velíšek and Davídek, (unpublished))

A frozen and freeze-dried sample of leafy vegetables (10 g) is crushed to a homogeneous mixture and extracted for 3 h in a Soxhlet extractor with hexane. The extract is concentrated to 10 ml on a rotary evaporator at 40°C and purified by chromatography on a 250×25 mm column prepared from a suspension of 50 g of alumina (neutral activity, grade 1) in hexane. The elution is carried out with the following solvents: hexane (200 ml), hexane–diethyl ether (98.5:1.5, v/v, 150 ml), hexane–diethyl ether (96:4, v/v, 300 ml). To the last of the eluants 100 μg of dotriacontane (internal standard) in 1 ml of heptane is added; the volume of this fraction is reduced to about 0.5 ml with a stream of nitrogen and the solution is analysed by GLC.

Gas–liquid chromatography
Detector: FID; column: glass, 2100×2 mm; stationary phase: 2.5% Dexsil 300GC; solid support: Chromosorb G AW DMCS (0.149–0.177 mm); temperatures: column 290°C, injector 290°C (Fig. 7.3).

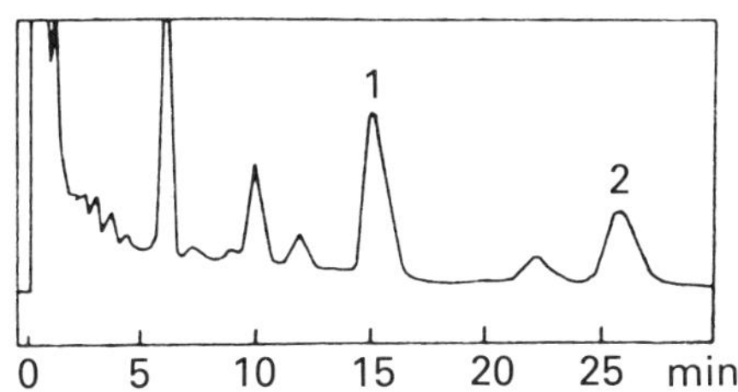

Fig. 7.3 — Determination of vitamin K_1 in spinach (according to Seifert (1979)). 1, dotriacontane (internal standard), 2, phylloquinone.

Notes
All the operations must be carried out away from air and direct daylight. Purified solvents free from peroxides (see p. 197) must be used.

7.2.5 Other active compounds

Ubiquinones (coenzymes Q) and other structurally similar and relatively non-volatile compounds (high molecular-weight benzoquinones) in pharmaceutical preparations and/or biological samples can be chromatographed without previous derivatization (as for vitamin K), employing very short columns packed with supports coated with low quantities of liquid stationary phases (mainly non-polar methylpolysiloxanes) at temperatures exceeding 300°C.

The first step in the evaluation of these biologically active compounds again includes a gentle isolation at temperatures not exceeding 60°C. Oxygen is to be avoided, as is any contact with alkaline reagents. The samples are most commonly extracted with lower primary alcohols and then re-extracted with lipophilic solvents such as hexane. The resulting extract can be further purified by chromatography on columns or thin layers of silica gel; lipids such as triacylglycerols and sterols still occurring in the purified extracts in small amounts do not interfere as they elute much sooner than ubiquinones (Dialameh and Olson, 1969; Morimoto and Imada, 1971; Dugan and Lundgren, 1964; Castello *et al.*, 1977; Yanotovskij *et al.*, 1968).

7.3 THE WATER-SOLUBLE VITAMINS

The procedures described for the preparation of aqueous extracts employing acid and/or enzymatic hydrolysis can be modified according to other existing convenient methodology (e.g. chemical and microbiological AOAC methods, (*Official Methods*, 1985)).

7.3.1 Thiamin

Since thiamin (vitamin B_1) is not volatile and is relatively heat-sensitive it does not lend itself to direct GLC analysis. Attempts have been made to chromatograph more volatile compounds of thiamin such as *O*-benzoylthiamin (Hagiwara *et al.*, 1963), trimethylsilyl (Amos and Neal, 1970) and trifluoroacetyl derivatives (Hilker and Mee, 1973) of thiamin and the corresponding substituted pyrimidine moiety of thiamin (Kawasaki and Iritani, 1968) which arises as a degradation product of thiamin under certain experimental conditions. The derivatives mentioned are not volatile enough at lower temperatures, whereas at temperatures above 250°C they readily decompose and are, thus, not suitable for GLC analysis. The methods have been applied, with limited success, to the analysis of the pure vitamin and to the determination of the vitamin in pharmaceutical multivitamin preparations and feed supplements.

Indirect methods in which thiamin is quantitatively split by sulphite treatment to produce the non-polar solvent-extractable 5-(2′-hydroxyethyl)-4-methyl-thiazole, which is then analysed by GLC as a free compound (Echols *et al.*, 1980) or as the corresponding trimethylsilyl ether (Dwivedi and Arnold, 1972) have also been developed for the determination of thiamin in commercial multivitamin and mineral preparations.

The only GLC method based on the same principle and applicable to foods has been developed for the analysis of thiamin in beans (Seifert and Miller, 1973). Thiamin, freed from its bound forms by acid hydrolysis and treatment with a proteolytic enzyme (pepsin) and phosphatase (takadiastase), is split in the presence of sulphite to the above mentioned thiazole, which is then extracted and directly analysed by GLC (Scheme 3).

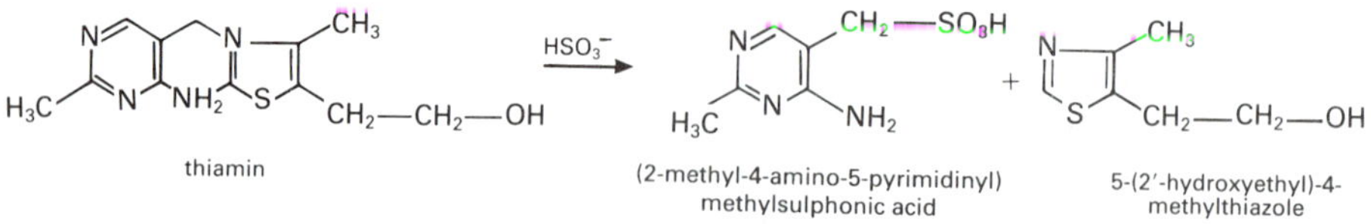

$$\text{thiamin} \xrightarrow{HSO_3^-} \text{(2-methyl-4-amino-5-pyrimidinyl)methylsulphonic acid} + \text{5-(2'-hydroxyethyl)-4-methylthiazole}$$

Scheme 3

A reliable and simple modification of this method for determining thiamin in many foods of both vegetable and animal origin has been recently developed (Echols *et al.*,1983, 1985; Velíšek *et al.*, 1986, 1987). Silicones, such as a 2:1 mixture of OV-7 and OV-22, may be employed as stationary phases for the separation of the trimethylsilyl ether of the substituted thiazole (Dwivedi and Arnold, 1972). The free base can be efficiently separated either on non-polar silicones (Echols *et al.*, 1980, 1983) or polar polyethyleneglycols (Seifert and Miller, 1973; Velíšek *et al.*, 1986, 1987). Owing to the relatively large amounts of thiamin in the majority of foods, an FID can be employed (Kawasaki and Iritani, 1968; Echols *et al.*, 1980; Dwivedi and Arnold, 1972; Seifert and Miller, 1973; Velíšek *et al.*, 1986). In cases where higher sensitivity (about 1000 times) and higher specificity are required, a specific detector such as a nitrogen–phosphorus detector (NPD) alkali flame-ionization detector, (AFID) (Echols *et al.*, 1980, 1983; Velíšek *et al.*, 1986) or a flame photometric detector (FPD) (sulphur) can be used (Velíšek *et al.*, 1986, 1987).

Procedure (according to Velíšek* et al., *(1987))

Depending on the thiamin content, 7–15 g of sample is transferred to a 100 ml volumetric flask, 75 ml of 0.05-M sulphuric acid is added and the material is hydrolysed (with occasional shaking) on a boiling water bath for 30 min. The flask is then cooled to room temperature and about 5 ml of 2.5M sodium acetate is added to adjust the pH of the mixture to 4.5. After the addition of 5 ml of 6% (m/m) takadiastase preparation, the flask is placed in a thermostat at 30°C for 14–16 h. Then 1 ml of 0.5% (m/m) cysteine solution is added, the flask heated at 37°C for 30 min, cooled to 20°C, the volume adjusted with water to 100 ml and the suspension filtered.

The filtrate (two 20-ml portions) is transferred into 250 ml beakers. Into one of the beakers 25 g of sodium sulphite (heptahydrate) is added and dissolved, and the pH of both of the solutions is adjusted to 4.9 with 20% (m/m) hydrochloric acid. The beakers, covered with a watch glass, are then heated on a steam bath for 1 h; 5 ml of 10% (m/m) trichloroacetic acid is then added, the solutions cooled to about 5°C, filtered and the precipitate on the filter washed with 25 ml of water. In the filtrate 10 g of sodium chloride is dissolved, the pH adjusted to 10–11 with sodium hydroxide solution (10 M) and the resulting 5-(2-hydroxyethyl)-4-methylthiazole extracted with four portions of chloroform, 50 ml each. The combined extracts are evaporated to dryness using a rotary evaporator. The residues are dissolved in 2 ml of chloroform, which contains 15 μg per 1 ml of 2-methylbenzothiazole (internal standard), and analysed by GLC. The content of 5-(2-hydroxyethyl)-4-methylthiazole found in sulphite-untreated sample (a measure of the spontaneous thiamin degradation) is subtracted from that found in sulphite-treated sample and the result is expressed as thiamin.

Gas–liquid chromatography

Detector: FPD (sulphur); column: glass, 2400×2 mm; stationary phase: 10% Carbowax 20M; solid support: Chromaton N-AW (0.125–0.16 mm); temperatures: column 230°C, injector 230°C (Fig. 7.4).

Notes

The stock solution of thiamin (thiamin content of 10 mg l^{-1}) is prepared by dissolving the corresponding amount of thiamin hydrochloride in 0.02M hydrochloric acid (thiamin hydrochloride must be previously dried by standing in a vacuum dessicator

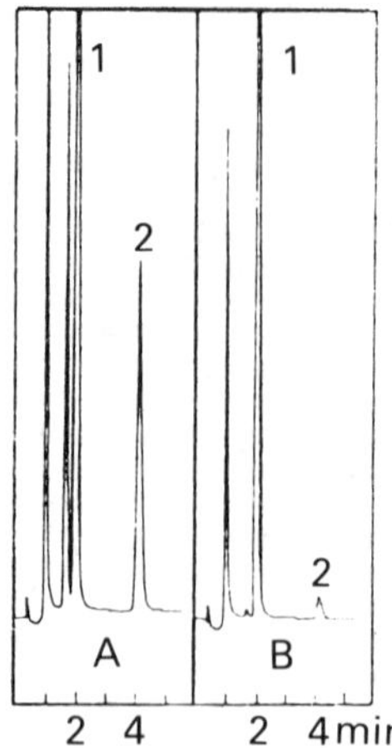

Fig. 7.4 — Determination of thiamin in pork meat (according to Velísek *et al.* (1987)). 1, 2-methylbenzothiazole (internal standard), 2, 5-(2′-hydroxyethyl)-4-methylthiazole; A, sodium sulphite-treated extract; B, untreated extract.

over phosphorus pentoxide for 24 h). This solution (100 ml) is treated with sodium sulphite according to the above procedure and the combined chloroform extracts are adjusted to 200 ml (stock solution of 5-(2-hydroxyethyl)-4-methylthiazole).

7.3.2 Nicotinic acid and nicotinamide

Nicotinic acid (vitamin B_3) has been analysed as the corresponding ethyl ester (Prosser and Sheppard, 1968; Sheppard and Prosser, 1971), trimethylsilyl ether (Janecke and Voege, 1968, 1971; Ponomaryev *et al.*, 1973) or *N*-ethylnicotinamide (Prosser and Sheppard, 1968; Sheppard and Prosser, 1971). Nicotinamide (vitamin PP) can be chromatographed either as a free compound (Tarli *et al.*, 1969; Ashby and Deavin, 1969) or, after derivatization, as the ethyl ester of nicotinic acid or *N*-ethyl-nicotinamide (Prosser and Sheppard, 1968; Sheppard and Prosser, 1971), trimethylsilyl derivative (Ponomaryev *et al.*, 1973) or nitrile of nicotinic acid (nicotinonitrile) (Vessman and Stromberg, 1975). By choosing the appropriate derivative, both of the two active compounds (the total vitamin content) can be determined simultaneously.

In foods, nicotinic acid, employed as an additive for the improvement of colour of meat products, has been determined as its trimethylsilylether (Bertling and Tietz, 1978). The only recently described method for the determination of the total natural vitamin content in foods (Davídek and Velíšek, 1987) is based upon acid hydrolysis of the sample, esterification of the free nicotinic acid with ethanol and subsequent determination employing a specific NPD.

Procedure (according to Davídek and Velísek (1987))

Sample (10 g) is hydrolysed in 50 ml of 0.5M sulphuric acid at 100°C for 1 h. After cooling to room temperature the pH of the suspension is adjusted to 4.5 by the addition of 10 ml of 2.5M sodium acetate and about 10 ml of 15% (m/m) sodium hydroxide, the volume made up to 100 ml with water and the suspension filtered. An aliquot of the filtrate (10 ml) in a 100-ml distillation flask is evaporated to dryness using a rotary evaporator, 2 ml of absolute ethanol containing 0.1 mg of α-picolinic acid are added to the residue (internal standard) followed by 0.5 ml of concentrated sulphuric acid and the solution is heated at 100°C for 90 min. The cooled solution is neutralized with several drops of concentrated aqueous ammonia, transferred to a separating funnel and extracted with five 3-ml portions of diethyl ether. The combined extracts are filtered through a layer of anhydrous sodium sulphate, the filter twice washed with 3 ml of the same solvent, the volume made approximately to 25 ml, and the filtrate mixed and analysed by GLC.

Gas–liquid chromatography
Detector: NPD; column: glass, 2400×2 mm; stationary phase: 10% Carbowax 20M; solid support: Chromaton N-AW (0.125–0.16 mm); temperatures: column 220°C, injector 240°C (Fig. 7.5).

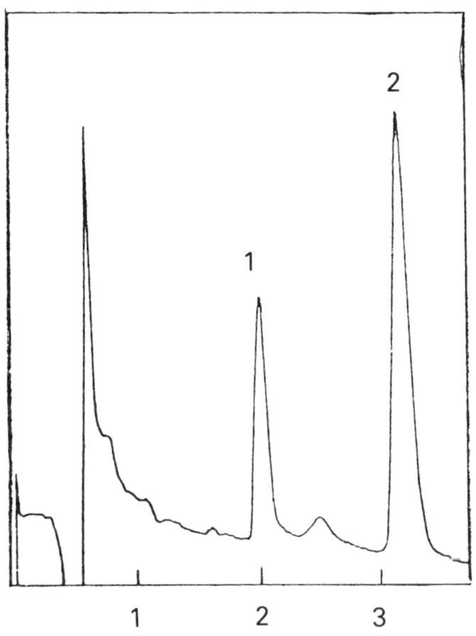

Fig. 7.5 — Determination of nicotinic acid in pork meat (according to Davídek *et al.* (1985)). 1, ethyl nicotinate; 2, ethyl α-picolinate (internal standard).

Notes
α-Picolinic acid may be replaced by quinolinic acid.

7.3.3. Pantothenic acid

All the biologically active compounds, i.e. pantothenic acid (vitamin B_5), its salts and panthenol can be assayed by GLC (Wisniewski, 1966; Prosser and Sheppard,

1969, 1971; Stone and Wright, 1971; Tarli *et al.*, 1971). The majority of GLC methods are, however, applicable only to relatively pure and simple samples such as pharmaceutical preparations.

Salts of pantothenic acid have been analysed after derivatization of the acid to the corresponding *N*-acetylethyl or *N*-trifluoroacetylethyl pantothenates (Prosser and Sheppard, 1969). Some other derivatives, including the trimethylsilyl derivative, have also been employed (Prosser and Sheppard, 1969, 1971).

A simpler and more convenient approach would seem to be the indirect method based upon the hydrolytic cleavage of pantothenic acid in acid medium and the analysis of the hydrolysis products, i.e. either β-alanine or pantoyl lactone (3,3-dimethyl-2-hydroxy-4-butanolide) (Scheme 4). β-Alanine can be chromatographed as the corresponding *N*-trifluoroacetylmethyl or *N*-trifluoroacetylbutyl ester (Tarli *et al.*, 1971); pantoyl lactone which is sufficiently volatile, may be analysed as the free lactone (Tarli *et al.*, 1971) or as the trimethylsilyl ether (Stone and Wright, 1971).

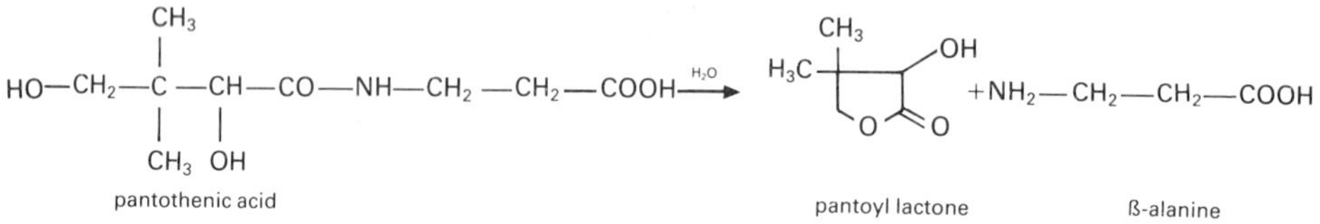

$$HO—CH_2—C(CH_3)_2—CH(OH)—CO—NH—CH_2—CH_2—COOH \xrightarrow{H_2O} \text{pantoyl lactone} + NH_2—CH_2—CH_2—COOH$$

pantothenic acid — pantoyl lactone — ß-alanine

Scheme 4

Gas chromatographic determination of free or bound pantothenic acid in foods (Tesmer *et al.*, 1980; Davídek *et al.*, 1985) is also based upon the hydrolysis of pantothenic acid by hydrochloric acid and separation of pantoyl lactone extracted into a non-polar solvent. The extract may be further purified by chromatography on a silica gel column (Tesmer *et al.*, 1980) or concentrated and directly analysed by GLC (Davídek *et al.*, 1985), employing polar stationary phases such as polyethyleneglycols. The procedure was evaluated by analysing various animal, vegetable and microbial samples, such as beef liver, spray-dried egg yolks, soyabean flour, whole-grain wheat flour, dried baker's yeasts, ready to eat meals, etc.

Procedure (according to Davídek* et al., *(1985))

According to the predetermined moisture content, an appropriate amount corresponding to 2 g of dry material is weighed into a reagent flask. Water and concentrated (36%, m/m) hydrochloric acid are added. The amount of water and hydrochloric acid is calculated to yield 20 ml of 25% (m/m) hydrochloric acid solution. The flask is closed, shaken until its contents are evenly distributed and immersed in a water bath and held at the temperature of 95–100°C for 5 h, with occasional stirring. The flask is then cooled to room temperature and the pH of the hydrolysate adjusted to 5.0 with 40% (m/m) sodium hydroxide. The neutralized hydrolysate is filtered under slight vacuum through a funnel equipped with a medium-porosity sintered glass and the filtrate collected in a separating funnel. The flask and the precipitate are washed with two 5-ml portions of water and the volume of the combined filtrate and washings adjusted to about 40 ml. The solution in the separating funnel is extracted with five 60-ml portions of dichloromethane and the extracts combined in a distillation flask containing 1 ml of ethyl laurate (ethyl dodecanoate, internal standard) solution in dichloromethane which contains 50 μg of the compound. The solution is evaporated to about 10 ml on a vacuum rotary evaporator and then to 100 μl by a gentle stream of nitrogen, and the concentrated solution is analysed by GLC.

Gas–liquid chromatography
Detector: FID; column: glass, 2400×2 mm; stationary phase: 10% Carbowax 20M; solid support: Chromaton N-AW (0.125–0.16 mm); temperatures: column 120–220°C (5°C min^{-1}), injector 220°C (Fig. 7.6).

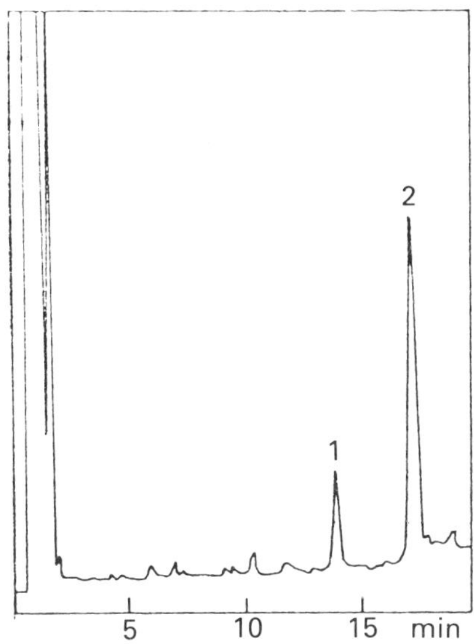

Fig. 7.6 — Determination of pantothenic acid in pork meat (according to Davídek *et al.* (1985)). 1, ethyl laurate (internal standard); 2, pantoyl lactone.

7.3.4 Pyridoxal, pyridoxol and pyridoxamine

The compounds in the free form are not sufficiently volatile for GLC, therefore acetates (Korytnyk, *et al.*, 1966; Prosser *et al.*, 1967; Sheppard and Prosser, 1970), trifluoroacetates (Korytnyk *et al.*, 1966; Ohniski *et al.*, 1967; Richter *et al.*, 1967; Sennello and Argoudelis, 1969; Patzer and Hilker, 1977; Lim *et al.*, 1982), heptafluorobutyrates (Williams, 1974), trimethylsilyl derivatives (Korytnyk *et al.*, 1966;

Ohniski *et al.*, Richeter *et al.*, 1967; Sennello and Argoudelis, 1969), benzyl and isopropylidene derivatives (Korytnyk *et al.*, 1966) have been employed for gas chromatographic separation and the determination of the three active forms of vitamin B_6, their phosphates and important metabolites in model solutions and pharmaceutical preparations.

During the last few years preliminary studies by Lim *et al.*, (1982) have demonstrated the usefulness of trifluoroacetates analysed on a GC system equipped with an ECD for the separation and quantitative evaluation of the naturally occurring active compounds in foods such as white bread, non-fat dry milk and sweet peas. Reaction with absolute ethanol is necessary to convert pyridoxal into its hemiacetal in order to distinguish it from pyridoxol after derivatization.

Procedure (according to Lim** et al., **(1982)

Aqueous slurry of the sample is prepared by homogenization of one part of the sample in two parts (m/m) of water. To 25–50 g of this slurry, 30 ml of 0.2M hydrochloric acid is added. The sample in a beaker is placed in a boiling water bath for 1 h with constant stirring, then cooled to room temperature and 2 ml of α-amylase solution (3 g in 2.5M sodium acetate), 2 ml of pepsin solution (3 g in 2.5M sodium acetate) and 1 ml of 1% (m/m) papain solution are added and the sample incubated at 37°C for 16 h in a shaker water bath. The sample is then filtered, washed with water and the combined filtrate and washings passed through a strongly acid cation exchange resin (the procedure adapted according to the AOAC procedure could be employed, (*Official Methods,* 1985)).

Derivatization

Aliquots of 150 μl of the food extracts (containing up to 100 ng μl^{-1} of B_6 compounds) in a vial are dried under a stream of nitrogen, absolute ethanol (50 μl) is added to convert pyridoxal to its hemiacetal, the closed vial is heated at 85°C for 30 min then cooled to room temperature and then ethanol is removed by nitrogen. *N*-Methylbis(trifluoroacetamide) (MBTFA) (50 μl) is added and the vial heated at 130°C for 20 min. After cooling to room temperature, 450 μl of ethyl acetate are added and the solution is analysed by GLC.

Gas–liquid chromatography

Detector: ECD, column: glass, 1540×2 mm; stationary phase: 10% SP-2100; solid support: Supelcoport (0.147–0.177 mm); temperatures: column 125°C, injector 205°C (Fig. 7.7).

Notes

Additional precision could be obtained, for example, by use of an internal standard such as deoxypyridoxine, omitting or reducing the enzymatic hydrolysis step, simplifying the purification procedure on cation exchangers, etc.

7.3.5 Ascorbic and dehydroascorbic acid

Application of GLC to the determination of L-ascorbic and L-dehydroascorbic acid (vitamin C) has been limited and only a few papers dealing with its application to

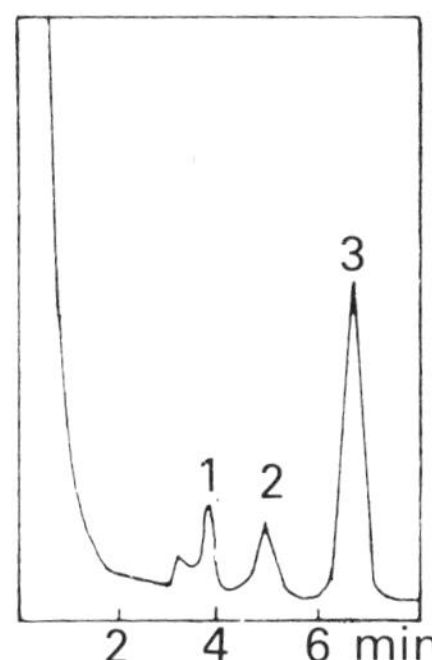

Fig. 7.7 — Determination of vitamin B_6 in milk (according to Lim *et al.* (1982)). 1, pyridoxol; 2, pyridoxal (hemiacetal with ethanol); 3, pyridoxamine (trifluoroacetates).

food have appeared (Gerstl and Ranfft, 1974; Schlack, 1974; Watanabe and Murota, 1981).

Ascorbic acid is an non-volatile compound but it yields a sufficiently volatile trimethylsilyl ether (similarly to sugars) which may be chromatographed on non-polar methylpolysiloxanes as stationary phases with an FID as detector.

According to Gerstl and Ranfft (1974), ascorbic acid may be extracted from food with metaphosphoric acid, interfering substances separated on a column of cellulose, and ascorbic acid in the dried eluate derivatized and analysed. In the method of Schlack (1974) ascorbic acid, together with the non-volatile organic acids, is precipitated from an ethanolic extract with lead acetate solution. It may be regenerated from its lead salt as the corresponding trimethylsilyl ether. A number of foods and food products were assayed for ascorbic acid content while employing the latter procedure.

The same derivatization procedure yields volatile trimethylsilyl derivatives from L-dehydroascorbic acid and 2,3-dioxo-L-gulonic acid from which the ascorbic acid derivative can be separated. Basic solvents, such as pyridine, must, however, be avoided during the determination of dehydroascorbic acid owing to its instability in alkaline media. An indirect determination of dehydroascorbic acid as ascorbic acid (reduced with hydrogen sulphide) is also possible (Gerstl and Ranfft, 1974).

The main problem encountered in the analysis of vitamin C in foods, i.e. elimination of various interfering substances, can thus be avoided by employing GLC methods as they combine separation and determination in one step. It is also well established that no single technique has been accepted as fully suitable for all foods and food products. Therefore, many points still remain to be improved in the methodology of the GLC determination of the two naturally occurring biologically active forms of vitamin C in foods, as well as of the related compounds employed as food additives (salts, esters and isomers of ascorbic acid).

Procedure (according to Velíšek and Davídek (unpublished), Schlack (1974))

Sample (15–20 g) is homogenized in about 55 ml of 95% (v/v) ethanol (the final ethanol concentration, depending on the moisture content of the sample, would be at least 90%) and the homogenate is transferred to a 100 ml volumetric flask using the same solvent. Liquid samples may be directly weighed into the volumetric flask. The flask is made up to the mark and kept (with occasional shaking) at room temperature for 1 h; the contents of the flask are filtered and 10 ml of the resulting filtrate are transferred to a centrifuge tube containing approximately 0.1 g of Celite 545. A solution (1 ml) of glutaric acid (internal standard) containing 1 mg of the acid in 95% ethanol, 1 ml of a saturated solution of lead acetate (neutral salt) and 30 ml of 95% ethanol are added, mixed and kept for 30 min at room temperature. Then the tube is centrifuged and the supernatant tested with one drop of lead acetate solution to confirm completeness of the precipitation of the acids. If this is incomplete 0.5 ml of lead acetate solution and 10 ml of 95% ethanol are added, the contents mixed, let stand for an additional 15 min and centrifuged again. The supernatant is decanted and discarded and the sediment washed with two 10-ml portions of 95% ethanol and 5 ml of diethyl ether. After each addition of solvent the precipitate of lead salts is well mixed, the suspension centrifuged and the solvent decanted and discarded. (Addition of 1 ml of citric acid solution of the same or even 10 times higher concentration than that of the internal standard solution is recommended during the analysis of samples with low ascorbic acid levels). Residual solvent is evaporated with a gentle stream of nitrogen; the precipitate is broken into a fine powder using a glass rod and dried for 1 h at 100°C.

Derivatization

About 0.1 g of freshly prepared anhydrous calcium sulphate and 1 ml of silylation reagent (pyridine–HMDS–TMCS, 10:9:6, v/v/v) are added, the mixture heated at 60°C for 1 h and then analysed by GLC.

Gas–liquid chromatography

Detector: FID; column: glass, 1540×2 mm; stationary phase: 3% SE-30; solid support: Varaport 30 (0.177–0.25 mm); temperatures: column 60–210°C (6°C min^{-1}), injector 250°C (Fig. 7.8).

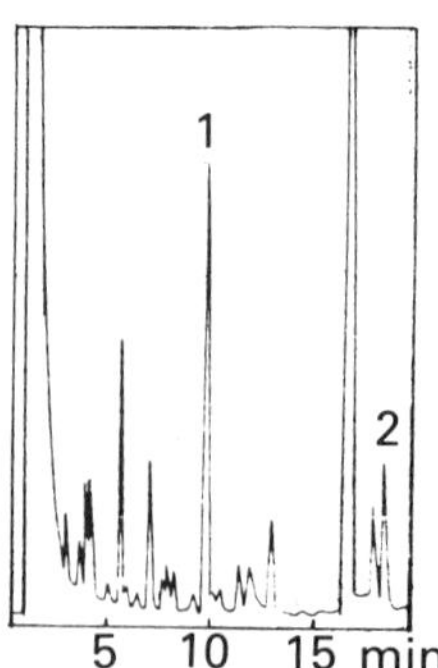

Fig. 7.8 — Determination of ascorbic acid in tomatoes (according to Gerstl and Ranfft (1974)). 1, glutaric acid (internal standard); 2, ascorbic acid (trimethylsilyl ethers).

Notes
Stock solution of L-ascorbic acid in 95% ethanol (1 mg of ascorbic acid per 1 ml) must be prepared daily.

7.3.6 Other active compounds

GLC methods for the separation and/or determination of the other water-soluble vitamins, such as riboflavin, biotin, folacin, corinoids (Janecke and Voege, 1968, 1971; Viswanathan *et al.*, 1970; Johnson and Boyden, 1977; Gaudiano *et al.*, 1978; Seifert, 1974; Levy *et al.*, 1964; Karayannis and Corwin, 1968) have been utilized to a small extent in the analysis of pharmaceutical preparations and model samples. In foods, however, these vitamins have not yet been evaluated.

A simple GLC method has been developed for the determination of *S*-methylmethionine (vitamin U) in vegetables (Bezzubov and Gessler, 1977, 1983). The method involves the degradation of the compound in an alkaline medium to homoserine and dimethylsulphide (Scheme 5), which is distilled off, trapped in toluene and the solution directly analysed.

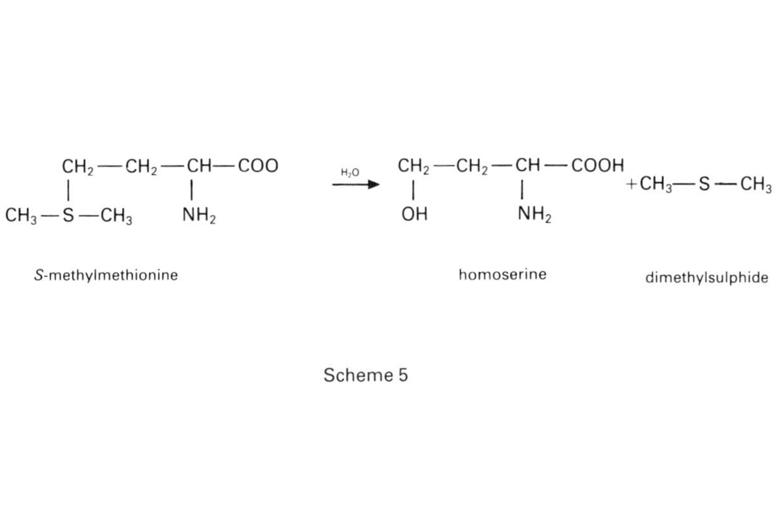

$$\underset{S\text{-methylmethionine}}{\underset{CH_3-\overset{|}{S}-CH_3\qquad\ NH_2}{CH_2-CH_2-\overset{}{CH}-COO}} \xrightarrow{H_2O} \underset{\text{homoserine}}{CH_2(OH)-CH_2-CH(NH_2)-COOH} + \underset{\text{dimethylsulphide}}{CH_3-S-CH_3}$$

Scheme 5

Procedure (according to Bezzubov and Gessler (1977))

The sample (100 g) is homogenized in 100 ml of borate buffer of pH 9.7–10.0 and the homogenate transferred to a three-necked (500-ml) distillation flask. The flask is connected via a side neck to a flask containing 100 ml of water and placed in a bath at 80°C. The distillation flask is connected via the second side neck to a series of two traps. The first one, containing 2 ml of 40% (m/m) sodium bisulphite, is cooled with ice. The second trap, with 2 ml of toluene containing 4 mg ml^{-1} of tetrachloromethane (internal standard), is cooled with solid carbon dioxide in acetone. Nitrogen at the rate of 50 ml min^{-1} is then bubbled into the flask containing the water and the distillation flask containing the sample is heated on a boiling water bath for 90 min. The toluene trap is then disconnected, allowed to stand till it reaches room temperature and the solution in the trap is analysed by GLC.

Gas–liquid chromatography
Detector: FID; column: stainless steel 3000×3 mm, stationary phase: 15% Carbowax 20M; solid support: Chromaton N-AW-DMCS (0.10–0.125 mm); temperatures: column 75°C, injector 130°C (Fig. 7.9).

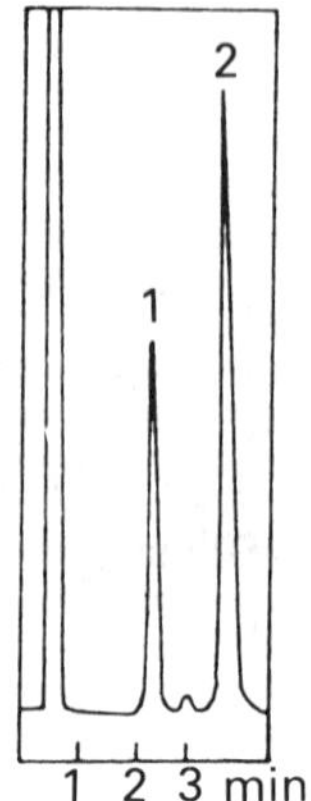

Fig. 7.9 — Determination of *S*-methylmethionine in tomatoes (according to Bezzubov and Gessler, (1977)). 1, dimethylsulphide; 2, tetrachloromethane (internal standard).

Notes
Stock solution of *S*-methylmethionine in borate buffer (pH 9.7–10.0) contains 50 mg of the compound per litre.

The reactivity of *myo*-inositol enables its gas chromatographic separation in the form either of volatile esters or of ethers, as in the case of other cyclitols and/or sugars. *Myo*-inositol has been determined as the corresponding hexa(trimethylsilyl)ether in yeast hydrolysates (Roberts *et al.*, 1965), multivitamin preparations (Sagara *et al.*, 1979), phospholipids (Flint *et al.*, 1965), animal tissues (Flint *et al.*, 1965; Narumi *et al.*, 1969), vegetables and other plant materials (McDonald and Newson, 1970; Reymakers, 1974). Some of the other biologically important inositols and other cyclitols (Lee and Baloou, 1965) as well as *myo*-inositol phosphate in mixtures with sugar phosphates (Sherma *et al.*, 1971) can also be separated and estimated by GLC as trimethylsilyl ethers. Methyl ethers are especially convenient for GLC–MS analysis (Binder and Haddon, 1984).

The hexaacetate has been employed for the determination of *myo*-inositol in white beans (Seifert, 1972). The described method seemed to be also applicable to other natural products. It had the advantage of simple sample preparation in comparison with the preparation of trimethylsilyl ethers. The method can be employed for the determination not only of free *myo*-inositol but also of the phytin content. Saccharides, if present in the sample, may be simultaneously determined. Other inositols may be analysed as well (Irving, 1981).

Procedures for the quantitative estimation of *myo*-inositol either in the form of the trimethylsilyl ether or acetate are similar to those employed in the analysis of sugars.

Several GLC procedures are also available for the determination of rutin. They always include, as in the case of other flavonols and other flavonoids, its prior separation by other chromatographic techniques. For example, rutin may be separated together with simple phenols, phenolic acids and other naturally occurring phenolic substances, such as flavonoid pigments. A convenient method applicable to beer and wine (Drawert *et al.*, 1977) is based on the separation of phenolic compounds from other substances (mainly organic acids, amino acids, saccharides) by adsorption on a polyvinylpolypyrrolidone (PVPP) column. After elution, concentration and silylation, the resulting trimethylsilyl derivatives of the individual phenolic compounds are separated and determined on either packed or capillary columns with non-polar silicones as stationary phases. Convenient methods have been also developed for the evaluation of rutin and other phenolics in orange juice and some fruits (Drawert *et al.*, 1980; Duggan, 1960).

Lipoic (thioctic) acid and 4-aminobenzoic acid may be analysed as their esters but the methods have not yet been developed for use on a routine basis (Huber *et al.*, 1966; Verga and Gnocchi, 1970).

REFERENCES

Amos, W. H. and Neal, R. A. (1970). Gas chromatography–mass spectrometry of the trimethylsilyl derivative of various thiamine metabolites, *Anal. Biochem.*, **36**, 332–337.

Ashby, J. D. and Deavin, J. C. (1969). Gas chromatographic determination of nicotinamide and thiamine in vitamin preparation, *J. Pharm. Pharmacol.*, **21** (Suppl.), 52–54.

Becktold, H., Sklein, F., Frenk, D. and Jöhnchen, E. (1984). Improved method for quantitative analysis of vitamin K_1 2,3-epoxide in human plasma by electron-capture gas–liquid capillary chromatography, *J. Chromatogr.*, **306**, 333–337.

Bell, J. G. and Christie, A. A. (1973). Gas–liquid chromatographic determination of vitamin D in cod-liver oil, *Analyst*, **98**, 268–273.

Bell, J. G. and Christie, A. A. (1973). Gas–liquid chromatographic determination of vitamin D_2 in fortified full-cream dried milk, *Analyst*, **99**, 385–396.

Bertling, L. and Tietz, I. (1978). Verwendung von Nikotinsäure in Fleischerzeugnissen, Fleischwirtschaft, **58**, 621–622.

Bezzubov, A. A. and Gessler, N. N. (1977). Gas chromatographic determination of *S*-methylmethionine (vitamin U) in plants, *Prikl. Biokhim. Mikrobiol.*, **13**, 301–308.

Bezzubov, A. A. and Gessler, N. N. (1983). Gas liquid column chromatography for studying vitamin U metabolism in humans and animals, *J. Chromatogr.*, **273**, 192–196.

Binder, R. G. and Haddon, W. F. (1984). Analysis of *O*-methyl inositols by gas–liquid chromatography–mass spectrometry, *Carbohydr. Res.*, **129**, 21–32.

Brubacher, G., Müller-Mulot, W. and Southgate, D. A. T. (1986). *Methods for the Determination of Vitamins in Food,* Elsevier Applied Science Publishers Ltd., London, New York.

Carrol, K. K. and Herting, D. C. (1964). Gas–liquid chromatography of fat-soluble vitamins, *J. Amer. Oil Chem. Soc.*, **41**, 473–464.

Castello, G., Bruschi, E., Ghelli, G. (1977). Gas chromatographic determination of the purity of vitamin K_3 (menadione). *J. Chromatogr.* **139**, 195–202.

Christie, A. A. (1975). The estimation of vitamin D and E in foods, *Chem. Ind.* 492–493.

Christie, A. A. and Wiggins, R. A. (1978). Developments in vitamin analysis. In *Developments in Food Analysis Techniques-1,* King, R. D. (ed.), Elsevier Applied Science Publishers Ltd., London, pp. 1–42.

Christie, A. A., Dean, A. C. and Millburn, B. A. (1973). The determination of vitamin E in foods by colorimetry and gas–liquid chromatography, *Analyst,* **98**, 161–167.

Dampert, W. V. and Beringer, H. (1976). Dünschicht- und Gas-chromatographie von Tocopherolen, Fette Seifen Anstrichm., **78**, 108–111.

Davídek, J., Velíšek, J., Černá, J. and Davídek, T. (1985). Gas chromatographic determination of pantothenic acid in foodstuffs, *J. Micronutr. Anal.*, **1**, 39–46.

Davídek, J. and Velíšek, J. (1986). Gas–liquid chromatography of vitamins in foods: the fat-soluble vitamins, *J. Micronutr. Anal.*, **2**, 81–96.

Davídek, J. and Velíšek, J. (1987). Gas–liquid chromatography of water-soluble vitamins in foods, *Proc. EURO FOOD CHEM IV,* Loen, Norway, 1987, pp. 101–105.

de Leenheer, A. P. and Cruyl, A. A. M. (1980). In *Methods in Enzymology, Vol. 67,* Part F, McCormick, D. B. and Wright, L. D. (eds), Academic Press, New York, pp. 335–343.

de Leenheer, A. P., Lambert, W. E. and de Ruyter, M. G. M. (1985). *Modern Chromatographic Analysis of the Vitamins,* Marcel Dekker, Inc., New York, Basel.

de Ritter, E. (1967). Newer analytical techniques—vitamins, *Ass. Food Drug Offic. US, Quart. Bull.*, **31**, 94–117.

De Vries, E. and Borsje, B. (1982). Analysis of fat-soluble vitamins, XXVII. High performance liquid chromatographic and gas–liquid chromatographic determination of vitamin D in fortified milk and milk powder: collaborative study, *J. Assoc. Offic. Anal. Chem.*, **65**, 1228–1234.

Dialameh, J. H. and Olson, R. E. (1969). Gas–liquid chromatography of phytyl ubiquinone, vitamin E, vitamin K_1 and homologues of vitamin K_2, *Anal. Biochem.*, **32**, 263–272.

Dickes, G. J. and Nicholas, P. V. (1976). *Gas chromatography in food analysis,* Butterworth, London, Boston.

Drawert, F., Lessing, V. and Leupold, G. (1977). Über die Abtrennung und gaschromatographische Bestimmung von niedermolekularen phenolischen Verbindungen, *Chem. Mikrobiol. Technol. Lebensm.*, **5**, 65–70.

Drawert, F., Leupold, G. and Pivernetz, H. (1980). Quantitative Bestimmung von Rutin, Hesperidin und Naringin in Orange Saft mittels Gas–Flüssigkeitchromatographie, *Chem. Mikrobiol. Technol. Lebensm.*, **8**, 189–191.

Dugan, P. and Lundgren, D. (1964). Microdetection of menadione, coenzyme Q_6 and certain related quinoid compounds in biological materials using electron capture–gas chromatography, *Anal. Biochem.*, **8**, 312–318.

Duggan, M. B. (1969). Methods for examination of flavonoids in fruits: application to flavonol glycosides and aglycones of apples, pears, and strawberries, *J. Ass. Offic. Anal. Chem.*, **52**, 1038–1043.

Dwivedi, B. K. and Arnold, R. G. (1972). Gas chromatographic estimation of thiamine, *J. Food Sci.*, **37**, 889–891.

Echols, R. E., Harrie, J. and Miller, R. H., Jr. (1980). Modified procedure for determining vitamin B_1 by gas chromatography, *J. Chromatogr.*, **193**, 470–475.

Echols, R. E., Miller, R. M., Winzer, W., Carmen, D. J. and Ireland, Y. R. (1983). Gas chromatographic determination of thiamine in meats, vegetables and cereals with a nitrogen–phosphorus detector, *J. Chromatogr.*, **262**, 257–263.

Echols, R. E., Miller, R. H. and Thompson, L. (1985). Evaluation of internal standards and extraction solvents in the gas chromatographic determination of thiamine, *J. Chromatogr.*, **347**, 89–97.

Fenton, T. W., Vogtmann, H. and Clandinin, D. R. (1973). Gas–liquid chromatography of vitamin-A_1 alcohol and vitamin A_1 acetate, *J. Chromatogr.*, **77**, 410–412.

Flint, D. R., Lee, T. C. and Huggins, C. G. (1965). A simple and rapid method for the determination of *myo*-inositol by gas–liquid chromatography, *J. Amer. Oil Chem. Soc.*, **42**, 1001–1002.

Gaudiano, A., Bellomonte, G., Sanzini, E., Gilardi, G. and Renzi, M. (1978). Colorimetric and gas-chromatographic method for determination of biotin in pharmaceutical preparations, *Ann. Ist. Super. Sanite*, **13**, 783–791.

Gerstl, R. and Ranfft, K. (1974). Gaschromatographische Methode zur Vitamin C-Bestimmung, *Z. Lebensm. Untersuch. Forsch.*, **154**, 12–17.

Hagiwara, H., Oka, Y., Hara, Y., Yurugi, S., Suzuki, J., Furuno, K. and Iida, T. (1963). Vitamin B_1 and related compounds. 104. New *S*-acylthiamine derivatives, *Ann. Rept. Takeda Res. Lab.*, **22**, 1–12.

Heftmann, E. (1976). *Chromatography of steroids*, Elsevier, Amsterdam.

Hilker, D. M. and Mee, J. M. L. (1973). Gas chromatography of thiamine and derivatives, *J. Chromatogr.*, **76**, 239–241.

Hommes, M. and van der Mijll Dekker, L. P. (1973). Gas chromatographic determination of vitamin D in cod liver oil, pharmaceuticals and veterinary preparations, *Internat. J. Vit. Nutr. Res.*, **43** 271–282.

Huber, J. D., Bingham, R. J. and Aurand, L. W. (1966). Isolation of thioctic acid in milk, *61st Annual Mtg., Amer. Dairy Sci. Ass.*, Corvalis, Ore, June 1966; *Abstr., J. Dairy Sci.*, **49,** (1966) 702.

Irwing, G. C. J. (1981). Gas chromatography of inositols as their hexakis-*O*-acetyl derivatives, *J. Chromatogr.*, **205**, 460–463.

Ishiguro, M. (1983). Determination of vitamin E by gas chromatography, *Kanzei Chue Bunsekishoho*, **24**, 45–49 (*Chem. Abstr.*, **100**, (1984) 119440s).

IUPAC (1981). Standard methods for the Analysis of Oils, Fats and Derivatives, 6th Edn, 1st Supplement: Part 4 (1981), Section II: Oils and Fats, 2.404. Identification and Determination of Tocopherols, pp. 234–241; *Pure and Applied Chem.*, **54**, (1982) 233–245.

Janecke, H. and Brendel, R. (1971). Zur Bestimmung von Vitamin D in biologischem Material (Milch), *Naturwissenschaften,* **58**, 54–55.

Janecke, A. and Voege, H. (1968). Zur Gaschromatographie von Vitamingemischen, *Naturwissenschaften,* **55**, 447–448.

Janecke, A. and Voege, H. (1971). Zur gaschromatographischen Bestimmung von silyliertem Biotin, seinem Oxydationsprodukten und anderen silylierten Vitaminen, *Z. Anal. Chem.*, **254**, 335–339.

Johnson, R. N. and Boyden, G. R. (1977). Characterization of biotin trimethylsilyl derivative, *J. Pharm. Sci.*, **66**, 1212–1213.

Karayannis, N. M. and Corwin, A. H. (1968). Hyperpressure gas chromatography. III. Gas chromatography of porphyrins and metalloporphyrins, *Anal. Biochem.*, **26**, 34–50.

Katsui, G. (1981). Assay methods for vitamin E. 3. Gas chromatographic method, *Vitamin,* **55**, 191–196.

Kawamoto, Y. and Mochida, Y. (1980). Study of methods for gas chromatographic determination of vitamin E in feeds, *Nihon Nosan Kogyo Kenkyu Nenpo,* **10**, 1–7.

Kawasaki, C. and Iritani, N. (1968). Biosynthesis of thiamine. II. Gas chromatography of pyrimidine moiety of thiamine, *Vitamin,* **38**, 38–40.

Kobayashi, T. (1980). In *Methods in Enzymology, Vol. 67,* Part F, McCormick, D. B. and Wright, L. D. (eds), Academic Press, New York, pp. 347–355.

Kobayashi, T., Adachi, A. and Furata, K. (1976). Studies on gas–liquid chromatographic determination of vitamin D. III. Gas–liquid chromatographic determination of vitamin D_3 in tuna liver and vitamin D_3 resin oils, *J. Nutr. Sci. Vitaminol.*, **22**, 215–224.

Korytnyk, W., Fricke, G. and Paul, B. (1966). Pyridoxine chemistry. XII. Gas chromatography of compounds in the vitamin B_6 group, *Anal. Biochem.*, **17**, 66–75.

Lee, Y. C. and Baloou, C. E. (1965). Gas chromatography of inositols as their trimethylsilyl derivatives, *J. Chromatogr.*, **18**, 147–149.

Levy, R. L., Gesser, H., Halevi, E. A. and Saidman, S. (1964). Pyrolysis gas chromatography of porphyrins, *J. Gas Chromatogr.* **2**, 254–255.

Lercker, G. and Caboni, M. F. (1985). Analysis of unsaponifiables by GLC, *Riv. Ital. Sostanze Grasse,* **62**, 193–198.

Lim, K. L., Young, R. W., Palmer, J. K. and Driskell, J. A. (1982). Quantitative separation of B-6 vitamins in selected foods by a gas–liquid chromatographic system equipped with an electron-capture detector, *J. Chromatogr.*, **250**, 86–89.

McDonald, R. E. and Newson, D. W. (1970). Extraction and gas–liquid chromatography of sweet potato sugars and inositol, *J. Am. Soc. Hort. Sci.*, **95**, 299–301.

Mancas, D. G. (1984). Gas-chromatographic determination of vitamin E acetate from premixes used in animal feeding, *Rev. Chim.*, **35**, 637–639 (*Chem. Abstr.*, **101**, 124 299x (1984)).

Mariani, G. and Fedeli, E. (1983). Rapid methods for the analysis of vegetable oils.

Note 1. Simultaneous determination of sterols and tocopherols, *Riv. Ital. Sostanze Grasse,* **59**, 557–565.

Mee, J. M. L., Brooks, C. C. and Yanagihara, K. H. (1975). Gas chromatographic –mass spectrometric investigation of the photoepoxidation of vitamin K_3, *J. Chromatogr.*, **110**, 178–181.

Meijboom, P. W. and Jongenotter, G. A. (1979). A quantitative determination of tocotrienols and tocopherols in palm oil by TLC–GLC, *J. Amer. Oil Chem. Soc.*, **56**, 33–35.

Mermet-Bouvier, R. (1972). Gas chromatography of the photochemical isomers of ergosterol, *J. Chromatogr. Sci.*, **10**, 733–736.

Mordret, F. X. and Laurent, A. M. (1978). Application of gas-phase chromatography on capillary glass column to the analysis of tocopherols, *Rev. Fr. Corps Gras,* **25**, 245–250.

Morimoto, H. and Imada, I. (1971). In *Methods in Enzymology, Vol. XVIII,* Part C, McCormick, D. B. and Wright, L. D. (eds), Academic Press, New York, pp. 169–179.

Nair, P. P., Sarlos, I. and Machiz, J. (1966). Microquantitative separation of isomeric dimethyltocols by gas-chomatography, *Arch. Biochem. Biophys.*, **114**, 488–493.

Nakata, Y. and Tsuchida, E. (1980). In *Methods in Enzymology, Vol. 67,* Part F, McCormick, D. B. and Wright, L. D. (eds), Academic Press, New York, pp. 148–155.

Nakata, Y., Mita, Y., and Khono, S. (1976). Studies on photodegradation of vitamin K_1 (phylloquinone). I. Determination of the vitamin by gas-chromatography, *J. Pharm. Soc. Japan,* **96**, 53–59.

Narumi, K., Arita, M., Kitagawa, M., Kumazawa, A. and Tsumita, T. (1969). Gas chromatographic analysis of free *myo*- and *scyllo*-inositols in animal tissues, *Jap. J. Exp. Med.*, **39**, 399–407.

Nelson, J. P., Milun, A. J. and Fisher, H. D. (1970). Gas chromatographic determination of tocopherols and sterols in soya sludges and residues—an improved method, *J. Amer. Oil Chem. Soc.*, **47**, 259–261.

Official Methods of Analysis of the Association of Official Analytical Chemists (1985), Horowitz, W. (ed.), 14th ed., Association of Official Analytical Chemists, Washington.

Ohniski, Y., Horii, Z. and Makita, M. (1967). Gas chromatography of vitamin B_6, *Pharm. Soc. Japan J.*, **87**, 747–748.

Panalaks, T. (1970). Gas chromatographic method for determination of vitamin D in fortified non-fat milk, *Analyst,* **95**, 862–867.

Patzer, E. M. and Hilker, D. M. (1977). New reagent for vitamin B_6 derivative formation in gas chromatography, *J. Chromatogr.*, **135**, 489–492.

Ponomaryev, A. M., Vselyubovskaya, O. A. and Yanotovskii, M. T. (1973). Gas chromatographic determination of nicotinic acid in the presence of nicotinamide and nicotinonitrile, *Zh. Analit. Khim.*, **28**, 2458–2463.

Prosser, A. R. and Sheppard, A. J. (1968). Gas–liquid chromatography of niacin and niacinamide, *J. Pharm. Sci.*, **57**, 1004–1006.

Prosser, A. R. and Sheppard, A. J. (1969). Gas–liquid chromatographic determination of pantothenates and panthenol, *J. Pharm. Sci.*, **58**, 718–721.

Prosser, A. R. and Sheppard, A. J. (1971). Gas–liquid chromatography of trimethylsilyl derivatives of pantothenyl alcohol and pantothenates, *J. Pharm. Sci.*, **60**, 909–913.

Prosser, A. R., Sheppard, A. J. and Libbey, D. A. (1967). Gas–liquid chromatography of vitamin B_6, *J. Assoc. Offic. Anal. Chem.*, **50**, 1348–1353.

Ranfft, K. (1973). Gas-chromatographic determination of added and naturally occurring vitamin E in feeding-stuffs, *Landw. Forsch. Sonderh.*, **28**, 146–152.

Rao, M. K. G. and Perkins, E. G. (1972). Identification and estimation of tocopherols and tocotrienols in vegetable oils using gas chromatography–mass spectrometry, *J. Agric. Food Chem.*, **20**, 240–245.

Reymakers, A. (1974). Extraction and purification of carbohydrates from plant material, *Pharm. Weekbl.*, **109**, 1229–1231.

Richter, W., Vecchi, M., Vetter, W. and Walther, W. (1967). Gas chromatographic and mass-spectrometric studies of trimethylsilyl derivatives of vitamin B_6, *Helv. Chim. Acta,* **50**, 364–376.

Riera, J. B. (1975). Analysis de los tocoferols de aceites vegetales por cromatografía en fase gaseosa, *Anal. Bromatol.*, **27**, 287–296.

Roberts, R. N., Johnston, J. A. and Fuhr, B. W. (1965). A method for the quantitative estimation of *myo*-inositol by gas–liquid chromatography, *Anal. Biochem.*, **10**, 282–289.

Sagara, K., Ikunaga, K., Kasuya, K., Kaito, T. and Anmo, T. (1979). Gas–liquid chromatographic determination of inositol in multivitamin preparation containing large amounts of sugars, *Chem. Pharm. Bull.*, **27**, 800–803.

Schlack, J. E. (1974). Quantitative determination of L-ascorbic acid by gas–liquid chromatography, *J. Assoc. Offic. Anal. Chem.*, **57**, 1346–1349.

Seifert, R. M. (1972). Analysis of *myo*-inositol in dry beans by gas chromatography of its hexa-acetate, *J. Ass. Offic. Anal. Chem.*, **55**, 1194–1198.

Seifert, R. M. (1974). An approach to the chemical analysis of folic acid: gas chromatography of a degradation product from synthetic folic acid samples, *J. Sci. Food Agric.*, **25**, 1509–1515.

Seifert, R. M. (1979). Analysis of vitamin K_1 in some green leafy vegetables by gas chromatography, *J. Agric. Food Chem.*, **27**, 1301–1304.

Seifert, R. M. and Miller, C. F. (1973). Analysis of thiamine in pinto beans by gas chromatography of the sulfite cleavage product 5-(2-hydroxyethyl)-4-methylthiazole, *J. Assoc. Offic. Anal. Chem.*, **56**, 1273–1276.

Sennello, L. T. and Argoudelis, C. J. (1969). Gas chromatographic procedure for the simultaneous determination of pyridoxine, ascorbic acid, and nicotinamide in vitamin capsules and tablets, *Anal. Chem.*, **4**, 171–173.

Sheppard, A. J. (1971). In *Methods in Enzymology, Vol. XVIII,* Part C, McCormick, D. B. and Wright, L. D. (eds), Academic Press, New York, pp. 461–464.

Sheppard, A. J. and Hubbard, W. D. (1971). In *Methods in Enzymology, Vol. XVIII,* Part C, McCormick, D. B. and Wright, L. D. (eds), Academic Press, New York and London, pp. 733–738.

Sheppard, A. J. and Prosser, A. R. (1970). In *Methods in Enzymology, Vol. XVIII,* Part A, McCormick, D. B. and Wright, L. D. (eds), Academic Press, New York, pp. 494–500

Sheppard, A. J. and Prosser, A. R. (1971). In *Methods in Enzymology, Vol. XVIII,* Part B, McCormick, D. B. and Wright, L. D. (eds), Academic Press, New York, pp. 17–20

Sheppard, A. J., Prosser, A. R. and Hubbard, W. D. (1971). In *Methods in Enzymology, Vol. XVIII,* Part C, McCormick, D. B. and Wright, L. D. (eds), Academic Press, New York, pp. 356–365

Sheppard, A. J., Prosser, A. R. and Hubbard, W. D. (1972). Gas chromatography of fat soluble vitamins: a review, *J. Amer. Oil Chem. Soc.*, **49**, 619–633.

Sherma, W. R., Goodwin, S. L. and Zinbo, M. (1971). Gas chromatography of some completely trimethylsilylated phosphates of inositol and their sugar/alcohols and sugars, *J. Chromatogr.* **9**, 363–367.

Sklan, D. (1980). In *Methods in Enzymology, Vol. 67,* Part F, McCormick, D. B. and Wright, L. D. (eds), Academic Press, New York, pp. 355–361.

Slover, H. T. (1971). Tocopherols in foods and fats. *Lipids,* **6**, 291–296.

Slover, H. T. and Thompson, R. H., Jr. (1981). Chromatographic separation of the stereoisomers of α-tocopherol, *Lipids,* **16**, 268–275.

Slover, H. T., Shelley, L. M. and Burks, T. L. (1967). Identification of and estimation of tocopherols by gas–liquid chromatography, *J. Amer. Oil Chem. Soc.*, **44**, 161–166.

Slover, H. T., Lehmann, J. and Valis, R. J. (1969). Vitamin E in foods. Determination of tocols and tocotrienols, *J. Amer. Oil Chem. Soc.*, **46**, 417–420.

Slover, H. T., Thompson, R. H., Jr. and Nerola, G. V. (1983). Determination of tocopherols and sterols by capillary gas chromatography, *J. Amer. Oil Chem. Soc.*, **60**, 1524–1528.

Spiegel, P. and Teply, O. (1984). Chromatographic determination of water-soluble vitamins (review article), Part 2. Gas chromatographic methods, *Sci. Pharm.*, **52**, 259–267.

Stone, J. C. and Wright, J. (1971). GLC analysis of the trimethylsilyl derivative of 3,3-dimethylbutyric acid gamma-lactone in pantothenyl alcohol, *J. Pharm. Sci.*, **60**, 163–164.

Takeuchi, A., Okano, T., Torii, M., Hataka, Y. and Kobayashi, T. (1985). Identification and determination of vitamin D_2 in various fungi, *Vitamin,* **59**, 195–205 (*Chem. Abstr.*, **103**, 84526g (1985)).

Tarli, P. Benocci, S. and Neri, P. (1969). Gas chromatographic determination of nicotinamide and lidocaine in pharmaceutical preparations, *Boll. Chim. Farm.*, **108**, 554–559.

Tarli, P., Benocci, S. and Neri, P. (1971). Gas chromatographic determination of pantothenates or panthenol in pharmaceutical preparations, *Anal. Biochem.*, **42**, 8–13.

Taylor, R. F. and Ikawa, M. (1980). In *Methods in Enzymology, Vol. 67,* Part F, McCormick, D. B. and Wright, L. D. (eds), Academic Press, New York, pp. 233–239.

Taylor, R. F. and Davies, B. H. (1975). Gas–liquid chromatography of carotenoids, and other terpenoids, *J. Chromatogr.*, **103,** 327–340.

Taylor, R. F. and Ikawa, M. (1971). Gas chromatography of carotenoids, *Anal. Biochem.*, **44,** 623–627.

Tesmer, E., Leinert, J. and Hötzel, D. (1980). Gas chromatographic determination of pantothenic acid in foods, *Nahrung,* **24**, 697–704.

Touw, H. M. D., Kröse, B. M. C. and Molenaar, H. M. (1972). Gas–liquid chromatographic determination of vitamin D_2 and D_3 in infant formulas and feeding preparations, *J. Assoc. Offic. Anal. Chem.*, **55**, 622–624.

Uebersax, P. (1970). Quantitative gaschromatographische Vitamin E-Bestimmung in Futtermitteln, *Mitt. Geb. Lebensmitteluntersuch. Hyg.*, **61**, 254–259.

Van Niekerk, P. J. (1982). Determination of Vitamins. In *HPLC in Food Analysis,* Macrae, R. (ed), Academic Press, London, pp. 187–226.

Vecchi, M., Vesely, J. and Oesterhelt, G. (1973). Application of high-pressure liquid chromatography and gas chromatography to problems in vitamin A analysis, *J. Chromatogr.*, **83**, 447–453.

Vecchi, M., Schmid, M., Walther, W. and Gerber, F. (1981). Determination of the diastereomers of vitamin K_1, *J. High Resolut. Chromatogr., Chromatogr. Commun.*, **4**, 257–259.

Velíšek, J. and Davídek, J. (1986). Gas–liquid chromatography of vitamins in foods: the water-soluble vitamins, *J. Micronutr. Anal.*, **2**, 25–42.

Velíšek, J., Davídek, J., Mňuková, J. and Pištěk, T. (1986). Gas chromatographic determination of thiamine in foods, *J. Micronutr. Anal.*, **2**, 73–80.

Velíšek, J., Davídek, J. and Pištěk, T. (1987). Determination of thiamine by gas chromatography. *Sbor. ÚVTIZ-Potravinářské Vědy,* **5**, 263–272.

Verga, R. and Gnocchi, A. (1970). Characterization and gas chromatographic determination of active principles of biological interest in pharmaceutical products (gastric mucoprotein and thioctic acid), *J. Chromatogr.,* **49**, 46–52.

Vetter, W., Vecchi, M., Gutmann, H., Rucegg, R., Walther, W. and Meyer, R. (1967). Gas chromatographische und massenspektrometrische Untersuchung von Phytylubichinon, Vitamin K_1 und Vitamin K_2, *Helv. Chim. Acta,* **50**, 1866–1879.

Vessman, J. and Stromberg, S. (1975). Gas–liquid chromatographic determination of nicotinamide in multivitamin formulations after conversion to nicotinonitrile, *J. Pharm. Sci.*, **64**, 311–313.

Viswanathan, V., Mahn, F. P., Venturella, V. S. and Senkowski, B. Z. (1970). Gas–liquid chromatography of d-biotin, *J. Pharm. Sci.*, **59**, 400–402.

Watanabe, T. and Murota, S. (1981). Gas chromatographic determination of L-ascorbic acid in food, *Kobe Joshi Daigaku Kiyo,* **12**, 87–88 (*Chem. Abstr.*, **96** (1982) 160 937w).

Wiggins, R. A. (1976). Replacement of biological methods by chemical and physical methods. Chemical analysis of vitamins A,D and E, *Proc. Anal. Div. Chem. Soc.*, **13**, 137–139.

Williams, A. K. (1974). Vitamin B_6: gas–liquid chromatography of pyridoxol, pyridoxal, and pyridoxamine, *J. Agric. Food Chem.* **22**, 107–109.

Wilson, P. W., Lawson, D. E. M. and Kodicek, E. (1969). Gas–liquid chromatography of ergocalciferol and cholecalciferol in nanogram quantities, *J. Chromatogr.*, **39**, 75–77.

Winkler, V. W. (1973). Collaborative study of a gas–liquid chromatographic method for the determination of water-soluble menadione (vitamin K_3) in feed premixes, *J. Assoc. Offic. Anal. Chem.*, **56**, 1277–1280.

Wisniewski, L. W. (1966). Trimethylsilyl derivatives in gas chromatographic analysis. Derivative formation with pantothenyl alcohol, *Facts Methods Sci. Res.*, **7**, 4–5.

Yanotovskij, M. T., Kozlov, E. I., Obolnikova, E. A., Volkova, O. I. and Samochvalov, G. I. (1968). Gas–liquid chromatography of a series of vitamin K and ubiquinones, *J. Gas Chromatogr.*, **6**, 520–522.

8

Gas chromatographic analysis of contaminants in food

John Gilbert

8.1 INTRODUCTION

Organic contamination of foods can occur at every stage of the food chain from atmospheric pollution of the environment, which can, for example, lead to contamination of growing vegetables, through to contamination of a highly processed finished food product, perhaps by migration from its plastic packaging material. Contamination usually occurs at trace levels but the level at which monitoring in the food supply is required will depend on what is known of the toxicity of the constituent and in some circumstances whether there are regulatory limits controlling its presence in foods. For highly toxic and carcinogenic contaminants methods of analysis may be required at sub-μg/kg levels, whereas for some less toxic constituents (perhaps a deliberate adulterant like diethylene glycol) monitoring at the mg/kg level may be sufficient. An overview of both the classes of different food contaminants and some generalized figures for levels at which monitoring is normally required is given in Table 8.1. It must be appreciated that this table is a generalization but it does illustrate the relative status of different contaminants and the differences in requirements for analysis.

Food contaminant analysis differs from other aspects of food analysis in two main respects — the level of concern, which in this instance is really very low even at the highest level of the indicated range, and the diversity of biological matrices that have to be handled. This means that methods for food contaminant analysis need to be sensitive, and because of the possibilities of interferences need to be specific. Sensitivity can be achieved by taking a large sample size and concentrating the analyte in a small final extract for analysis or by having a very sensitive end-detector. Similarly, specificity can be achieved by either a lengthy sample preparation and clean-up to separate the analyte from potential interferences, or using a less selective work-up but utilizing a more specific detector. Most analytical methods balance these two factors to arrive at a method that is not too time-consuming in sample preparation but is neither too extreme in its requirements for analyte detection.

Table 8.1 — Overview of applications of GC in food contaminant analysis

Class of contaminant	Examples	Matrix	Derivative	Required Limit of detection (μg/kg)	Detector	Alternative methods
Environmental	PCBs	Fish	None	10–50	ECD	None
	PCDDs	Milk; meat	None	0.01	HRMS	None
	PAHs	Vegetables, oils	None	0.1	FID; MS	HPLC (FL)
Pesticides	Organochlorine	Milk; eggs	None	100	ECD	None
	Organophosphorus	edible crops	None	100	FPD	None
Fumigants	EDB	Cereals	None	30	ECD	None
Mycotoxins	Trichothecenes	Cereals	TMS; HFB	50	ECD; MS	TLC; HPLC (UV)
	Patulin	Fruit juice	TMS	5–10	FID; MS	HPLC (UV)
Veterinary drugs	Anabolics	Animal tissue	TMS	1	MS	RIA
	Sulphonamides	Milk, eggs	Me; HFB	100	ECD	ELISA; HPLC
Nitrosamines	NDMA	Meat; beer	None	1–10	TEA	None
Packaging	Monomers	Processed foods:	None	10	FID; AFID	None
	Plasticizers	dairy products	None	100	FID; MS	HPLC
		confectionary				
Adulterants	DEG; Methanol	wine	None	10 mg/kg	FID	None

PCBs — polychlorinated biphenyls; PCDDs — polychlorinated dibenzodioxins; PAHs — Polyaromatic hydrocarbons; NDMA — *N*-Nitrosodimethylamine; DEG — diethyleneglycol; TMS— trimethylsilyl; HFB — heptafluorobutyryl; HRMS — high-resolution selected-ion monitoring; FID — flame ionization detector; ECD — electron-capture detector; FPD — flame photometric detector; MS — selected ion monitoring; TEA — thermal-energy analyser; AFID — nitrogen specific detector; HPLC (FL) — fluorescence detection; HPLC (UV) — ultraviolet detection; RIA — radioimmunoassay; ELISA — enzyme-linked immunosorbant assay; EDB — ethylene dibromide.

Hence, the most highly specific gas chromatographic (GC) detector — high resolution mass spectrometry (MS), which obviously involves expensive analytical instrumentation and high operating costs, is only chosen to be used in circumstances when there are no alternative detection methods available to meet the requirements of the analysis — for example isomer-specific tetrachlorodibenzodioxin analysis.

Most currently available analytical techniques are used for the analysis of food contaminants although principally thin-layer chromatography (TLC), high-performance liquid chromatography (HPLC) and GC are used as separation methods prior to end-measurement. Also, and of increasing importance there is the use of immunological based methods such as enzyme-linked immunosorbent assay (ELISA) and these are now available for determining some of the veterinary drug residues and for a limited number of mycotoxins. Faced with a choice of methods the volatile analytes tend to be analysed exclusively by GC (in particular headspace GC) and the higher molecular weight nonvolatile constituents by HPLC, provided they contain a suitable chromophore for detection. Some intermediate compounds, particularly those which are heat-labile but for which flame ionization detection (FID) is nevertheless the appropriate detection method, may be analysed by supercritical fluid chromatography. This is, however, still a relatively new technique and is not yet established for routine food contaminant monitoring, although its potential has already been demonstrated in the pesticide (carbamates) and mycotoxin fields (macrocyclic trichothecenes).

A number of contaminants do, however, fall between the extremes of volatility and for these the choice, usually of GC or HPLC, is more difficult. This is frequently the case for analytes that require derivatization prior to GC analysis (which can be a disadvantage) or for analytes that can be chemically degraded to smaller species suitable for GC monitoring. The principal advantage of a GC method is the ease with which confirmation can be carried out by mass spectrometry (MS) either by full scanning or by selected-ion monitoring (SIM). For food contaminant analysis, rigorous confirmation is important and therefore the ability to carry out GC–MS confirmation must weigh as an important factor in selection of a method.

In this chapter is is not possible to exhaustively review all food contaminants for which GC is applicable and therefore a selective approach has been adopted looking at the area where GC is really the only appropriate technique — for the very volatile contaminants — and looking at the areas where there are choices available between GC and HPLC and highlighting the particular advantages of using GC. Important areas, such as GC analysis of pesticides, have not been covered as they are discussed in Chapter 9 and elsewhere (Roseboom, 1984); *N*-nitrosamines have been covered in detail by Scanlan (1984) and Sen (1984) and are rather specialised in relation to the GC detection method unique to this class of compound.

In view of the need for confidence in correct identification of food contaminants MS is frequently used as a primary detection method, and therefore in this chapter GC–MS (SIM) will be regarded as a selective GC detection method.

8.2 VOLATILE CONTAMINANTS — HEADSPACE GC

There are two basic approaches to the headspace analysis of volatile contaminants in foods — the static headspace method, which involves sampling at equilibrium from the headspace gas above the heated food, and the dynamic or 'purge and trap' approach. The static approach has probably been used most frequently for quantitative monitoring of food contaminants and is suitable for relatively large numbers of samples, for example for use in surveys. The dynamic approach is particularly useful for achieving very low levels of detection as the analyte is effectively concentrated. This method is appropriate for identification of contaminants where larger amounts of material may be required for initial full spectroscopic characterization. Both approaches have been dealt with in specialist texts (Hachenberg & Schmidt, 1977; Kolb, 1980, 1984) and here only an outline of the principles are given with an emphasis on practical detail. Only recent applications are reviewed and for earlier references the reader is referred to other texts (Gilbert, 1984).

8.2.1 Principles of headspace GC

Headspace GC is an attractive technique for food contaminant analysis because it is simple, involves minimal sample preparation and is amenable to automation. The principle of the technique is as follows.

The sample in liquid form or slurried with water is sealed in a vial with a rubber septum, heated until equilibrium partition is established between analyte and gaseous headspace, and then a fixed volume of headspace gas is withdrawn for analysis. The key word is 'equilibration' and this is essential for quantification. The gaseous sample (usually of the order of 1 ml or so) is injected into the GC and analysed in the normal way by packed or capillary column (for capillary GC, split injection is necessary). For quantification a calibration curve can be constructed by spiking analyte into uncontaminated material or, where control material is not available, the method of standard addition can be used.

Some practical points about headspace analysis are not always obvious from the literature and are worth emphasizing:

(1) Partition equilibrium must be demonstrated and conditions must be established for each particular compound of interest and each different matrix, and this involves choice of both appropriate temperature and time. For most low-boiling contaminants to be determined in solution, then 1 h at 79–90°C is normally sufficient for equilibrium partition. For some high-boiling components or compounds present in solid matrices equilibration may take several hours. This is obviously impractical and in this situation one must either change to another method of analysis or use an alternative approach to make headspace suitable.

(2) It is worth remembering that when working in a laboratory environment there are often relatively high levels of solvent vapours and other potential interferences present and therefore care is needed to seal 'clean' air in the headspace of the vials. This is sometimes best carried out away from the laboratory atmosphere.

(3) Some rubber septa employed for sealing the vials can be a source of interferences that are released when the vial is heated. This can be overcome by switching to PTFE (polytetrafluoroethylene)-faced septa, but these are more prone to leaks, particularly during manual sampling. It is always necessary to analyse a number of blank vials, vials containing only solvent, as well as the normal control vials to check for interferences.

(4) Some very reactive contaminants, such as ethylene oxide and methyl bromide, could be lost through chemical reaction with components in the matrix at high temperatures during sample equilibration. The effect is evidenced by poor reproducibility and by a decrease in apparent concentration with an increase in sample equilibration time. In these circumstances a cold solvent extraction method might be preferable to the headspace approach.
(5) If sample preparation involves grinding and mixing then aeration can lead to analyte losses, which will be particularly apparent for very volatile contaminants. This can be minimized by sample handling, for example grinding, at low temperature, such as under liquid nitrogen.

Sampling is potentially a problem with all food contaminant assays but becomes exaggerated as sample sizes become smaller. For the headspace approach frequently sample sizes are only 2 g, and therefore very thorough mixing and grinding of larger samples is essential before sub-samples are taken for headspace analysis. In view of the small sample size it is worth analysing a greater number of sub-samples than normal to test the homogeneity of mixing of the initially taken sample.

When manual headspace sampling as opposed to dedicated automated techniques is employed, there can be particular problems associated with the injection. Condensation of sample vapour may be a problem with aqueous samples but can be overcome by using a heated syringe. 'Blow-back' when injecting large sample volumes into the gas chromatograph can be a problem resulting in difficulties in obtaining good quantification — syringes are available allowing compression of the sample before injection, which helps this problem to some extent. Carry-over from one injection to the next can be a source of error — vaccum syringe cleaners are useful to minimize this effect.

The normal range of GC detectors have been used for headspace GC analysis chosen as appropriate for the analyte, and in addition MS has been employed either for confirmation or as a primary detection method.

8.2.2 Automated headspace GC–MS

GC–MS systems usually have a permanently fixed GC instrument interfaced through heated transfer lines. It would therefore be inconvenient to remove the GC and reinterface to another headspace GC whenever this mode of analysis was required. This problem has been overcome by mounting the headspace GC on a trolley and simply wheeling it to the MS whenever needed. Coupling is achieved by passing the GC column to a stainless steel transfer line connected right through the existing GC and directly coupling to the MS (Gilbert & Startin, 1981). With fused-silica columns this is very simple as the column end can be directly connected to the MS and thus one can avoid all connectors with the disadvantages inherent of deadspace volumes. No other interfacing is required and, although pressure pulses do disturb the MS response during injection, this does not cause any problems and the system rapidly re-stabilizes. Using automated headspace equipment the instrument can be allowed to run unattended, and with the MS operated in the SIM mode it is sufficiently stable not to require tuning during the course of a day's work.

The use of the MS as a headspace GC detector does offer some very specific attractions:

(1) Achievable sensitivity is better than can be attained with other detectors, which means that headspace analysis can be utilized for low-volatility compounds with unfavourable partition. Thus the headspace approach can be extended to less-volatile analytes, which on equilibrium strongly favour remaining in solution rather than in the gaseous headspace.
(2) As there is no sample clean-up using the headspace approach the specificity is achieved by GC alone. Using the MS as a detector, a higher specificity is achieved than with other GC detectors. It is possible, by choice of appropriate ions and by switching ions during the course of a GC run, to carry out multi-compound assays. This high confidence in the specificity of monitoring allows for rapid method development.

8.2.3 Monomer analysis in foods

Foods packaged in plastics can be contaminated by migration of residual monomers from the plastic container into the food. An important use of headspace GC is for the analysis of residual monomer levels in the plastics and of contamination levels in the foods. The principal monomers of interest have to date been vinyl chloride, acrylonitrile, butadiene, styrene, vinylidene chloride and terephthalic acid. Interest was initiated in the monomers with the discovery that vinyl chloride was a liver carcinogen and that it was present at surprisingly high residual levels in PVC (polyvinyl chloride) and also migrated into some foods, notably alcoholic beverages contained in airline spirit miniature bottles. Interest in the other monomers followed, based on their toxicity where known, their structural relationship to vinyl chloride and the extent of use of the parent polymer for food packaging. It should be noted that, although most of the work to date has concerned only these six monomers, the proposed EEC Positive list contains 450 named monomers and other starting substances (Rossi, 1988) so there still remains considerable work to be done in this area in terms of assessment of migration.

Plastics for food packaging for which monomer analysis has been undertaken find a wide range of applications in the United Kingdom. PVC in rigid form is used for bottles, principally orange squash and cooking oil, and is also used in thinner but still rigid form for nestings for cakes, biscuits and chocolates. PVC flexible film is used in the home for covering foods and in supermarkets as a overwrap, usually for an expanded polystyrene tray for packaging fresh and cooked meat, fruit and vegetables. Polystyrene and acrylonitrile butadiene styrene (ABS) are used in the form of tubs, ABS principally for soft margarine, whilst polystyrene is used for dairy and desert-type products such as cream, yoghurt, cottage cheese, chocolate desserts, etc. Poly(vinylidene chloride) (PVDC) is used as a film coating for biscuits and snack products, and as a co-polymer in the form of 'chub' packs for paté, processed cheese and cooked meats. The use of poly(ethylene terephthalate) (PET) has grown significantly and this material in the form of bottles is now widely used for carbonated beverages as well as for alcoholic drinks, such as beer, cider and for some spirit miniatures. The use of all these plastics also extends beyond food packaging and many materials are sold as containers for use in cooking, particularly PET which is used in the form of trays for microwave applications, and this area of migration must also be included in any assessment. Thus, methods of analysis for monomers must be sufficiently versatile to handle a wide range of different food types. Often processed foods may contain a number of added volatile constituents, such as flavourings, that are potential sources of interferences in the monomer analysis, and the GC approach must be able to cope with these potential problems.

For the analysis of monomers in both plastics and foods the simple approach of headspace GC can be employed for their quantitative determination. This involves placing a sample of the plastic (usually dissolved in a solvent such as dimethylacetamide or dimethylformamide) in a glass vial or placing an aqueous slurry of the food in a vial and then proceeding as described above. For GC analysis an FID is usually employed for vinyl chloride, and vinylidene chloride or a nitrogen-specific detector in the case of acrylonitrile. In some instances for the analysis of less volatile constituents, such as styrene monomer, the injection of water into the vial can enhance sensitivity by altering the analyte partition in favour of the headspace (Steichen, 1976). Headspace GC–MS is advised for confirmation, particularly for monomer analysis of foodstuffs, which frequently contain volatile interferences.

Calibration for headspace GC is by spiking known amounts of the monomer in a suitable solvent into vials containing samples of the dissolved blank plastic (i.e. with known zero residual monomer levels) or into vials containing blank food samples. Calibration curves can then be constructed, although it is recommended that the analysis of calibration standards be randomly interspersed with those of the actual survey. Where blank material is unavailable then the method of standard addition can be employed, but this is very time-consuming, particularly for large-scale surveys. The analysis of frequent blank samples is essential as a check on the cleanliness of the procedure. Frequent sources of contamination are the headspace air sealed into the vial and the rubber septum.

Typical residual levels of monomers in the plastics vary from low levels of below 1.0 mg/kg for vinyl chloride and vinylidene chloride, to several hundreds of mg/kg for styrene monomer, where the high levels are caused by depolymerization during fabrication of the finished article. Typical levels of monomers found in the foods do not directly correlate with the residual levels in the plastics, as the partition coefficient (solubility) of the monomer in the food and monomer volatility will exert an important influence on migration. Vinyl chloride monomer is currently not detectable in the UK in foods packaged in PVC; acrylonitrile levels are below 20 μg/kg; vinylidene chloride cannot be generally detected; and styrene levels range from 5 to 40 μg/kg depending on the food product. Styrene contamination, however, is in some ways self-limiting as levels above the taste threshold will lead to detectable taint problems (Miltz *et al.*, 1980).

Residual acrylonitrile monomer levels have been determined in ABS plastic packaging materials by dissolution in either dichlorobenzene (Gilbert & Startin, 1982), dimethylsulphoxide (McNeal and Breder, 1981), or methyl isobutyl ketone (Page & Charbonneau, 1983) and either automated or manual headspace analysis using a nitrogen/phosphorus specific detector. The choice of solvent for dissolution of the plastic seems largely a matter of personal preference, although greater responses for acrylonitrile were observed for solution in propylene carbonate or dichlorobenzene (McNeal & Breder, 1981). Impurities in the solvents can present problems in headspace analysis of plastics, but these can often be removed by sparging with an inert gas prior to use of the solvent. Analysis of cellulose and polyolefin films (biscuit and snack food wrappers which had a vinylidene chloride/acrylonitrile co-polymer coating) for levels of acrylonitrile monomer has been carried out by direct headspace analysis (Gilbert & Startin, 1982). This involved heating the film at 120°C for 30 min prior to manual sampling. A limit of detection of 10 μg/kg residual acrylonitrile in the film was obtained using GC/MS SIM monitoring.

Acrylonitrile has been determined in 3% acetic acid, used as a food simulant for carbonated beverages, and with the addition of sodium sulphate to the solution to alter the equilibrium partition in favour of the headspace a limit of detection of 0.07 μg/kg was achieved (McNeal and Breder, 1981). For the analysis of acrylonitrile in soft margarine by automated headspace with a nitrogen detector a limit of detection of 5 μg/kg was obtained. Propionitrile was used as an internal standard (Gilbert & Startin, 1982). A similar procedure was employed for margarine, honey, butter, cheese and peanut butter but with the addition of sodium chloride to the food sample to improve headspace partition. Limits of detection were from 4 to 10 μg/kg, and the CV (coefficient of variation) of the procedure ranged from 7.5 to 10.2% using manual headspace sampling with propionitrile as internal standard (Page & Charbonneau, 1983). This latter procedure was tested by collaborative trial and recommended as the AOAC official first action method for determining acrylonitrile above 15 μg/kg in specified foods (Page, 1985). A problem observed with determining acrylonitrile in proteinaceous foods by headspace analysis was the progressive loss of monomer with respect to internal standard on protracted heating. A chemical reaction with food components was demonstrated which could be overcome by carrying out headspace analysis in an acidified medium. When food samples were prepared with 10% phosphoric acid and sodium chloride in a 1:1:0.5 slurry, headspace analysis was satisfactory for luncheon meat (CV=6.6–17.5%) and peanut butter (Page & Charbonneau, 1985).

Styrene monomer has been determined in polystyrene and related food packaging materials by headspace GC using an FID. The plastics were dissolved in dimethylacetamide in sealed headspace vials and immediately prior to equilibration the polymer was precipitated by the addition of water — this had the effect of altering the equilibrium partition in favour of the headspace and gave a limit of detection of 1 mg/kg in the polymer (Gilbert & Startin, 1983). Concern that heating of a polymer solution might lead to depolymerization and artificially high residual styrene monomer levels has led some workers to prefer to use HPLC with UV detection for plastics analysis (Varner & Breder, 1981a). Polystyrene cups have been tested for migration of styrene into hot coffee, tea, water and 8% ethanol, and with manual headspace sampling limits of detection from 3 to 10 μg/kg have been reported (Varner & Breder, 1981b). For the analysis of fatty foods such as margarine for styrene contamination the limit of detection by headspace GC with an FID is too high to be of practical use (350 μg/kg). To overcome this problem GC–MS has been employed using SIM, monitoring m/z 104 (the base peak and molecular ion for styrene), whereby this limit can be improved to 15 μg/kg (Gilbert & Startin, 1981). An alternative approach has been to carry out azeotropic distillation of the food sample and then to analyse the aqueous distillate by headspace GC (Varner *et al.*,

1983). This procedure is rather cumbersome, particularly if there are large numbers of samples to be analysed, but it was possible to achieve a limit of detection of 25 μg/kg for styrene in margarine. A large survey of retail foods in the UK including yoghurts, creams, salads, coleslaws, soft cheeses, margarines, hot and cold beverages from dispensing machines, spreads, fresh and cooked meats, candied fruits and take-away 'fast foods' was carried out for styrene contamination using headspace GC–MS (Gilbert & Startin, 1983). Similar surveys have been reported for retail products in Chile (Santa Maria *et al.*, 1986) and in Australia (Flanjak & Sharrad, 1984), the latter work showing a good agreement between results for styrene levels determined in food samples by both the techniques of headspace GC and HPLC.

Residual 1,3-butadiene has been determined in food contact plastics, chewing gum, oils and yoghurt using manual headspace GC with an FID. Confirmation of the μg/kg levels detected was, however, recommended owing to the non-specific nature of the detection method and the low levels (McNeal & Breder, 1987). Automated GC–MS using SIM for m/z 54 (the molecular ion and base peak of 1,3-butadiene) indicated levels below 0.2 μg/kg in retail samples of soft margarine (Startin & Gilbert, 1984). Vinylidene chloride was monitored at levels of 5 μg/kg by electron-capture headspace GC, but was not detected in retail samples of biscuits, cakes and snack food products (Gilbert *et al.*, 1980). The insolubility of PET in solvents usually chosen for headspace analysis necessitated a solid headspace approach to determining residual acetaldehyde in PET carbonated beverage bottles. Samples of the plastic were ground to a fine powder under liquid nitrogen, heated at 130°C for 1 h to release residual acetaldehyde, and then equilibrated in the headspace waterbath at 92°C before analysis (Dong *et al.*, 1980). A taint problem in samples of maple syrup packaged in plastic tubs was attributed to the presence of methyl methacrylate, toluene and styrene. These were identified by headspace GC–MS and quantified by manual headspace (FID) using the method of standard addition (Hollifield *et al.*, 1980).

The technique of multiple headspace extraction, where repeated headspace analysis is undertaken with gas extraction of the sample in between, enables concentrations to be determined by the exponential decrease in analyte, without the necessity of calibration or standard addition. This approach has been employed for determining residual solvents, such as ethyl acetate, methanol, ethanol and ethylene glycol, in printed aluminium film, and for monomer analysis. Using automated headspace multiple extraction with glass capillary GC analysis a CV of 3.4% was obtained for determining vinyl chloride monomer in PVC resin sample (Kolb *et al.*, 1981).

8.2.4 Other volatile contaminants

Ethylene dibromide (EDB) is mixed with other organic compounds, such as carbon tetrachloride and carbon disulphide, to make commercial fumigant formulations. These mixtures have in the past been used to spray stored grain, citrus fruits and milling machinery to control insect infestations. EDB has also been used by farmers to eliminate groundworms in soil before planting, There has therefore been concern about possible residue levels in foods and a need to carry out monitoring. In the USA tolerance levels have been set ranging from 0.9 mg/kg for raw grains to 0.03 mg/kg for finished food products.

Analysis for fumigant residues has conventionally been based on cold solvent extraction of grains or a water/azeotropic distillation of fruit. These methods work reasonably but have the severe disadvantages of being time-consuming in the clean-up stage and necessitating extensive analyses of spiked samples as a constant check on recovery. Surprisingly headspace GC has been little used for fumigant analysis despite its obvious attractions. Automated headspace GC–MS has been used for determining EDB in fresh fruits monitoring the base peak at *m/z* 107 (M-Br) together with the bromine isotope ion at *m/z* 109 (Gilbert *et al.*, 1985a). Preliminary spiking of EDB into water at various levels and monitoring *m/z* 107 in the headspace provided good linearity for calibration over the concentration range 0–100 μg/kg. Solutions were heated for different times and in a random order to avoid systematic errors. The linearity of calibration gave good confidence in the chosen equilibration conditions of 1 h at 70°C. The GC analysis employed a 25 m fused-silica capillary GC column (0.33 mm i.d.) operated isothermally at 60°C. A typical SIM chromatogram for the analysis of a sample of lemons is shown in Fig. 8.1. The EDB peak was found to elute

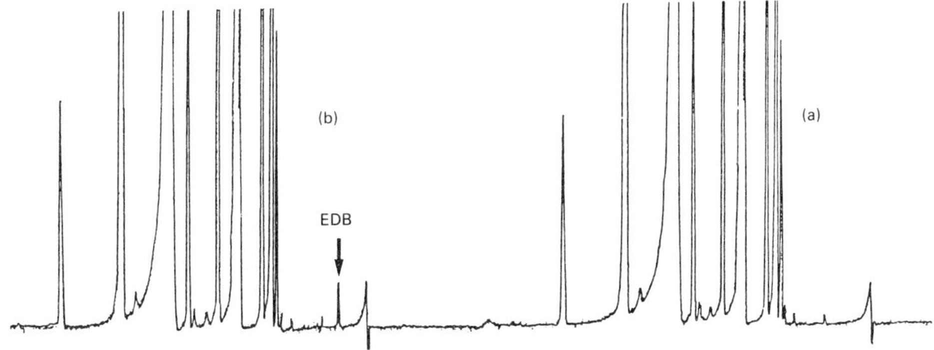

Fig. 8.1 — Headspace GC analysis of ethylene dibromide (EDB) fumigant residues in a sample of lemon fruit. Detection by GC–MS monitoring *m/z* 107. GC conditions — column (25 m×0.3 mm i.d.) fused silica CP SIL 5CB operated at 10 p.s.i. (6.9×10^4 Pa) helium carrier-gas flow isothermally at 60°C. Split injection (20:1) with an injection time of 1 s, using an automated headspace analyser (Perkin Elmer F42). Chromatograms: (a) control lemon sample with below 1 μg/kg contamination with EDB; (b) spiked lemon sample containing 10 μg/kg EDB. (Reproduced with permission from Gilbert *et al.*, 1985.)

1 min after the air peak. No EDB was detected in control samples of fruit. In this analysis the ion at *m/z* 107 was clearly not a very specific ion to monitor and the additional peaks in the chromatogram are believed to originate from terpenoid compounds present in the peel of the fruit (Gilbert *et al.*, 1985a). The limit of detection for citrus fruits, mangoes, papayas, and green peppers was established at 1 μg/kg and the limit of quantification at 5 μg/kg (S/N=10:1). The CV was 6.8% for fruit samples spiked at 20 μg/kg (n=12) and the assay was linear over the range 0–1.0 mg/kg.

For validation of the headspace approach, spiked matrices were produced by addition of EDB in propanol to aqueous slurries of fruit from which samples of 100 g were taken for distillation and 4 g for headspace analysis. Commercially fumigated samples would have been preferable to allow for any fumigant binding to the matrix but it is difficult to obtain samples with the required target residue levels. Spiked samples were, however, left for 24 h before analysis. Headspace GC–MS of EDB in citrus fruits gave good agreement with the spiking levels but analysis by distillation of the same samples gave slightly lower results. For fumigant analysis the headspace approach was between 5 and 18 times faster than conventional distillation techniques. This does not even allow for the additional samples that need to be analysed by the distillation approach to establish recovery and for replication.

Methyl bromide is frequently used as a fumigant for nuts, grain and cocoa beans. For the headspace analysis of these materials, water was added to the sample contained in a vial, which was then sealed with a PTFE-faced septum. After equilibration at 32°C for periods of 1–6 h analysis was by automated headspace GC with a 30-m thick-film fused-silica capillary column and electron-capture detection (ECD) (DeVries *et al.*, 1985). Somewhat surprisingly the samples were not ground to a powder, which together with their higher fat content explains the long equilibration time required for the larger particle-sized peanuts. The limit of detection of the method was 0.4 μg/kg and CVs ranging from 6.5 to 11.6% were obtained depending on the sample type.

For certain solid food matrices there is advantage in digesting the food sample in acid to produce a slurry for headspace analysis of halocarbons (Entz & Hollifield, 1982). This of course is provided that the analytes are unaffected by the acid treatment. On treatment of such foods as fish and meat in a sealed vial with 10M sulphuric acid the protein was digested, but the lipid, though sometimes charred, floated around as a layer on the digestion medium. No measurable degradation was observed for chloroform, 1,1,1-trichloroethane, carbon tetrachloride, trichloroethylene, and tetrachloroethylene. Equilibration of the multi-phase system was established at 90°C for 1 h and automated headspace analysis of these volatile halocarbons was by capillary GC with electron-capture detection. Digestion can produce volatile non-halogenated compounds from the food, but with the specific detection of electron-capture this does not present a problem. For the analysis of a wide range of foods, CVs of 20% or less were achieved; the large variations were probably where matrix inhomogeneity was a problem. Limits of detection were sub-μg/kg for aqueous foods and in the range 10–50 μg/kg for lipid-containing matrices. Fish and several processed food samples including jelly, chocolate sauce, ice cream, and mayonnaise were analysed for halocarbons using this headspace technique (Entz & Hollifield, 1982). The same technique was later applied to the analysis of composite total diet samples for the presence of seven volatile brominated and chlorinated constituents (Entz *et al.*, 1982). Margarine samples were analysed for the presence of tetrachloroethylene by automated headspace GC with electron-capture detection (Entz & Diachenko, 1988), and although the solvent was detected in the adhesives from the wrappers, the highest levels were found in samples collected from a supermarket located next to a dry cleaner. Contamination of samples of olive oil with tetra- and trichloroethylene has also been detected by capillary headspace GC–MS monitoring two characteristic ions for each of the two chlorinated constituents (Startin, 1988; unpublished results). Typical chromatograms illustrating blanks and oilve oil samples contaminated with low-mg/kg amounts of these solvent residues are shown in Fig. 8.2. Methylene chloride residues have been detected at mg/kg levels in decaffeinated tea and coffee by manual headspace sampling onto a packed GC column using a Coulson electrolytic conductivity detector (Page & Charbonneau, 1984). Sodium sulphate was added to samples to improve headspace partition, and with methyl bromide as an internal standard CVs of 10% or less were obtained.

8.2.5 Purge and trap analysis

The principle of the 'purge and trap' method is that the sample in a liquid slurry is sealed in a flask and whilst being stirred is sparged with an inert gas. The gas passes through a trap containing a small amount of adsorbent, such as Tenax, which retains the volatile analyte. When the total amount of volatile analyte has been purged from the food matrix, the trap is removed, the analyte is eluted with a small volume of solvent, such as hexane, and then analysed by GC. Sparging usually takes from 0.5 to 1.0 h for complete purging of the sample, and recoveries from fortified samples are generally better than 90%. This approach has the following advantages:

(a) It is non-labour-intensive (purging can be carried out unattended).
(b) The apparatus is simple and readily available in most analytical laboratories.
(c) The extract is concentrated and relatively clean, which means that the limits of detection that can be achieved are much lower than by solvent extraction or by static headspace analysis.

The method was developed initially for the analysis of EDB in grains (Heikes, 1985a) and then for milled grain products, intermediate grain-based foods and animal feeds (Heikes, 1985b). Using both ECD and Hall electrolytic conductivity detection, EDB could be quantified at a limit of 0.5 μg/kg and gave good agreement with solvent extraction methods. When the trap was modified to include Amberlite XAD resin as well as Tenax, the technique was extended to the analysis of carbon disulphide (using flame photometric detection), methylene chloride, chloroform, 1,2-dichloroethane, methyl chloroform, carbon tetrachloride, trichloroethylene, EDB, and tetrachloroethylene (Heikes & Hopper, 1986). The method was also applied to the analysis of table-ready foods, such as cheese, margarine, frozen dinners, peanut butter, canned vegetables, etc. Recoveries of all nine analytes from fortified samples ranged from 83 to 104%, and chromatograms were very clean despite the complex food matrices analysed. Limitations of detection were in the low to sub-μg/kg for the halogenated analytes and around 12 μg/kg for carbon disulphide (Heikes, 1987a). For decaffeinated instant coffees and decaffeinated ground coffees the same approach was applied for the analysis of several chlorinated solvents, including methylene chloride. Chromatograms were very clean and limits of detection for methylene chloride were 0.4 μg/kg in the solid coffee, and 10 μg/kg in the coffee brew (Heikes, 1987b).

The analysis of residual organic solvents in samples of laminated food packaging materials was carried out by applying a vacuum to bags and drawing air from the interior through a Tenax trap (Eiceman & Karasek, 1981). Various volatile organic compounds, including methanol, 1-ethoxy-2-propanol, 1-propanol, propyl acetate, 2-methyl-2-propanol, *tert*-butanol and a C_9 hydrocarbon, were identified in lami-

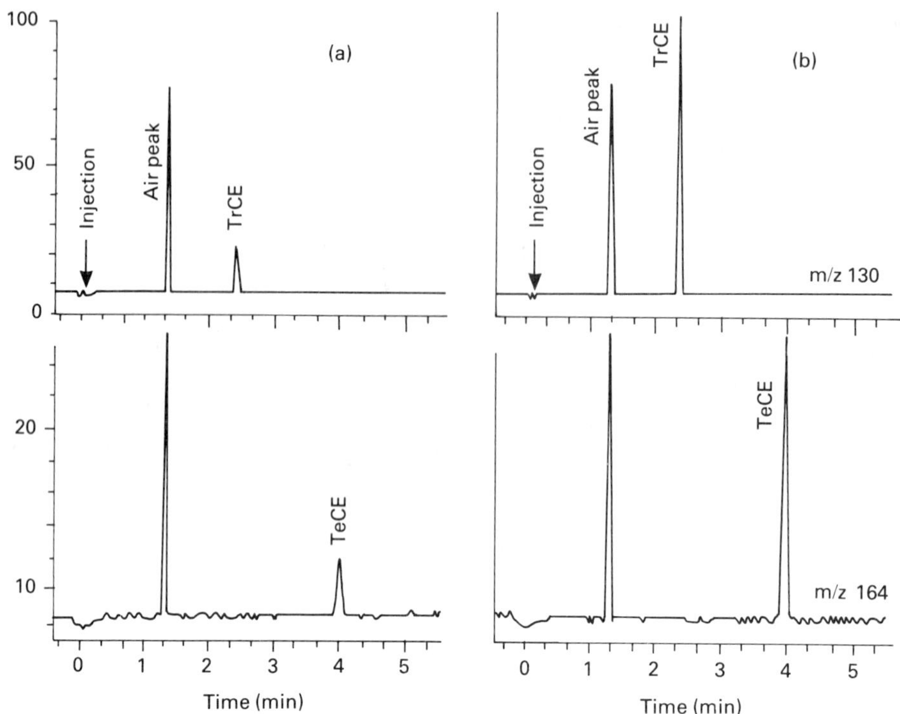

Fig. 8.2 — Headspace GC analysis of tri (TrCE) and tetra-chloroethylene (TeCE) contamination of a sample of olive oil. Detection by GC–MS monitoring *m/z* 130 for trichloroethylene and *m/z* 164 for tetrachloroethylene. GC conditions — column (25 m×0.3 mm i.d.) thick film (1 μm) fused-silica BPl operated at 0.8 bar (8×10^4 Pa) helium carrier-gas flow isothermally at 60°C. Split injection (20:1) with an injection time of 2 s, using an automated headspace analyser (Perkin Elmer F42). Chromatograms: (a) contaminated olive oil sample; (b) contaminated olive oil sample spiked with 1 mg/kg of both TrCE and TeCE. (Startin, 1988; unpublished results.)

nated plastic material, and benzene was detected by GC–MS in a commercial potato crisp package.

8.3 MYCOTOXINS

Mycotoxins are toxic secondary fungal metabolites which can arise in nuts and cereals during growth, production and storage, or in animal products by transfer when contaminated feedstuffs are used. The main foods that are affected are nuts and nut products, cereals, fruit and dairy products, such as milk. A large number of mycotoxins of diverse chemical structures have been identified as being produced by fungi (Cole & Cox, 1981), but only a relatively small number have been shown to occur naturally in foods and only a limited number are routinely monitored. Most mycotoxins are polar and of relatively high molecular weight and therefore TLC and HPLC have tended to be the analytical methods of choice. GC is used in some instances for mycotoxin analysis as an alternative to HPLC, but the toxins generally then require derivatization prior to chromatography. The only advantage that GC has to offer over HPLC for mycotoxin analysis is the ability to carry out confirmation by GC–MS with relative ease, and the greater separating power which is achievable by capillary GC compared to HPLC. Derivatization for GC can, however, often bring its own problems and the selectivity offered by UV or fluorescence detection by HPLC is usually preferable to the non-specificity of an FID.

8.3.1 Aflatoxins

HPLC analysis of aflatoxins can be carried out routinely using fluorescence detection, and sub-μg/kg sensitivity is easily achieved in food samples. The chromatography is generally clean, and the characteristic pattern of peaks makes confirmation in products such as nuts and cereals as a rule unnecessary, as the presence of the toxins in these products is usually by no means unexpected. Confirmation of HPLC results cannot be carried out routinely by LC (liquid chromatography)–MS as it is difficult with this technique to achieve the desired sensitivity. Some attempts have therefore been made to carry out GC–MS analysis of aflatoxins.

The high polarity and low volatility of these compounds has meant that analysis can only be carried out on short fused-silica capillary columns (15 m) and the columns must have a low activity (less than 1 mg/kg metals present) and an absence of active sites that might cause losses by adsorption. Friedli (1981) first carried out GC analysis of aflatoxin B_1 standard, and this was later followed by the GC–MS analysis of the aflatoxin in peanut samples (Rosen *et al.*, 1984). A conventional sample clean-up was employed, together with a further silica Sep-Pak stage and on-column injection with a bonded-phase capillary column operated with a temperature programme from 200 to 290°C. Monitoring the molecular ion for aflatoxin B_1 (m/z 312) at a mass spectrometer resolution of 1000 was insufficient to resolve the toxin from interferences present in the peanuts. A resolution of 3000 was required (monitoring m/z 312.0633) and a sensitivity of 0.1 μg/kg was achieved (Rosen *et al.*, 1984). A similar confirmation was used by Trucksess *et al.* (1984a) for maize and peanut butter, using a slightly different clean-up procedure and a 6-m fused-silica GC column. High-resolution SIM was avoided by choosing to use the more selective negative-ion chemical ionization mode of operation whereby a full spectrum was obtained for confirmation rather than attempting SIM quantification. Aflatoxin G_1 and G_2 could not be chromatographed, and there are no reports of GC analysis of aflatoxin M_1 — the aflatoxin metabolite found in dairy products.

8.3.2 *Fusarium* toxins

The *Fusarium* fungi occur naturally on cereals and can give rise to a number of mycotoxins of which the group of compounds known as the trichothecenes together with zearalenone are the most important. The trichothecenes comprise about 100 structurally related compounds based on a characteristic trycyclic ring system and each containing a 12,13-epoxide group. The structures of the trichothecenes differ in the position, number and type of substituents in the ring systems and vary significantly in both molecular weight and polarity (Ueno, 1983). The trichothecenes that have been found to occur naturally are 4-deoxynivalenol (DON or 'vomitoxin'), nivalenol, T-2 toxin, diacetoxyscirpenol, HT-2 toxin and neosolaniol. More than one trichothecene may be present in the same cereal sample and the oestrogenic compound zearalenone is often found to co-occur with the trichothecenes. Simultaneous monitoring is therefore normally required for a number of these *Fusarium* toxins in the same food sample. The occurrence, particularly of DON, is worldwide and this toxin is most frequently found in wheat and maize (Yoshizawa & Hosokawa, 1983; Scott, 1984; Karppanen *et al.*, 1985; Hagler *et al.*, 1984). Monitoring is not required at the same sensitivity as for the aflatoxins and generally measurement below 0.1 mg/kg is not of much interest.

The trichothecenes provide an example where TLC, HPLC and GC are all currently used for analysis of contaminated cereals and therefore a direct comparison can be made of the advantages of each approach (Scott, 1982). TLC is the simplest and most rapid approach and utilizes specific visualization reagents, for example for the detection of the -enone function aluminium chloride is used (Kamimura *et al.*, 1981), and for detection of the epoxide group *p*-nitrobenzylpyridine is employed (Takitani *et al.*, 1979). The disadvantage of TLC is that it is prone to interferences and has a relatively high limit of detection. It is nevertheless appropriate as a rapid screening method for trichothecenes, particularly DON (Trucksess *et al.*, 1984b, 1987). Trichothecenes do not contain a chromophore and therefore for HPLC analysis detection is a problem, although low-wavelength UV has been used for DON and some other trichothecenes (Lauren & Greenhalgh, 1987). Low-wavelength UV is not, however, suitable for detecting T-2 toxin. Again, sensitivity can be a disadvantage in using HPLC and also quantification may be difficult as the trichothecenes are often eluted on the tail of the solvent peak. For GC analysis the trichothecenes need to be derivatized, which adds an extra stage to the procedure and can be a source of difficulty both in achieving complete derivatization where there are multiple functional groups, and where the derivatives may be prone to decomposition on subsequent analysis. GC detection has been by FID, ECD or MS, depending on personal preference and on the nature of the trichothecene derivative that has been formed.

There have been a large number of GC methods published for the analysis of trichothecenes and these differ in that some are aimed at determining only the polar trichothecenes such as DON and nivalenol, whilst others aim to cover a wide range, extending from nivalenol to the relatively non-polar T2-toxin. The conditions for extraction from the cereal and the subsequent sample clean-up differ for the two extremes of compound type, and the multi-trichothecene assays inevitably compromise on some aspects of the analysis. No one aspect of any of the analytical methods can be adopted in isolation and the suitability of a particular GC derivative is intrinsically linked with that of the sample clean-up. A change in the sample clean-up to one recommended for a different derivative may give rise to interferences not previously seen.

Scott *et al.* (1981) published the first method, which has been reasonably well-tested, for the analysis of DON alone in cereals using packed-column GC with electron-capture detection of the heptafluorobutyryl (HFB) ester derivative. A similar procedure using the HFB derivative for capillary GC was reported (Cohen & Lapointe, 1982) but with a modified sample preparation. Both these methods have the drawback of the sample preparation being time-consuming and attempts have been made subsequently for more rapid clean-up using commercially available prepacked cartridge columns (Terhune *et al.*, 1984). This latter method, again using the

HFB derivative with a GC packed column and an ECD, gave a limit of detection of 20 μg/kg and a CV of 9.4% at the 500 μg/kg level. Mulders and van Impelen-Peek (1986) noted poor reproducibility with HFB derivatives of DON, which was attributed to adsorption onto the surface of glass vials and could be overcome by deactivation of the glass by silylation. Steinmeyer *et al.* (1985) have changed the sample extraction conditions to give a method that recovers DON and nivalenol, and report limits of detection of 20–30 μg/kg for DON and 100–300 μg/kg for nivalenol using capillary GC (ECD) of the HFB derivative. The non-polar trichothecenes T-2 toxin, HT-2 toxin and diacetoxyscirpenol have also been determined in cereals by capillary GC of the HFB derivatives and have been confirmed by GC–MS (SIM). Formation of the HFB derivative is time-consuming, but complete derivatization seems relatively easy to achieve and the derivative once formed is stable for at least several days. Confirmation of results by GC–MS has been successfully carried out for HFB derivatives, but the relatively high mass of the derivative (e.g molecular weight of 884 for DON) can be a disadvantage for some mass spectrometers, for which this may be towards the limit of the mass range.

T-2 toxin, HT-2 toxin, neosolaniol and zearalenone have been analysed in grain samples as trimethylsilyl (TMS) ether derivatives by capillary GC (FID) with alkanes used as internal standards (Ilus *et al.*, 1981). These four toxins, together with DON and fusarenon-X, have been determined in cereal samples by a similar approach, forming the TMS derivative by reaction at 60°C with *N,O*,-bis(trimethylsilyl)trifluoroacetamide (Bata *et al.*, 1983). This derivatization procedure has, however, been shown to lead to partial derivatization with DON (Gilbert *et al.*, 1985), and for trichothecenes containing the 8-enone functionality, derivatization with TMS-imidazole was demonstrated as necessary. The TMS derivative has the advantage of being electron-capturing and recent methods have been published with improved sample clean-up for simultaneous monitoring of DON and nivalenol using the ECD (Tanaka *et al.*, 1985; Scott *et al.*, 1986; Sugimoto *et al.*, 1987). These methods report high recoveries (80–90%) and low μg/kg sensitivity. TMS derivatives have proved useful for GC–MS confirmation using multiple ion monitoring for DON and nivalenol (Tanaka *et al.*, 1985) or single ion monitoring the molecular ion (*m/z* 512) for TMS–DON (Gilbert *et al.*, 1983), although the precision in GC–MS has not been as good as would normally be expected for SIM operation. Scott and Kanhere (1986) have compiled retention time data for 10 different trichothecenes as both TMS and HFB derivatives on six fused-silica columns of different polarities which forms a useful basis for multi-trichothecene method development.

A source of particular difficulty in the quantification of trichothecenes has been the lack of a good internal standard. An appropriate internal standard should have a similar structure to the trichothecenes, similar functionality for derivatization and a close GC retention time to the analyte. Compounds such as methoxychlor (Romer *et al.*, 1978), hexachlorobiphenyl (Blaas *et al.*, 1984), and alkanes (Ilus *et al.*, 1981) have been employed but are rather too different in structure. The deuterated TMS derivative of T-2 toxin has been used successfully for SIM GC–MS (Rosen & Rosen, 1984) although there is a risk of deuterium exchange with this type of derivative, and it can only be used as an internal standard at the end of the assay. The best internal standard employed for trichothecene analysis has been an isomer of T-2 toxin which can be chemically synthesized (Stahr, 1981) and meets all the criteria for internal standards. For the future a wider range of non-naturally occurring trichothecenes need to be synthesized for use in trichothecene assays.

The trifluoroacetyl (TFA) derivatives of the trichothecenes formed by reaction with trifluoroacetic anhydride in combination with sodium bicarbonate have been found to be useful GC derivatives (Kientz & Verweij, 1986) (Fig. 8.3). Care is,

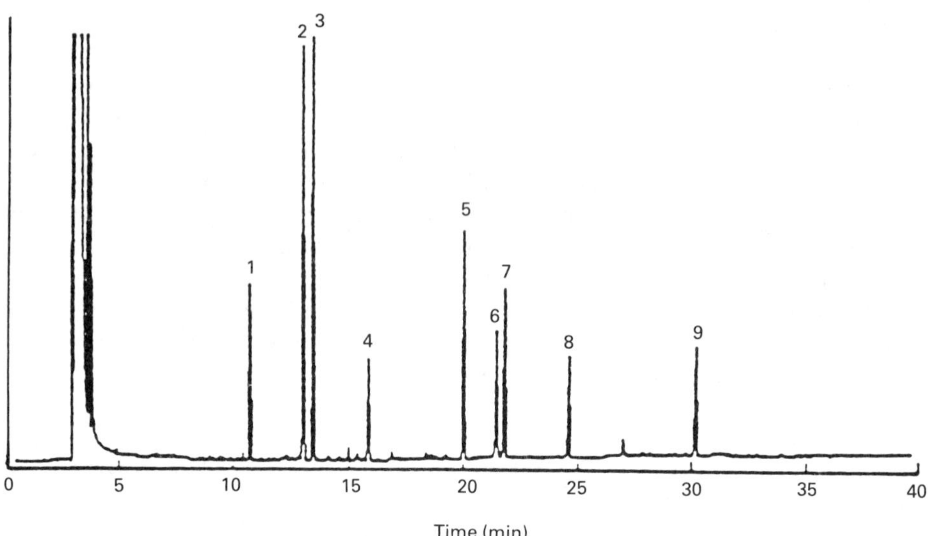

Fig. 8.3 — Analysis of a standard mixture of trichothecene mycotoxins as TFA derivatives. Detection by flame ionization. GC conditions — column (50 m×0.3 mm i.d.) fused-silica CP-SIL 8CB operated with split injection (100:1). Peak identification: (1) nivalenol; (2) 4-deoxynivalenol; (3) T-2 tetraol; (4) fusarenone X; (5) Docosane internal standard; (6) Neosolaniol; (7) diacetoxyscirpenol; (8) HT-2 toxin; (9) T-2 toxin. (Reproduced with permission from Kientz and Verweij, 1986.)

however, required to remove excess derivatizing reagent and to remove residual water from the sample to avoid derivative decomposition on GC injection (Kientz & Verwij, 1987). Even with these precautions the TFA derivative of neosolaniol was unstable during GC temperature-programming. Pentafluoropropionyl and TFA derivatives of trichothecenes were particularly useful for GC–MS since characteristic ions were produced, and high sensitivity could be achieved in the negative-ion chemical ionization mode (Krishnamurthy and Sarver, 1986).

Other related trichothecenes have been analysed in foods by GC methods, for example the de-epoxidized metabolite of DON called DOM-1 has been determined in milk as both its HFB and TMS derivative at a limit of 1 μg/kg (Swanson *et al.*, 1986). Also, only by capillary GC was an isomer of DON (iso-DON) separated from DON itself and identified as its TMS derivative. Subsequent work showed iso-DON to be formed on heating of DON, and to be present in bread and breakfast cereals prepared from naturally DON-contaminated flour (Greenhalgh *et al.*, 1984).

Stachybotrytoxicosis is a problem associated with fungal toxins from *Stachybotrys*, usually in relation to hay, straw or grain that has been infected and causes mortality of farm animals ingesting the material. The toxins that have been identified as being implicated are the macrocyclic trichothecenes. These are 4,15-esters of verrucarol, of which Cole and Cox (1981) list 12 diesters and a further six triesters. Data on the incidence of the macrocyclic trichothecenes is limited as they are difficult to analyse, having a low volatility, and minimal fluorescence or UV absorbing properties. Some GC–MS work has, however, been carried out with the TMS derivatives of macrocyclic trichothecenes on very short fused-silica columns of 7 m in length with on-column injection and programmed from 200 to 340°C (Rosen *et al.*, 1986). The molecular weights of the six derivatized macrocyclics analysed ranged from 484 to 632, but each gave interpretable spectra and molecular ion information indicating chromatography without decomposition. Krishnamurthy *et al.* (1987) have preferred the approach of alkaline hydrolysis of the ester ring system of macrocyclics to yield verrucarols, which are then analysed as HFB derivatives by capillary GC–MS. Using negative-ion chemical ionization and SIM, low-μg/kg detection was demonstrated for the analysis of macrocyclic trichothecenes in environmental and plant samples.

8.3.3 Other volatile mycotoxins

Very few other mycotoxins are sufficiently volatile for GC analysis and in most cases HPLC is by far the preferred approach. The mycotoxin patulin (4-hydroxy-4H-furo(3,2)pyran-2(6*H*)-one) is produced by a number of organisms and particularly brown rot fungi, which affect apples and other fruit. This is not a mycotoxin of much toxicological significance but it is regarded as a useful indicator of food quality, and is frequently monitored in apple and grape juices. GC is more often chosen only for confirmation whilst HPLC is preferred for routine monitoring, and when GC is used the acetate, chloroacetate or TMS derivatives are employed. Confirmation of the presence of patulin has been carried out by full scanning GC–MS of the underivatized compound (Cerutti *et al.*, 1984) or by SIM of the TMS derivative (Mortimer *et al.*, 1985) or acetate derivative (Ralls & Lane, 1977). Price (1979) has used high resolution SIM (10 000 resolution) and with a deuterated TMS internal standard has achieved a precision of 9% and a limit of detection of 0.2 μg/l.

8.4 VETERINARY DRUG RESIDUES IN ANIMAL PRODUCTS

A wide range of veterinary drugs may be administered to farm animals for therapeutic purposes, to improve feed conversion or to regulate reproduction. Antibiotics, anthelmintics, glucocorticoids, prostaglandins, vitamins, hormones, tranquilizers, vaccines, trace minerals and inorganic substances may all be given therapeutically, whilst anabolic agents (growth promoters), antibiotics, ionophores (rumen active anaboles) and coccidiostats can be used to improve performance in meat production. The use of these drugs may represent a potential hazard to human health, because of the possibility of harmful residues occurring in meat and meat products. Strict regulations govern the use of these compounds and the observance of correct withdrawal periods prior to slaughter of treated animals should, however, militate against the occurrence of detectable residues. Monitoring for residues of veterinary drugs is, however, essential in the interests of food safety, to check for compliance with tolerance limits where these exist, and also to generate surveillance data on which to assess likely human exposure. Immunological methods are widely used for screening for drug residues, and many drugs are more suitable for HPLC than GC analysis; GC is nevertheless the prime method of choice for the more volatile and derivatizable components and in the GC–MS mode is frequently used where confirmation is necessary.

8.4.1 Anabolic steroids

Methodology for detecting the presence of both synthetic anabolic and natural hormones has been developed for monitoring these compounds in urine, blood, plasma and bile as well as in animal tissues. An extensive review by Ryan (1976) demonstrates the wide range of different hormones that require monitoring and the choices of approach available to the analyst. For example, for diethylstilboestrol (DES) alone there are papers reporting the use of seven different derivatives and four different modes of GC detection. To achieve adequate specificity and the required sensitivity the use of the ECD is preferred, with formation of a suitable halogen-containing derivative. Laitem *et al.* (1978) have formed the perfluoro-esters of DES, dienoestrol and hexoestrol and achieved sub-μg/kg sensitivity in animal tissue. For hexoestrol the heptafluorobutyryl ester derivative has been used (Tobioka and Kawashima, 1981), again enabling μg/kg sensitivity. Extensive sample clean-up is frequently necessary for veterinary drug analysis in animal tissue and it may often be difficult with the necessary elaborate sample preparation to achieve consistent and satisfactory recoveries of analyte. For this reason the use of stable isotope-labelled internal standards, incorporated at the beginning of the assay, and the use of GC–MS for monitoring have many advantages and this is frequently the approach of choice in this area (Van Peteghem *et al.*, 1987).

A particular problem with DES is the existence of *cis*-and *trans*-isomers, which are well separated by GC. Covey *et al.* (1985) noted that isomerization took place during derivatization, which was variable and was strongly matrix-dependent. This perhaps argues for quantitative results for DES to be reported on the basis of a total *cis* plus *trans* concentration.

Anabolic steroids (10 in total) of various types related to testosterone, methanedienone, and norethandrolone together with their respective metabolites have been detected in urine by capillary GC–MS after formation of their TMS derivatives. Limits of detection of around 1 μg/kg were reported for monitoring of three selected ions (Cartoni *et al.*, 1983). The simultaneous detection of DES, dienoestrol, hexoestrol, methyltestosterone, ethinyloestradiol, zeranol, and trenbolone in urine have been similarly carried out by GC–MS multiple-ion monitoring (Tuinstra *et al.*, 1983). The separation of the individual compounds of interest in multi-residue procedures can be a problem, particularly for DES, dienoestrol and hexoestrol, but these difficulties can be overcome by the use of both the TMS and heptafluorobutyryl derivatives, as well as combinations of capillary columns of different polarities. Durbeck and Buker (1983) have determined DES, dienoestrol, and hexoestrol in animal tissue using the TMS derivatives and accurately recording retention indices by incorporation of standard mixtures of alkanes. Quantification was by standard addition to a parallel sample in each case and correlation from the statistically confirmed peak areas with the corresponding calibration curve. Bis-trifluoroacetyl derivatives have been employed for the monitoring of DES and its phosphorylated precursors using dimethylstilbestrol as an internal standard (Abramson and Lutz, 1985).

An approach intended to avoid GC–MS but still enabling adequate specificity and sensitivity to be achieved is the use of multi-dimensional coupled HPLC–GC (Grob *et al.*, 1986). A system comprising a size-exclusion HPLC column coupled to a capillary GC using a retention gap has been used with both FID and ECD for the analysis of DES in urine. In the ECD mode the pentafluorobenzyl ether derivative of DES was formed in the urine sample prior to HPLC. A sample cut from the HPLC (about 300 μL) containing the DES peak was directly switched to the capillary GC, utilizing the retention gap principle to ensure sharp injection without peak broadening. With relatively little sample preparation it was demonstrated that a detection limit of 0.1 to 0.3 μg/kg could be achieved for DES in urine by this approach using the ECD.

8.4.2 Sulphonamides

Veterinary drugs of the sulphonamide class are frequently employed as antimicrobial agents for the treatment of poultry, cattle and pigs. As there are tolerance limits for residues in some countries (e.g. in the USA there is a tolerance limit of 100 μg/kg in uncooked edible tissue), these compounds have received considerable attention in relation both to preliminary screening methods and to rigorous confirmation and quantification of data. Analytical methods for the analysis of sulphonamides have been reviewed by Horwitz (1981) and are amongst the veterinary drugs considered in some detail in the methodological overview paper by Petz (1984). There are at least 20 different sulphonamides, and although sulphadimidine (sulphamethazine) appears to be the most widely used, assay methods should have the capability of simultaneous monitoring of a number of individual residues from this class of compound. For GC analysis, derivatization of sulphonamides is required. The most frequently described procedures involve either single-stage N^1-methylation with diazomethane, i.e. methylation of the nitrogen adjacent to the SO_2 (Manuel and Steller, 1981; Stout *et al.*, 1984), or N^1-methylation followed by acylation of the NH_2 function with pentafluoropropionic anhydride or other appropriate reagent (Simpson *et al.*, 1985). Subsequent GC analysis is normally then undertaken using specific nitrogen detection, ECD or GC–MS. For the derivatization of sulphonamides there is some evidence of side-products formed with diazomethane (Gilbert *et al.*, 1986), and this problem, which can be common to all derivatization procedures, must be guarded against in development of quantitative assays.

The potential of capillary GC to offer the capability for multi-residue analysis is illustrated in Fig. 8.4, where the separation is shown of a standard mixture of 17 sulphonamides and three nitrofurans. Two pairs of compounds could not be separated on this particular column, although a change in stationary phase or GC derivative could be used to achieve the desired resolution of critical pairs. For the analysis of these constituents in tissue matrix the ability of the sample preparation and clean-up to remove interferences from the extract would need to be demonstrated for each and every analyte.

Manuel and Stellar (1981) have determined seven sulphonamides by packed

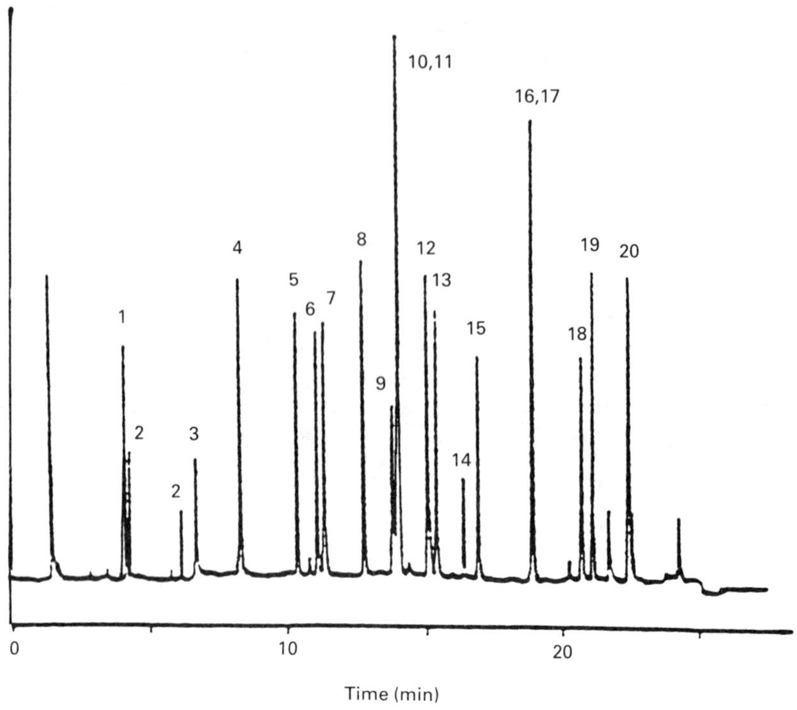

Fig. 8.4 — Analysis of a standard mixture of sulphonamide and nitrofuran veterinary drugs. Nitrogen-specific detection (alkali FID). GC conditions — column (30 m×0.26 mm i.d.) fused-silica Durabond DB-1 operated with a nitrogen carrier-gas flow of 2 ml/min with a temperature programme from 180°C at 3°C/min to 260°C. Split injection (10:1). Peak identification as methyl derivatives with noted exceptions (1) sulphanilamide (underivatized); (2) sulphamoxole; (3) furazolidone (underivatized); (4) nitrofuranthoin; (5) sulphamethoxazole; (6) sulphpyridine; (7) sulphazole; (8) sulphadiazine; (9) sulphachlorpyrazine; (10) sulphamethoxypyrazine; (11) sulphamerazine; (12) sulphadimidine; (13) sulphamethoxypyridazine; (14) sulphachloropyridazine; (15) sulphaethoxypyridazine; (16) sulphadozine; (17) sulphadimethoxine; (18) furaltadone (underivatized); (19) sulphaphenazole; (20) sulphaquinoxaline. (Reproduced with permission from Petz, 1984.)

column (ECD) analysis of a methylated sample extract with a limit of detection of 100 μg/kg, whilst Suhre *et al.* (1981) reported a stable isotope GC–MS method for determining sulphadimidine employing a ^{13}C-labelled internal standard. More recently this same group of workers (Simpson *et al.*, 1985) have extended the methodology to the simultaneous assay of five sulphonamides (sulphadimidine, sulphadimethoxine, sulphabromethazine, sulphathiazole, and sulphaquinoxaline) in animal tissue. The methods differ in that the latter employs a dual derivatization forming the *N*-1 methyl and *N*-4 acyl compound, which has the advantage of shortening the GC run time and also negating the effect of a background interference for sulphadimidine. Quantification in each case was against one of the five respective stable isotope internal standards. Garland *et al.* (1980) have also used GC–MS but preferred to use isobutane chemical ionization for the estimation of sulphadimethoxine in liver and kidney samples, and employed the d_6-internal standard. Using sample extracts prepared for preliminary ECD screening, Stout *et al.* (1984) used GC–MS for confirmation of sulphadimidine in animal tissues, but without the inclusion of a stable isotope-labelled internal standard. Monitoring three ions in positive chemical ionization (CI) was used as the basis for confirmation at the 100-μg/kg level.

Most methods concentrated only on monitoring the drugs as administered although in a few cases additionally the analysis of their metabolites has been included (Matusik *et al.*, 1987). Some extraction procedures for animal tissue do include a deconjugation stage, but mostly the aspect of the residues being in a conjugated form in the sample matrix has been rather neglected.

8.4.3 Other veterinary drugs

There is considerable interest in monitoring many other veterinary drugs of widely differing structures and types that potentially could be present as trace residues in animal tissues. HPLC is frequently the most appropriate method, although in some instances GC is an alternative or the drugs can be chemically degraded to smaller species suitable for GC determination.

GC methods for determining the broad-spectrum antibiotic chloramphenicol in animal tissue, milk and eggs have been reviewed by Allen (1985) and contrasted with TLC and HPLC approaches. Most workers choosing GC have used TMS or heptafluorobutyryl derivatization and with ECD generally have achieved better limits of detection (of the order of 1–10 μg/kg) than the alternative approaches. GC has also been used to confirm the effectiveness of radioimmunoassay as a screening method for chloroamphenicol (Arnold and Somogyi, 1985) — interestingly these workers used chloroamphenicol *m*-isomer as an internal standard for GC, which improved the precision for milk analysis.

Nelson *et al.* (1983) used thiamphenicol as internal standard, and using the TMS derivative with an ECD achieved limits of detection of 5 μg/kg chloramphenicol in liver and chicken muscle tissue. As with other drugs GC–MS has been used both as a primary detection method and for confirmation of chloramphenicol, for example in chicken tissue, with a limit of detection of 15 μg/kg when three ions were monitored (Bories *et al.*, 1983). Another antibiotic, lincomycin A, is a tetrahydroxy compound of significantly different structure to chloramphenicol that can be analysed by GC as its TMS derivative. McMurray *et al.* (1984) carried out single-ion monitoring in EI using m/z 126 for confirmation of lincomycin A in animal feedstuffs and no interferences were experienced with a limit of detection of 0.1 mg/kg. However, it was not possible to distinguish lincomycin A from clindamycin, another antibiotic of the same family, which co-chromatographed. These compounds could however be easily distinguished in the CI mode of operation and clearly this would be the better approach for monitoring residues in animal tissues, which might also give rise to greater problems of interferences than feedstuffs when monitoring low-mass ions.

Methods have been reviewed for the analysis of the polyether antibiotics in animal tissues (Weiss & McDonald, 1985). These high molecular weight multifunctional compounds such as lasalocid, salinomycin and monensin are difficult to analyse by gas chromatography and therefore TLC and HPLC methods tend to predominate. A confirmatory method has however been reported for lasalocid in beef liver (Weiss *et al.*, 1983) involving silylation of a purified extract and then pyrolysis GC–MS with monitoring of four characteristic ions. This confirmation was shown to be effective at the 50-μg/kg level in spiked liver samples. HPLC was used for determining monensin (Takatsuki *et al.*, 1986) but for positive samples the methyl ester tris-TMS ether derivative was formed and confirmation was carried out by full scanning GC–MS but at the relatively high limit of detection of 5 mg/kg. For other macrolide antibiotics (oleandomycin, kitasamycin, spiramycin and tylosin) in beef and pork the approach has involved degradation prior to GC–MS determination of a derived part of the original compound (Takatsuki *et al.*, 1987a). The procedure involved hydrogenation of the carbon–carbon double bonds in the antibiotics, acid hydrolysis, followed by acylation to form unique products capable of GC analysis. Limits of detection of between 0.1 and 0.5 mg/kg were demonstrated. For erythromycin in beef and pork (Takatusuki *et al.*, 1987b) acid hydrolysis and acylation were employed prior to analysis by GC–MS. A similar degradative approach has been used to determine the antibiotic avilamycin (molecular weight 1400) which after hydrolysis and methylation yields 3,5-dichloro-4,6-dimethoxy-2-methoxybenzoic acid methyl ester (molecular weight 279) which can be determined by electron-capture GC (Formica and Giannone, 1986).

Grohmann *et al.* (1983) have reported the analysis of four tranquilizers (azaperone, diaxepam, acepromazine and xylazine) in pork liver samples, from animals treated with these compounds to reduce stress during transport to slaughterhouses. Capillary GC with specific nitrogen detection (NDP) and GC–MS for confirmation were shown as potentially viable approaches for monitoring these four compounds together with a metabolite of azaperone. Glycerol formal (a mixture of the two isomeric compounds 5-hydroxyl-1,3-dioxane and 4-(hydroxymethyl)-1,3-dioxolane) can be employed as a solvent for administering animal health products and may give rise to residues in the animal tissues. A GC–MS method has been reported (VandenHeuvel *et al.*, 1983) with a limit of detection of 50 μg/kg in edible tissues (fat, kidney, liver and muscle). An internal standard of d_2-glycerol was employed and a packed-column GC method used to measure total glycerol formal without separation of isomers.

Pyrantel and its analogue morantel are anthelmintic agents which could give rise to residues in animal tissue and milk. Alkaline hydrolysis of, for example, morantel produces 3(3-methyl-2-thienyl)acrylic acid and *N*-methyl-1,3-propanediamine as reaction products. After methylation GC–MS was carried out to monitor the acid fragment achieving a limit of detection in tissue below 50 μg/kg (Lynch and Bartolucci, 1982). For milk samples the other fragment (the diamine) was monitored by GC (ECD) after conversion to a *N,N*-bis-(2-nitro-4-trifluoromethylphenyl) derivative (Lynch *et al.*, 1986).

A GC–MS procedure has been reported for determining ipronidazole, the active compound in medicated pre-mix feedstuff used to control histomoniasis in turkeys (Garland *et al.*, 1980). GC–MS was recommended for confirmation of ipronidazole and for when co-eluting interferences were apparent in HPLC analysis (Morriss *et al.*, 1987). The coccidiostat clopidol has been determined in chicken tissues by esterification with propionic anhydride in alkaline solution and electron-capture GC at a limit of 0.5 μg/kg (Ekstrom and Kuivinen, 1984). Xylonidine is of interest as a forced moulting agent which might be administered to poultry to improve egg production. A method was developed (Taylor *et al.*, 1982) to monitor possible residues of the compound in egg yolk and egg albumen, based on GC–MS stable isotope dilution, using d_4-xylonidine as internal standard.

8.5 CONTAMINATION FROM PACKAGING MATERIALS

Plastic materials for packaging and other food contact applications are constantly changing in type as new materials are developed and constantly increasing in extent of usage. The following classes of compound may be present in different plastic food packaging materials and may in each case provide a source of material for migration and thus may ultimately lead to contamination of food: monomers, oligomers, heat stabilizers, plasticizers, antioxidants, UV absorbers, processing aids, colorants, adhesives and printing inks. Each group on this list may comprise in itself a large number of components, and single ingredients are themselves normally of commercial quality and therefore may contain impurities. Faced with this complexity, assessment of this source of food contamination normally starts with analysis of the packaging material, and then when levels of 'targetted' components have been measured in the packaging, work can begin on foodstuff analysis.

As GC is very widely used in this area of work for monitoring contamination of foods from packaging, it is not possible exhaustively to cover this topic in this chapter. Instead selected areas of current importance have been reviewed, chosen particularly where there has been recent progress in the development of new methodology for determining these trace constituents in foods.

8.5.1 Plasticizers

Plasticizers are used to provide flexibility to films and to coatings, and a variety of types are used for different food contact applications. For PVC film ('clingfilm') the principal plasticizer in the UK has in the past been di(2-ethylhexyl)adipate (DEHA), although this is now being partially replaced in the film by so-called polymeric plasticizers. This type of 'clingfilm' is widely used in the home for wrapping and covering foods, particularly for subsequent storage in the refrigerator. The film is also used in supermarkets for covering fresh and cooked meat (usually overwrapping the food on an expanded polystyrene tray) as well as for cheese, fruit, and some vegetables. PVC/PVDC (Saran film) is a flexible film sold specifically for covering foods during cooking in a microwave oven and this is plasticized with acetyltributyl citrate. The same material is also used in the form of 'chub' packs (sausage-shaped packaging with lead end-seals) for pates, black pudding and smoked cheeses. Phthalates (mainly dibutyl-, butylbenzyl-, and dicyclohexyl-) are used to provide flexibility to the nitrocellulose coating of regenerated cellulose film (RCF). RCF finds very wide application for the packaging of confectionary (twist-wrapped candies, toffees and fudges), for cooked meat pies and pasties, and for cakes and biscuits. Epoxidized soyabean oil (ESBO) is used in conjunction with other plasticizers in both PVC and PVDC, and serves not only as a plasticizer but also as a secondary heat stabilizer. ESBO is thus used in all the above applications and additionally is used at high incorporation levels in PVC gasket seals, used for sealing the lids of jars for baby foods, jams and pickles as well as the 'crown caps' used for bottled beer and other beverages.

8.5.2 Analysis of the plasticizers in the food packaging films

For analysis of the plasticizers in films, chloroform extraction is carried out by allowing the film to stand overnight in the solvent. No single analysis is adequate for detecting all plasticizers, but a knowledge of the film type is helpful in deciding on the most fruitful approach. Probe analysis by MS can indicate whether ESBO and/or monomeric/polymeric plasticizers are present — this is not quantitative and mixed spectra are obtained.

Capillary GC analysis of the chloroform extract containing one or more internal standards is, however, the best method for quantifying individual monomeric plasticizers. A typical FID chromatogram showing a test mixture of plasticizer standards used for characterizing films is illustrated in Fig. 8.5. Identification of the plasticizers can usually be made on the basis of retention times compared with standards. This approach cannot, however, be used for determining polymeric plasticizers, which need to be analysed separately. The GC analysis can also be difficult for quantification of the complex mixtures of isomeric plasticizers, such as those encountered for example for di(2-ethylhexyl)phthalate. DEHA and diphenyl-2-ethylhexyl phosphate also co-elute under the normal GC conditions and therefore GC–MS is required to distinguish between these two components.

ESBO has been determined in rigid plastics, films and gaskets by extraction into hexane, transmethylation and then derivatization of the epoxides to form 1,3-dioxolanes (Castle *et al.*, 1988a). Quantification by capillary GC is based on the level of monoepoxidized methyl octadecanoate relative to the amount of monoepoxidized methyl eicosanoate internal standard.

8.5.3 Plasticizers in foods

8.5.3.1 Phthalates and adipates

Monitoring of plasticizer contamination both in the environment and in foods has been of interest for many years, and has been the subject of a recent comprehensive review (Giam and Wong, 1987). Most plasticizers are lipid-soluble and therefore analytical schemes normally involve initial fat extraction from the food. Separation of the higher molecular weight lipids from the lower molecular weight plasticizers can be achieved by size exclusion chromatography (SEC) (Shepherd, 1984). The SEC stage of the analysis has the advantage of being amenable to automation. Analysis of the cleaned-up extract is by packed-column or, for preference, capillary-column GC. GC analysis using an FID can be effective for determining relatively high levels of plasticizer contamination in comparatively simple foods (Sandberg *et al.*, 1982), and an example of DEHA contamination in a sample of fresh meat is shown in Fig. 8.6 illustrating the 'cleanliness' of the sample extract after SEC clean-up. However, for more complex foods, for example confectionary products and some complete cooked meals, interference peaks will be present and may obscure the region of the chromatogram where the plasticizers are expected to elute. In these

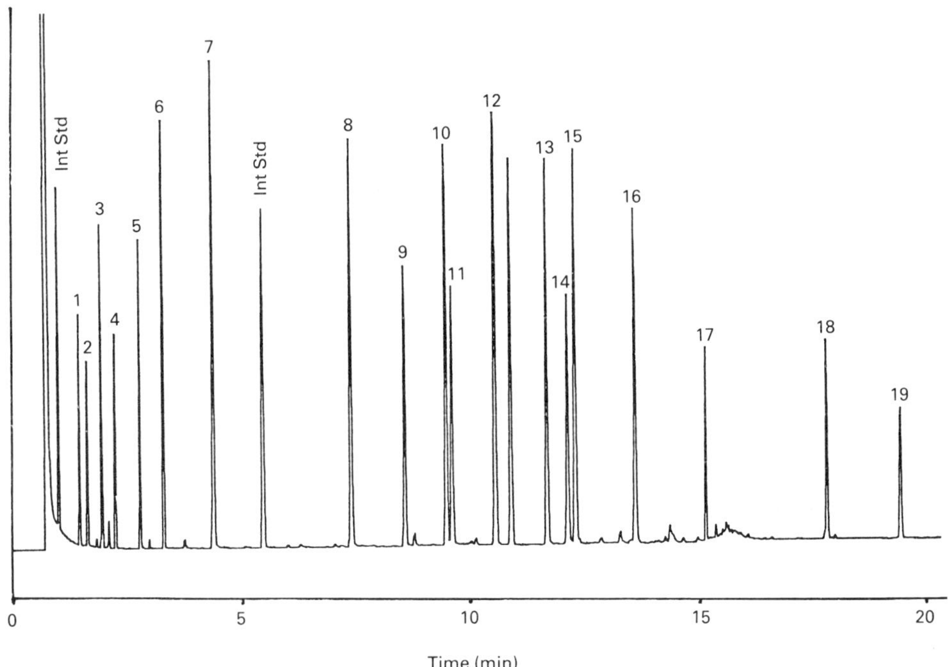

Fig. 8.5 — Analysis of a typical test mixture of plasticizer standards used for the identification of plasticizers in retail samples. GC conditions — fused-silica capillary GC (CP SIL 5CB) operated at 160°C then programmed at 20°C/min to 300°C. Flame ionization detection. Peak identification: (1) dimethyl phthalate; (2) dimethyl terephthalate; (3) diethyl phthalate; (4) di-isopropyl phthalate; (5) dipropyl phthalate; (6) di-isobutyl phthalate; (7) dibutyl phthalate; (8) dibutyl sebacate; (9) acetyltributyl citrate; (10) butylbenzene phthalate; (11) dihexyl phthalate; (13) di-(2-ethylhexyl)adipate/diphenyl-(2-ethylhexyl)phthalate; (14) dicyclohexyl phthalate; (15) di-(2-ethylhexyl)phthalate; (16) dinonyl phthalate; (17) di-(2)-ethylhexylsebacate; (18) diundecyl phthalate; (19) didodecyl phthalate.

instances, it is advisable to employ a mass spectrometer as the detector and carry out selected ion monitoring. The use of GC–MS has the added advantage that stable isotope-labelled internal standards can then be employed, and the assay becomes one of isotope dilution. Although the use of the mass spectrometer does require a high level of sophistication of instrumentation, for large numbers of survey samples data can be generated quickly, with a high precision and with a high confidence in correct identification of the plasticizers being monitored. Precisions of the order of 2% have been routinely obtained and the approach has been applied to a diversity of food types from cheese, fruit, chocolate to complete cooked meals (Startin *et al.*, 1987a).

Typical stable isotope-labelled internal standards which have been used are: d_4-dibutylphthalate and d_4-dicyclohexylphthalate (synthesized from the d_4-phthalic acid), d_4-DEHA (synthesized from d_4-adipic acid) (Startin *et al.*, 1987a; Castel *et al.*, 1988b). The unsymmetrical phthalate esters could not be readily synthesized and for these the corresponding symmetrical ester was employed as internal standard.

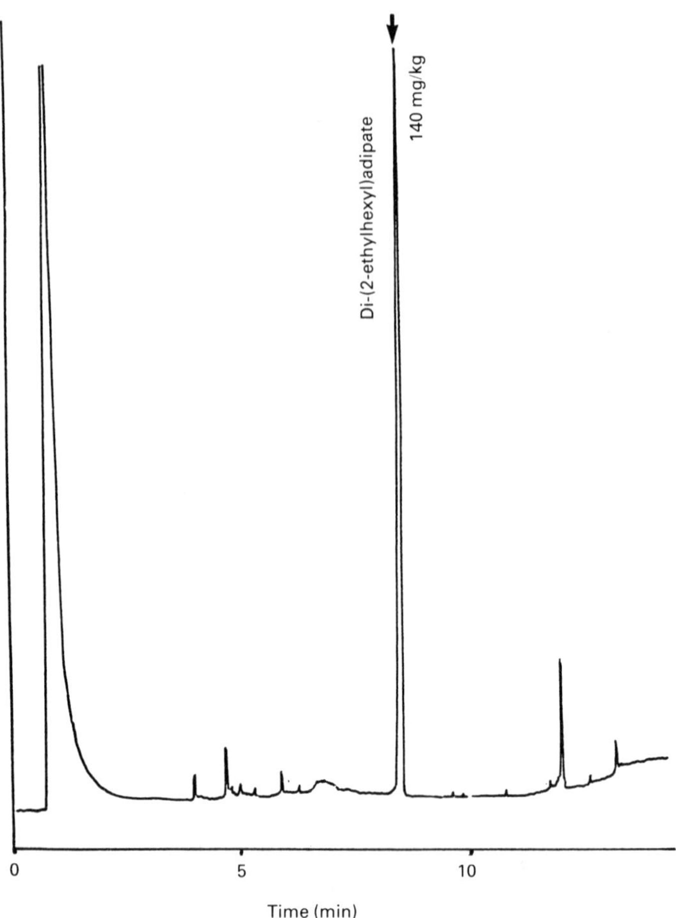

Fig. 8.6 — Analysis of the plasticizer di-(2-ethylhexyl)adipate in a retail sample of fresh meat (mince) wrapped in PVC film. Sample extract cleaned-up by size-exclusion chromatography. GC conditions — fused-silica capillary column (CP SIL 5CB) operated at 160°C then programmed at 20°C/min to 300°C. Flame ionization detection.

In surveys of retail foods and in experiments to study the influence of various factors on migration, several hundred plasticizer determinations have been carried out (Startin *et al.*, 1987b; Castle *et al.*, 1988b). In no instance has there been any evidence of interference with any of the food types, and thus there has been a high degree of confidence both in the adequacy of the clean-up and in the specificity of the end determination. Compared with other organic contaminants found in foods, plasticizer migration occurs at a relatively high level and the limit of detection employed for these measurements was therefore arbitrarily set at 0.1 mg/kg, a level below which contamination would be regarded as being of no significance. The mass spectrometric system has, however, for this analysis an intrinsic capability of detecting far lower levels.

8.5.3.2 Acetyltributyl citrate

Acetyltributyl citrate (ATBC) is a monomeric plasticizer generally used at levels of between 4 and 5% in PVC/PVDC co-polymer films. These films are used for wrapping retail foods such as cheese, and for 'chub' packs as well as being specifically recommended by the manufacturers for domestic use for covering foods during microwave cooking.

To determine ATBC migration into foods by isotope-dilution GC–MS analysis, the deuterated internal standard was synthesized containing from one to four deuterium atoms (Castle *et al.*, 1988c). The analytical method was essentially identical to that developed for other plasticizers and was shown to have a limit of detection of 0.1 mg/kg (it is quite possible but unnecessary to achieve greater sensitivity). The precision of the assay was within the general range expected of this type of method, giving an RSD of around 1.5%. This method has been employed to measure migration of ATBC from plasticized film during microwave cooking of soups, cakes, puddings, and meat dishes forming composite meals (Castle *et al.*, 1988d).

8.5.3.3 Epoxidized soyabean oil

Epoxidized soyabean oil (ESBO) is used as a plasticizer and secondary heat stabilizer in flexible PVC applications at levels of between 3 and 7%, as well as in other plastics where it might, for example, function as a lubricant. The approximate component fatty acid composition of ESBO is 50% di-epoxidized, 26% mono-epoxidized and 11% triepoxidized components with about 13% saturated fatty acids. Direct GC analysis of ESBO is not feasible owing to its polydisperse nature and low volatility. A GC approach to ESBO analysis is, however, possible, based on degradation of the triglyceride to its constituent fatty acid components using an epoxidized fatty acid ester as an internal standard (Castle *et al.*, 1988e). Transmethylation under basic conditions converts the triglyceride to its fatty acid esters. These fatty acid methyl esters do not receive further clean-up from the co-extracted food lipids but are instead derivatized to form 1,3-dioxolanes, which are then amenable to direct GC analysis. The formation of the dioxolane derivative increases the molecular weight of the epoxidized components sufficiently to separate them from food-constituent fatty acid esters. Additionally the dioxolanes have characteristic mass spectra that make them suitable candidates for GC–MS SIM analysis. Analysis and quantification is based on the C_{18}-diepoxidized fatty acid methyl ester component from ESBO using a C_{20}-diepoxidized fatty acid ester internal standard. This assay has a limit of detection of 1 mg/kg of ESBO and a relative standard deviation of 7%. The method has been applied to a wide range of food types. However, owing to the possibility of the presence of naturally occurring epoxy-constituents in foods it is essential that blank values are obtained for food samples not having had prior contact with ESBO-plasticized material.

8.5.4 Glycol softeners

Glycol softeners are used in regenerated cellulose film to provide flexibility, and are typically used at levels up to 27% w/w in the film. Prior to 1986 monoethylene glycol (MEG) and diethylene glycol (DEG) were widely employed, but have now been replaced by alternative softeners. The softeners now most frequently used are combinations of two or more of the following: propylene glycol, triethylene glycol, poly(ethylene glycol), glycerol and urea. Although MEG and DEG are still permitted as RCF softeners, recent EEC regulations now restrict the total level of MEG plus DEG in the food as a result of migration not to exceed a limit of 50 mg/kg. This effectively eliminates the use of MEG and DEG as softeners in their own right, but allows for impurities in the alternative softeners, and allows for the fact that small amounts of MEG and DEG will inevitably be minor components of poly(ethylene glycol). Typical applications of RCF in the UK are the wrapping of confectionery products (particularly for twist-wrapped boiled sweets, toffees, fudge and chocolates), and for cooked 'ready-to-eat' products, such as pies, cakes and biscuits.

Glycol softeners can be analysed in RCF packaging materials by extraction into acetone or acetonitrile, with the inclusion of 1,4-butanediol as an internal standard, derivatization with silylating reagent and then analysis by GC. All the glycols can be quantified by this analysis except for poly(ethylene glycol), where only its lower molecular weight components can be detected, although these could be used as a basis for the determination.

The analysis of glycols in foods is particularly difficult because of problems of efficient extraction from the food matrix, the similarity in behaviour of the glycols to other polar molecules present in foods, and the lack of a suitable chromophore or other 'handle' for detection. Pongratz and Bauer (1985) have proposed two alternative sample clean-ups to be selected on the basis of the fat and sugar content of the foodstuff, prior to GC analysis of the underivatized DEG. These workers particularly emphasize the dangers of contamination arising during the sample work-up from DEG which is frequently present as an impurity in laboratory reagents as well as for example in filter papers. A method has also been reported for the analysis of DEG in sausages wrapped in RCF material (Tannert and Schulze-Feldmann, 1986). This involves a fairly simple clean-up and again GC analysis with an FID of the underivatized glycol in the sample extract. An effective method has been developed, which although rather lengthy, can when applied with care give good results for the determination of MEG and DEG in chocolate and other confectionary products (Castle *et al.*, 1988f). The method involves dissolution of the sample and extraction into hot water, followed by defatting with hexane and precipitation of the sugars. The TMS derivative of the glycol is formed and the extract analysed by capillary GC with FID. Derivatization of the glycols was preferred, as poor peak shapes, due to chromatographic adsorption, and the elution of late-running peaks can cause problems with the analysis of underivatized sample extracts. Confirmation of identification, particularly where the method is to be used for regulatory control of MEG and DEG levels in foods, is recommended to be carried out by full scanning MS. Most problems in this analysis are caused by contamination occurring in solvents, reagents or from glassware and particular care is necessary in this aspect of sample preparation and quality assurance is essential in the form of frequent analyses of blank samples.

A slightly modified approach to the above method for MEG and DEG can also be applied to the analysis of the replacement softeners currently employed in RCF formulations (Castle *et al.*, 1988g) A problem with analysing for propylene glycol in retail survey samples is that this compound is widely used as a permitted solvent for the addition of additives to foods. Thus, although the methods may be effective for estimating levels of propylene glycol in foods, for survey samples there is no way of knowing whether the glycol has originated from the wrapping material or is present as a deliberate food additive. For the determination of triethylene glycol in foods, co-extracted components can interfere with the analysis using the GC flame ionization detector, and thus for the analysis of this glycol GC–MS SIM was preferred (Castle *et al.*, 1988g).

8.5.5 Other contaminants from packaging

GC is widely used for the analysis of packaging materials and films that are suspected as having given rise to taints or off-odours in foods. Although there are a few general accounts of applications in this area (Wyatt, 1986), most of the work is undertaken by industry in relation to specific complaints or disputes and is rarely published. However, a good illustrative example of the application of GC in this area was the identification of a musty component present in a food packaging film (McGorrin *et al.*, 1987). By use of capillary GC with odour appraisal of the peaks, the component with the taint could be identified, and with subsequent headspace GC–MS, GC–MS–MS and GC–FT–IR some speculation could be made as to its possible structure. Synthesis of possible candidate components showed the taint compound to be 4,4,6-trimethyl-1,3-dioxane, derived by reaction of 2-methyl-2,4-pentanediol solvent in the film with formaldehyde of an unknown source.

Other applications of GC in the analysis of food contaminants by migration from plastic materials are varied and range from determining minor decomposition products present during polymer production to measuring leaching of major intrinsic components of the finished plastic articles. The following provide examples of these extremes. Tetramethylsuccinonitrile (TMSN) is a decomposition product of forming agent or catalyst used in PVC production. Using an NPD a packed-column GC method with nitrobenzene internal standard has been reported for determining TMSN in PVC samples (Ishiwata *et al.*, 1986a). Migration of TMSN into food-simulating solvents was measured with a limit of detection of 1 μg/kg (Ishiwata *et al.*, 1987). In contrast melamine is a major constituent of melamine–formaldehyde polymer used for dishes and cups for repeat-use applications. The migration of melamine into food simulating solvents has been carried out by packed column GC using an NPD (Ishiwata *et al.*, 1986b). Interestingly, migration into 3% acetic acid increased with repeated use of the containers, which was coincident with an increase in surface roughness of the material.

8.6 CONCLUSION

This chapter has been deliberately very selective in the areas chosen to exemplify the use of GC for food contaminant analysis. The three main areas discussed in detail are of current topical interest and typically illustrate some of the approaches to the use of GC, and the reasons for selection of GC as opposed to alternative chromatographic techniques. For the future, although undoubtedly many GC analyses of food contaminants will continue to be undertaken as described, some indications of future developments are already apparent — firstly, the move towards routine use of MS for GC detection is likely to increase, owing at least in part to reductions in the relative capital costs of mass spectrometers and the development of new instrumentation, such as the ion-trap detector. Secondly, the use of multi-dimensional chromatography to enable sample preparation to be reduced and to automate analysis may well become more widely adopted.

This latter approach can take the form of GC–GC or LC–GC where in either case the second separation technique receives only a pre-selected portion of eluate from the first. The two techniques have complementary separating ability, thereby improving overall selectivity. Relatively crude sample extracts can be handled by the first technique as the bulk of analysed material is switched to waste and never passes to the end-detector. GC–GC with two capillary columns of different polarity coupled in series has been used for the analysis of ethyl carbamate in wine, sherry, liqueur, brandy, whisky and beer (van Ingen *et al.*, 1987) where it was shown possible to monitor down to contamination levels of 10 μg/kg with direct analysis of a concentrated dichloromethane extract of the alcoholic beverage. Similarly, the advantages of GC–GC have been shown for the analysis of polychlorinated biphenyl (PCB) mixtures (Duinker *et al.*, 1988) where complete separation and improved sensitivity (ECD) could be achieved that was not possible with a single GC column separation. A similar approach using two GC columns but switching from the first column into a cold trap, and then flash evaporation into the second column (Ligon and May, 1984) has been applied to PCB, nitrosamine and polychlorinated dibenzofuran analysis.

An alternative to GC–GC that of coupling HPLC with capillary GC has been pioneered by Grob *et al.* (1984). This approach has been demonstrated as effective for the analysis of wax esters in olive oil (Grob and Laubli, 1986), the determination of *p*-hydroxyphenyl-l-butanone-3 in raspberry sauce (Grob and Stoll, 1986) and the analysis of PCBs in fish samples (Grob *et al.*, 1987). In the future the number of reported applications of these multi-dimensional techniques can be expected to increase as full advantage is taken of the discriminating power of combined chromatographic approaches. These methods can be seen as an alternative to the use of MS for detection, and an alternative that offers the added attraction of being amenable to full automation from the handling of relatively crude sample extracts to the end-measurement of the food contaminant.

REFERENCES

Abramson, F. P. and Lutz, M. P. (1985). Application of a capillary gas chromatographic–selected-ion recording mass spectrometric technique to the analysis of diethylstilbestrol and its phosphorylated precursors in plasma and tissues. *J. Chromatogr.*, **339**, 87–95.

Allen E. H. (1985). Review of chromatographic methods for chloramphenicol residues in milk, eggs, and tissues from food-producing animals. *J. Assoc. Off. Anal. Chem.*, **68**, 990–999.

Arnold, D. and Somogyi, A. (1985). Trace analysis of chloramphenicol residues in eggs, milk, and meat: comparison of gas chromatography and radioimmunoassay. *J. Assoc. Off. Anal. Chem.*, **68**, 984–990.

Bata, A., Vanyi, A. and Lasztity, R. (1983). Simultaneous detection of some Fusariotoxins by gas–liquid chromatography. *J. Assoc. Off. Anal. Chem.*, **66**, 577–581.

Blaas, W., Kellert, M., Steinmeyer, S., Tiebach, R. and Weber, R. (1984).

Untersuchung von cerealien auf deoxynivalenol und nivalenol im unteren μg/kg-bereich. *Z. Lebensm. Unters. Forsch.*, **179**, 104–108.

Bories, G. S. F., Peleran, J-C. and Wal, J-M. (1983). Liquid chromatographic determination and mass spectrometric confirmation of chloramphenicol residues in animal tissues. *J. Assoc. Off. Anal. Chem.*, **66**, 1521–1526.

Cartoni, G. P., Ciardi, M., Giarrusso, A. and Rosati, F. (1983). Capillary gas chromatographic mass spectrometric detection of anabolic steroids. *J. Chromatogr.*, **279**, 515–522.

Castle, L., Sharman, M. and Gilbert, J. (1988a). Analysis of epoxidised soya bean oil additive in plastics by gas chromatography. *J. Chromatogr.*, **437**, 274–280.

Castle, L., Mercer, A. J., Startin, J. R. and Gilbert, J. (1988b). Migration from plasticized films into foods 3. Migration of phthalate, sebacate, citrate and phosphate esters from films used for retail food packaging. *Food Addit. Contam.*, **5**, 9–20.

Castle, L., Gilbert, J., Jickells, S. M. and Gramshaw, J. W. (1988c). Analysis of the plasticiser acetyltributyl citrate in foods by stable isotope dilution gas chromatography–mass spectrometry. *J. Chromatogr.*, **437**, 281–286.

Castle, L., Jickells, S. M., Sharman, M., Gramshaw, J. W. and Gilbert, J. (1988d). Migration of the plasticizer acetyltributyl citrate from plastic film into foods during microwave cooking and other domestic use. *J. Food Protect.*, **51**, 916–919.

Castle, L., Sharman, M. and Gilbert, J. (1988e). Gas chromatographic–mass spectrometric determination of epoxidised soya bean oil contamination of foods by migration from plastic packaging. *J. Assoc. Off. Anal. Chem.*, **71**, 1183–1186.

Castle, L., Cloke, H. R., Startin, J. R. and Gilbert, J. (1988f). Gas chromatographic determination of monoethylene glycol and diethylene glycol in chocolate packaged in regenerated cellulose film. *J. Assoc. Off. Anal. Chem.*, **71**, 499–503.

Castle, L., Cloke, H., Crews, C. and Gilbert, J. (1988g). The migration from regenerated cellulose film into foods of propylene-, mono-, di-, and triethylene glycol. *Z. Lebensm. Unters. Forsch.*, **187**, 463–467.

Cerutti, G., Finoli, C., Vecchio, A. and Volonterio, A. G. (1984). formazione e presenza di patulina in mele e loro prodotti. *Estratto da Technologie Alimentari Conservazione degli Alimenti*, **19**, 1–8.

Cohen, H. and Lapointe, M. (1982). Capillary gas–liquid chromatographic determination of vomitoxin in cereal grains. *J. Assoc. Offic. Anal. Chem.*, **65**, 1429–1434.

Cohen, H. and Lapointe, M. (1984). Capillary gas–liquid chromatographic determination of T-2 toxin, HT-2 toxin, and diacetoxyscirpenol in cereal grains. *J. Assoc. Offic. Anal. Chem.*, **67**, 1105–1107.

Cole, R. J. and Cox, R. H. (1981). *Handbook of Toxic Fungal Metabolites*. Academic Press, New York.

Covey, T., Maylin, G. and Henion, J. (1985). Quantitative secondary ion monitoring gas chromatography/mass spectrometry of diethylstilbestrol in bovine liver. *Biomed. Mass Spectrom.*, **12**, 274–282.

DeVries, J. W., Broge, J. M., Schroeder, J. P., Bowers, R. H., Larson, P. A. and Burns, N. M. (1985). Headspace gas chromatographic method for determination of methyl bromide in food ingredients. *J. Assoc. Off. Anal. Chem.*, **68**, 1112–1116.

Dong, M., DiEdwardo, A. H. and Zitomer, F. (1980). Determination of residual acetaldehyde in polyethylene terephthalate bottles, preforms, and resins by automated headspace gas chromatography. *J. Chromatogr. Sci.*, **18**, 242–246.

Duinker, J. C., Schulz, D. E. and Petrick, G. (1988). Multidimensional gas chromatography with electron capture detection for the determination of toxic congeners in polychlorinated biphenyl mixtures. *Anal. Chem.*, **60**, 478–482.

Durbeck, H. W. and Buker, I. (1983). A simple GC/MS method for the determination of diethylstilbestrol and related stilbene derivatives in meat and meat products. *Fresenius Z. Anal. Chem.*, **315**, 479–485.

Eiceman, G. A. and Karasek, F. W. (1981). Identification of residual organic compounds in food packages. *J. Chromatogr.*, **210**, 93–103.

Ekstrom, L-G. and Kuivinen, J. (1984). Gas chromatographic determination of clopidol in chicken tissues. *J. Assoc. Off. Anal. Chem.*, **67**, 955–957.

Entz, R. C. and Diachenko, G. W. (1988). Residues of volatile halocarbons in margarines. *Food Add. Contamin.*, **5**, 267–276.

Entz, R. C. and Hollifield, H. C. (1982) Headspace gas chromatographic analysis of foods for volatile halocarbons. *J. Agric. Food Chem.*, **30**, 84–88.

Entz, R. C., Thomas, K. W. and Diachenko, G. W. (1982). Residues of volatile halocarbons in foods using headspace gas chromatography. *J. Agric. Food Chem.*, **30**, 846–849.

Flanjak, J. and Sharrad, J. (1984). Quantitative analysis of styrene monomer in foods. A limited Fast Australian survey. *J. Sci. Food Agric.* **35**, 457–462.

Formica, G. and Giannone, C. (1986). Gas chromatographic determination of avilamycin total residues in pig tissues, fat, blood, faeces, and urine. *J. Assoc. Off. Anal. Chem.*, **69**, 763–766.

Garland, W. A., Hodshon, B. J., Chen, G., Weiss, G., Felicito, N. R. and MacDonald, A. (1980). Determination of ipronidazole and its principal metabolite in Turkey skin and muscle by combined gas chromatography–negative chemical ionization mass spectrometry–stable isotope dilution. *J. Agric. Food Chem.*, **28**, 273–277.

Giam, C. S. and Wong, M. K. (1987). Plasticizers in food. *J. Food Protect.*, **57**, 769–782.

Gilbert, J. (1984) Confirmation and quantification of trace organic food contaminants by mass spectrometry–selected ion monitoring. In *The Analysis of Food Contaminants* Gilbert, J. (ed.) Elsevier Applied Science, London.

Gilbert, J. and Startin, J. R. (1981). Single-ion monitoring of styrene in foods by coupled mass spectrometry automatic headspace gas chromatography. *J. Chromatogr.*, **205**, 434–437.

Gilbert, J. and Startin, J. R. (1982). Determination of acrylonitrile monomer in food packaging materials and in foods. *Food Chem.*, **9**, 243–252.

Gilbert, J. and Startin, J. R. (1983). A survey of styrene monomer levels in foods and plastic packaging by coupled mass spectrometry–automatic headspace gas chromatography. *J. Sci. Food Agric.*, **34**, 647–652.

Gilbert, J., Shepherd, M. J., Startin, J. R. and McWeeny, D. J. (1980). Gas chromatographic determination of vinylidene chloride monomer in packaging films and foods. *J. Chromatogr.*, **197**, 71–78.

Gilbert, J., Shepherd, M. J. and Startin, J. R. (1983). A survey of the occurrence of the trichothecene mycotoxin deoxynivalenol (vomitoxin) in UK grown barley and imported maize by GC–MS. *J. Sci. Fd. Agric.*, **34**, 86–92.

Gilbert, J., Startin, J. R. and Crews, C. (1985). Automated headspace GC/MS analysis of ethylene dibromide fumigant residues in fresh fruits. *Food Add. Contamin.*, **2**, 55–61.

Gilbert, J., Startin, J. R. and Crews, C. (1985). Optimisation of conditions for trimethylation of trichothecene mycotoxins. *J. Chromatogr.*, **319**, 376–381.

Gilbert, J., Startin, J. R., Shepherd, M. J. and Mitchell, J. C. (1986). Identification of a novel side-product formed during the methylation of sulphapyridine prior to gas chromatographic analysis. *J. Chromatogr.*, **356**, 206–211.

Greenhalgh, R., Gilbert, J., King, R. R., Blackwell, B. A., Startin, J. R. and Shepherd, M. J. (1984). Synthesis, characterization, and occurrence in bread and cereal products of an isomer of 4-deoxynivalenol (vomitoxin). *J. Agric. Food Chem.*, **32**, 1416–1420.

Grob, K. and Laubli, T. (1986). Determination of wax esters in olive oil by coupled HPLC–HRGC. *J. High Res. Chrom. & Chrom. Comm.*, **9**, 593–594.

Grob, K. and Stoll, J.-M. (1986). Loop type interface for concurrent solvent evaporation in coupled HPLC–GC. Analysis of raspberry ketone in a raspberry sauce as an example. *J. High Res. Chrom. & Chrom. Comm.*, **9**, 518–523.

Grob, K., Frohlich, D., Schilling, B., Neukom, H. P. and Nageli, P. (1984). Coupling of high performance liquid chromatography with capillary gas chromatography. *J. Chromatogr.*, **295**, 55–61.

Grob, K., Neukom, H. P. and Etter, R. (1986). Coupled high-performance liquid chromatography–capillary gas chromatography as a replacement for gas chromatography–mass spectrometry in the determination of diethylstilbestrol in bovine urine. *J. Chromatogr.*, **357**, 416–422.

Grob, K., Muller, E. and Meier, W. (1987). Coupled HPLC–GC for determining PCB's in fish. *J. High Res. Chrom. & Chrom. Comm.*, **10**, 416–417.

Grohmann, H. G., Scheutwinkel-Reich, M., Preiss, A. M. and Stan, H-J. (1983). Gas chromatographic and gas chromatographic–mass spectrometric analysis of psychopharmaceutics used in meat production employing various ionisation techniques for mass spectrometry. In *Recent Developments in Mass Spectrometry in Biochemistry, Medicine and Environmental Research*. Edited by A. Frigerio. Elsevier, Amsterdam, pp. 117–127.

Hachenberg, H. and Schmidt, A. P. (1977) *Gas Chromatographic Headspace Analysis*. Heyden, London.

Hagler, W. M., Tyczkowska, K. and Hamilton, P. B. (1984). Simultaneous occurrence of deoxynivalenol, zearalenone, and aflatoxin in 1982 scabby wheat from midwestern United States. *Applied Environ. Microbiol.*, **47**, 151–154.

Heikes, D. L. (1985a). Purge and trap method for determination of ethylene dibromide in table-ready foods. *J. Assoc. Off. Anal. Chem.*, **68**, 431–436.

Heikes, D. L. (1985b). Purge and trap method for determination of ethylene dibromide in whole grains, milled grain products, intermediate grain-based foods, and animal feeds. *J. Assoc. Off. Anal. Chem.*, **68**, 1108–1111.

Heikes, D. L. (1987a). Purge and trap method for determination of volatile halocarbons and carbon disulphide in table-ready foods. *J. Assoc. Off. Anal. Chem.*, **70**, 215–226.

Heikes, D. L. (1987b). Determination of residual chlorinated solvents in decaffeinated coffee by using purge and trap procedure. *J. Assoc. Off. Anal. Chem.*, **70**, 176–180.

Heikes, D. L. and Hopper, M. L. (1986). Purge and trap method for determination of fumigants in whole grains, milled grain products, and intermediate grain-based foods. *J. Assoc. Off. Anal. Chem.*, **69**, 990–998.

Hollifield, H. C., Breder, C. V., Dennison, J. L., Roach, J. A. G. and Adams, W. S. (1980). Container-derived contamination of maple syrup with methyl methacrylate, toluene, and styrene as determined by headspace gas–liquid chromatography. *J. Assoc. Off. Anal. Chem.*, **63**, 173–177.

Horwitz, W. (1981). Analytical methods for sulphonamides in foods and feeds. I. Review of methodology. *J. Assoc. Off. Anal. Chem.*, **64**, 104–130.

Ilus, T., Niku-Paavola, M-L. and Enari, T-M., (1981). Chromatographic analysis of *Fusarium* toxins in grain samples. *Eur. J. Appl. Microbiol. Biotechnol.*, **11**, 244–247.

Ingen, van R. H. M., Nijssen, L. M., Van den Berg, F. and Maarse, H. (1987). Determination of ethyl carbamate in alcoholic beverages by two-dimensional gas chromatography. *J. High Res. Chrom. & Chrom. Commun.*, **10**, 151–152.

Ishiwata, H., Inoue, T. and Yoshihira, K. (1986a). Gas chromatographic determination of tetramethylsuccinonitrile in poly(vinyl chloride) products in contact with food. *J. Chromatogr.*, **370**, 275–279.

Ishiwata, H., Inoue, T. and Tanimura, A. (1986b). Migration of melamine and formaldehyde from tableware made of melamine resin. *Food Addit. Contamin.*, **3**, 63–70.

Ishiwata, H., Inoue, T. and Yoshihira, K. (1987). Tetramethylsuccinonitrile in polyvinyl chloride products for food and its release into food simulating solvents. *Z. Lebensm. Unters Forsch.*, **185**, 39–42.

Kamimura, H., Nishijima, M., Yasuda, K., Saito, K., Ibe, A., Nagayama, T., Ushiyama, H. and Naoi, Y. (1981). Simultaneous detection of several *Fusarium* mycotoxins in cereals, grains, and foodstuffs. *J. Assoc. Off. Anal. Chem.*, **64**, 1067–1073.

Karppanen, E., Rizzo, A., Berg, S., Lindfors, E. and Aho, R. (1985). Fusarium mycotoxins as a problem in Finnish feeds and cereals. *J. of Agric. Sci in Finland*, **57**, 195–206.

Kientz, C. E. and Verweij, A. (1986). Trimethylsilylation and trifluoroacetylation of a number of trichothecenes followed by gas chromatographic analysis on fused silica capillary columns. *J. Chromatogr.*, **355**, 299–240.

Kientz, C. E. and Verweij, A. (1987). Decomposition of trifluoroacetyl derivatives of some trichothecenes on fused-silica capillary columns during gas chromatographic analysis. *J. Chromatogr.*, **407**, 340–342.

Kolb, B. (1980). *Applied Headspace Gas Chromatography*. Heyden, London.

Kolb, B. (1984). Analysis of food contaminants by headspace gas chromatography. In *Analysis of Food Contaminants* Gilbert, J. (ed.), Elsevier Applied Science, London, pp. 117–154.

Kolb, B., Popisil, P. and Auer, M. (1981). Quantitative analysis of residual solvents in food packaging printed films by capillary gas chromatography with multiple headspace extraction. *J. Chromatogr.*, **204**, 371–376.

Krishnamurthy, T. and Sarver, E. W. (1986). Mass spectral investigations on trichothecene mycotoxins. III. Synthesis, characterization and applications of pentafluoropropionyl and trifluoroacetyl esters of simple trichothecenes. *J. Chromatogr.*, **355**, 253–264.

Krishnamurthy, T., Sarver, E. W., Greene, S. L. and Jarvis, B. B. (1987). Mass spectral investigations on trichothecene mycotoxins. II. Detection and quantitation of macrocyclic trichothecenes by gas chromatography/negative ion chemical ionization mass spectrometry. *J. Assoc. Off. Anal. Chem.*, **70**, 132–140.

Laitem, L., Gaspar, P. and Bello, I. (1978). Stable derivatives for the gas chromatographic determination of synthetic anabolic stilbene residues (diethylstilbestrol, dienestrol, and hexestrol) in meat and organs of treated cattle at the sub-parts per billion level. *J. Chromatogr.*, **156**, 267–273.

Lauren, D. R. and Greenhalgh, R. (1987). Simultaneous analysis of nivalenol and deoxynivalenol in cereals by liquid chromatography. *J. Assoc. Off. Anal. Chem.*, **70**, 479–483.

Ligon, W. V. and May, R. J. (1984). Target compound analysis by two-dimensional gas chromatography–mass spectrometry. *J. Chromatogr.*, **294**, 77–86.

Lynch, J. M. and Bartolucci, S. R. (1982). Identification and confirmation of pyrantel- and morantel-related residues in liver by gas chromatography–mass spectrometry with selected ion monitoring. *J. Assoc. Off. Anal. Chem.*, **65**, 640–646.

Lynch, M. J., Burnett, D. M., Mosher, F. R., Dimmock, M. E. and Bartolucci, S. R. (1986). Determination of morantel-related residues in bovine milk by electron capture gas chromatography. *J. Assoc. Off. Anal. Chem.*, **69**, 646–651.

McGorrin, R. J., Pofahl, T. R. and Croasmun, W. R. (1987). Identification of the musty component from an off-odor packaging film. *Anal. Chem.*, **59**, 1109–1112.

McMurray, C. H., Blanchflower, W. J. and Rice, D. A. (1984). Gas chromatographic–mass spectrometric detection and quantification of lincomycin in animal feedstuffs. *J. Assoc. Off. Anal. Chem.*, **67**, 582–588.

McNeal, T. and Breder, C. V. (1981). Headspace sampling and gas–solid chromatographic determination of residual acrylonitrile in acrylonitrile copolymer solutions. *J. Assoc. Off. Anal. Chem.*, **64**, 270–275.

McNeal, T. and Breder, C. V. (1982). Manual headspace gas–solid chromatographic determination of sub-parts per trillion levels of acrylonitrile in 3% acetic acid. *J. Assoc. Off. Anal. Chem.*, **65**, 184–187.

McNeal, T. and Breder, C. V. (1987). Headspace gas chromatographic determination of residual 1,3-butadiene in rubber-modified plastics and its migration from plastic containers into selected foods. *J. Assoc. Off. Anal. Chem.*, **70**, 18–21.

Manuel, A. J. and Steller, W. A. (1981). Gas–liquid chromatographic determination of sulphamethazine in swine and cattle tissues. *J. Assoc. Off. Anal. Chem.*, **64**, 794–799.

Matusik, J. E., Guyer, C. G., Geleta, J. N. and Barnes, C. J. (1987). Determination of desaminosulphamethazine, sulfamethazine, and N^{14}-acetylsulfamethazine by gas chromatography with electron capture detection and confirmation by gas chromatography–chemical ionization mass spectrometry. *J. Assoc. Off. Anal. Chem.*, **70**, 546–553.

Miltz, J., Elisha, C. and Mannheim, C. H. (1980). Sensory threshold of styrene and the monomer migration from polystyrene food packages. *J. Food Process. and Preserv.*, **4**, 281–289.

Morris, W. J., Nandrea, G. J., Roybal, J. E., Munns, R. K., Shimoda, W. and Skinner, H. R. (1987). Quantitative confirmation of dimetridazole and ipronidazole in swine feed by capillary gas chromatography/mass spectrometry with multiple ion detection. *J. Assoc. Off. Anal. Chem.*, **70**, 630–634.

Mortimer, D. N., Parker, I., Shepherd, M. J. and Gilbert, J. (1985). A limited survey of retail apple and grape juices for the mycotoxin patulin. *Food Addit. Contam.* **2**, 165–170.

Mulders, E. J. and van Impelen-Peek, H. A. M. (1986). Gas chromatographic determination of deoxynivalenol in cereals. *Z. Lebensm. Unters Forsch.*, **183**, 406–409.

Nelson, J. R., Copeland, K. F. T., Forster, R. J., Campbell, D. J. and Black, W. D. (1983). Sensitive gas–liquid chromatographic method for chloramphenicol in animal tissues using electron-capture detection. *J. Chromatogr.*, **276**, 438–444.

Page, B. D. (1985). Determination of acrylonitrile in foods by headspace–gas chromatography with nitrogen-sensitive detection: collaborative study. *J. Assoc. Off. Anal. Chem.*, **68**, 776–782.

Page, B. D. and Charbonneau, C. F. (1983). Determination of acrylonitrile in foods by headspace gas–liquid chromatography with nitrogen–phosphorus detection. *J. Assoc. Off. Anal. Chem.*, **66**, 1096–1105.

Page, B. D. and Charbonneau, C. F. (1984). Headspace gas chromatographic determination of methylene chloride in decaffeinated tea and coffee, with electrolytic conductivity detection. *J. Assoc. Off. Anal. Chem.*, **67**, 757–761.

Page, B. D. and Charbonneau, C. F. (1985). Improved procedure for determination of acrylonitrile in foods and its application to meat. *J. Assoc. Off. Anal. Chem.*, **68**, 606–608.

Petz, M. (1984). Chemical residue analysis of veterinary drugs in food. Part I. General methodology and gas chromatographic procedures. *Z. Lebensm. Unters Forsch.*, **180**, 267–279.

Pongratz, K-P. and Bauer, E. (1985). The working-up of solid foodstuffs for the determination of diethylene glycol. *Ernahrung/Nutrition* **9**, 865–866.

Price, K. R. (1979). A comparison of two quantitative mass spectrometric methods for the analysis of patulin in apple juice. *Biomed. Mass Spectrom.*, **6**, 573–574.

Ralls, J. W. and Lane, R. M. (1977). Examination of cider vinegar for patulin using mass spectrometry. *J. Fd. Sci.*, **42**, 1117–1119.

Romer, T. R., Boling, T. M. and McDonald, J. L. (1978). Gas–liquid chromatographic determination of T-2 toxin and diacetoxyscirpenol in corn and mixed feeds. *J. Assoc. Off. Anal. Chem.*, **61**, 801–808.

Roseboom, H. (1984). Recent advances in pesticide residue analysis. In *Food Constituents and Food Residues*. Lawrence, J. L. (ed.) Marcel Dekker, New York, pp. 489–533.

Rosen, R. T. and Rosen, J. D. (1984). Quantification and confirmation of four

Fusarium mycotoxins in corn by gas chromatography–mass spectrometry–selected ion monitoring. *J. Chromatogr.*, **283**, 223–230.

Rosen, R., Rosen, J. D. and DiProssimo, V. P. (1984). Confirmation of aflatoxin B_1 in peanuts by gas chromatography/mass spectrometry/selected ion monitoring. *J. Agric. Food Chem.*, **32**, 276–278.

Rosen, J. D., Rosen, R. T. and Hartman, T. G. (1986). Capillary gas chromatography–mass spectrometry of several macrocyclic trichothecenes. *J. Chromatogr.*, **355**, 241–251.

Rossi, L. (1988). Activities of the Commission of the European Communities concerning materials and articles intended to come into contact with foodstuffs — a review. *Food Add. Contamin.*, **5**, 21–31.

Ryan, J. J. (1976). Chromatographic analysis of hormone residues in food. *J. Chromatogr.*, **27**, 53–89.

Sandberg, E., Vax, R., Albanus, L., Mattsson, P. and Nilsson, K. (1982). Contamination of foodstuffs by plasticizers from plastic film. *Var Foda*, **34**, 470–482.

Santa Maria, I., Carmi, J. D. and Ober, A. G. (1986). Residual styrene monomer in Chilean foods by headspace gas chromatography. *Bull. Environ., Contam. Toxicol.*, **37**, 207–212.

Scanlan, R. A. (1984). Chemiluminescence for measurement of *N*-nitrosamines in foods. In *Analysis of Food Contaminants*. Gilbert, J. (ed.), Elsevier Applied Science Publisher (Barking, U.K.), pp. 321–369.

Scott, P. M. (1982). Assessment of quantitative methods for determination of trichothecenes in grains and grain products. *J. Assoc. Off. Anal. Chem.*, **65**, 876–883.

Scott, P. M. (1984). The occurrence of vomitoxin (deoxynivalenol, DON) in Canadian grains. In *Toxigenic Fungi–Their Toxins and Health Hazard*, edited by H. Kurata and Y. Ueno, Kodanasha Ltd. (Tokyo) and Elsevier (Amsterdam), pp. 182–189.

Scott, P. M. and Kanhere, S. R. (1986). Comparison of column phases for separation of derivatized trichothecenes by capillary gas chromatography. *J. Chromatogr.*, **368**, 374–380.

Scott, P. M., Lau, P.-Y. and Kanhere, S. R. (1981). Gas chromatography with electron capture detection and mass spectrometric detection of deoxynivalenol in wheat and other grains. *J. Assoc. Off. Anal. Chem.*, **64**, 1364–1370.

Scott, P. M., Kanhere, S. R. and Tarter, E. J. (1986). Determination of nivalenol and deoxynivalenol in cereals by electron-capture gas chromatography. *J. Assoc. Off. Anal. Chem.*, **69**, 889–893.

Sen, N. P. (1984). Recent advances in the analysis of nitrosamines in food. In *Food Constituents and Food Residues*. Lawrence, J. L. (ed.), Marcel Dekker, New York, pp. 453–489.

Shepherd, M. J. (1984). Size exclusion and gel chromatography; theory, methodology and applications to the clean-up of food samples for contaminant analysis. In *Analysis of Food Contaminants*, Gilbert, J. (ed.), Elsevier Applied Science Publishers (Barking, UK), pp. 1–66.

Simpson, R. M., Suhre, F. B. and Shafer, J. W. (1985). Quantitative gas chromatographic–mass spectrometric assay of five sulfonamide residues in animal tissue. *J. Assoc. Off. Anal. Chem.*, **68**, 23–26.

Stahr, H. M., Hyde, W., Lederdal, D. and Pfeiffer, R. (1981). Trichothecene mycotoxin analysis for veterinary diagnostic toxicology. *Abst. 95th Annual Meeting of the AOAC,* **191**, 65.

Startin, J. R. and Gilbert, J. (1984). Single ion monitoring of butadiene in plastics and foods by coupled mass spectrometry–automatic headspace gas chromatography. *J. Chromatogr.*, **294**, 427–430.

Startin, J. R., Parker, I., Sharman, M. and Gilbert, J. (1987a). Analysis of di-(2-ethylhexyl)adipate plasticiser in foods by stable isotope dilution gas chromatography–mass spectrometry. *J. Chromatogr.*, **387**, 509–514.

Startin, J. R., Sharman, M., Rose, M. D., Parker, I., Mercer, A. J., Castle, L. and Gilbert, J. (1987b). Migration from plasticized films into floods. 1. Migration of di-(2)-ethylhexyl)adipate from PVC films during home-use and microwave cooking. *Food Addit. Contam.*, **4**, 385–398.

Steichen, R. J. (1976). Modified solution approach for the gas chromatographic determination of residual monomers by head-space analysis. *Anal. Chem.*, **48**, 1398–1402.

Steinmeyer, S., Tiebach, R. and Weber, R. (1985). Determination of deoxynivalenol and nivalenol in cereals by gas chromatography of the heptafluorobutyrates. *Z. Lebensm. Unters Forsch.*, **181**, 198–199.

Stout, S. J., Steller, W. A., Manuel, A. J., Poeppel, M. O. and daCunha, A. R. (1984). Confirmatory method for sulfamethazine residues in cattle and swine tissues, using gas chromatography–chemical ionization mass spectrometry. *J. Assoc. Off. Anal. Chem.*, **67**, 142–144.

Sugimoto, T., Yamamoto, K. and Minamisawa, M. (1987). Efficient method for the determination of deoxynivalenol and nivalenol in cereals by gas chromatography. *Proc. Jpn. Assoc. Mycotoxicol.*, **25**, 37–39.

Suhre, F. B., Simpson, R. M. and Shafer, J. W. (1981). Qualitative/quantitative determination of sulphamethazine in swine tissue by gas chromatographic/electron impact mass spectrometry using a stable isotope labeled internal standard. *J. Agric. Food Chem.*, **29**, 727–729.

Swanson, S. P., Dahlem, A. M., Rood, H. D., Cote, L-M., Buck, W. B. and Yoshizawa, T. (1986). Gas chromatographic analysis of milk for deoxynivalenol and its metabolite DOM-1. *J. Assoc. Off. Anal. Chem.*, **69**, 41–43.

Takatsuki, K., Suzuki, S. and Ushizawa, I. (1986). Liquid chromatographic determination of monensin in chicken tissues with fluorometric detection and confirmation by gas chromatography–mass spectrometry. *J. Assoc. Off. Anal. Chem.*, **69**, 443–448.

Takatsuki, K., Ushizawa, I. and Shoji, T. (1987a). Gas chromatographic–mass spectrometric determination of macrolide antibiotics in beef and pork using single ion monitoring. *J. Chromatogr.*, **391**, 207–217.

Takatsuki, K., Suzuki, S., Sato, N., Ushizawa, I. and Shoji, T. (1987b). Gas chromatographic/mass spectrometric determination of erythromycin in beef and pork. *J. Assoc. Off. Anal. Chem.*, **70**, 708–713.

Takitani, S., Asabe, Y., Kato, T., Suzuki, M., Ueno, Y. (1974). Spectrodensitometric determination of trichothecene mycotoxins with 4-(p-nitrobenzyl) pyridine. *J. Chromatogr.* **172**, 335–342.

Tanaka, T., Hasegawa, A., Matsuki, Y., Ishii, K. and Ueno, Y. (1985). Improved methodology for simultaneous detection of trichothecene mycotoxins deoxynivalenol and nivalenol in cereals. *Food Addit. Contam.*, **2**, 125–137.

Tannert, U. and Schulze-Feldmann, B. (1986). Migration behaviour of diethylene glycol from regenerated cellulose. *Deut. Lebensm-Rund.*, **82**, 86–88.

Taylor, J. E., Iderstine, A. V., Weppelman, R. M., Olson, G., Walker, R. W., Mertel, H. E. and VandenHeuvel, W. J. A. (1982). Residues of the forced molting agent xylonidine in chicken egg yolk and albumen as determined by combined gas–liquid chromatography–mass spectrometry. *J. Agric. Food Chem.*, **30**, 858–861.

Terhune, S. J., Nguyen, N. V., Baxter, J. A., Pryde, D. H. and Qureshi, S. A. (1984). Improved gas chromatographic method for quantitation of deoxynivalenol in wheat, corn and feed. *J. Assoc. Off. Anal. Chem.*, **67**, 1102–1104.

Tobioka, H. and Kawashima, R. (1981). Gas–liquid chromatographic determination of hexestrol residues in adipose tissue. *J. Assoc. Off. Anal. Chem.*, **64,** 709–713.

Trucksess, M. W., Brumley, W. C. and Nesheim, S. (1984a). Rapid quantitation and confirmation of aflatoxins in corn and peanut butter, using a disposable silica gel column, thin layer chromatography, and gas chromatography/mass spectrometry. *J. Assoc. Off. Anal. Chem.*, **67**, 973–975.

Trucksess, M. W., Neisheim, S. and Eppley, R. M. (1984b). Thin layer chromatographic determination and deoxynivalenol in wheat and corn. *J. Assoc. Off. Anal. Chem.*, **67**, 40–43.

Trucksess, M. W., Flood, M. T., Mossoba, M. M. and Page, S. W. (1987). High-performance thin-layer chromatographic determination of deoxynivalenol, fusarenon-X, and nivalenol in barley, corn, and wheat. *J. Assoc. Off. Anal. Chem.*, **35**, 445–448.

Tuinstra, L. G. M. Th., Traag, W. A., Keukens, H. J. and Mazijk, R. J. (1983). Procedure for the gas chromatographic/mass spectrometric confirmation of some exogenous growth-promoting compounds in the urine of cattle. *J. Chromatogr.*, **279**, 533–542.

Ueno, Y. (1983). *Trichothecenes — Chemical, biochemical and toxicological aspects. Developments in food science 4.* Kodanasha Ltd (Tokyo) and Elsevier (Amsterdam).

VandenHeuval, W. J. A., Baylis, F. P., Brown, J. E., Wallace, D. H., Minsker, D. H., Robertson, R. T., Rosegay, A. and Walker, R. W. (1983). Gas–liquid chromatography–chemical ionization selected ion monitoring assay for glycerol formal in animal tissues. *J. Agric. Food Chem.*, **31**, 548–553.

Van Peteghem, C. H., Lefevere, M. F., Van Haver, G. M. and De Leenheer, A. P. (1987). Quantification of diethylstilbestrol residues in meat samples by gas chromatography–isotope dilution mass spectrometry. *J. Agric. Food Chem.*, **35**, 228–231.

Varner, S. L. and Breder, C. V. (1981a). Liquid chromatographic determination of residual styrene in polystyrene food packaging. *J. Assoc. Off. Anal. Chem.*, **64**, 647–652.

Varner, S. L. and Breder, C. V. (1981b). Headspace sampling and gas chromatographic determination of styrene migration from food-contact polystyrene cups into beverages and food simulants. *J. Assoc. Off. Anal. Chem.*, **64**, 1122–1130.

Varner, S. L., Breder, C. V. and Fazio, T. (1983). Determination of styrene migration from food-contact polymers into margarine, using azeotropic distillation and headspace gas chromatography. *J. Assoc. Off. Anal. Chem.*, **66**, 1067–1073.

Weiss, G. and MacDonald, A. (1985). Methods for determination of ionophore-type antibiotic residues in animal tissues. *J. Assoc. Off. Anal. Chem.*, **68**, 971–980.

Weiss, G., Kaykaty, M. and Miwa, B. (1983). A pyrolysis gas chromatographic –mass spectrometry confirmatory method for lasolocid sodium in bovine liver. *J. Agric. Food Chem.*, **31**, 78–81.

Wyatt, D. M. (1986). Analytical analysis of tastes and odors imparted to foods by packaging materials. *J. Plastic Film and Sheeting*, **2**, 144–152.

Yoshizawa, T. and Hosokawa, H. (1983). Natural occurrence of deoxynivalenol and nivalenol, trichothecene mycotoxins, in commercial foods. *J. Food Hygienic Soc. Japan*, **24**, 413–415.

9

Pesticides

Hans-Juergen Stan

9.1 APPLICATION OF PESTICIDES IN AGRICULTURAL PRODUCTION AND FOOD STORAGE

Pesticides may be defined as chemical substances intended for preventing, destroying, repelling or mitigating the effects of pests. In general, substances used as plant growth regulators and defoliants are also included in this term. In 1980, some 530 000 tons (4.8×10^5 t) of pesticides were used in the production of food as well as clothing and other industrial goods in the United States for the more than 270 million inhabitants. That means an annual input of 2 kg of pesticides per person into our environment. Pesticides are an integral part of world agriculture and under present conditions are considered to be indispensable. Plants that supply our main source of food are susceptible to 80 000 to 100 000 diseases caused by fungi, viruses, bacteria and other types of microorganisms. Some 3000 nematodes and 10 000 species of insects attack crop plants. They compete furthermore with 30 000 species of weed, of which about 1800 cause economic losses worldwide. In the developed countries, crop losses due to pests are about 30% despite the use of pesticides and other sophisticated control methods (Ware, 1983).

Studies conducted between 1976 and 1980 compared the yields from test plots, where insecticides were applied, to adjacent plots without treatment to control the insects. The data suggested for the major crops corn, wheat, potatoes and cotton an average loss of 50% to insects (see Table 9.1).

Table 9.1 — Comparison of percentage losses caused by insects (Data from Ware, 1983)

Commodity	With treatment	Without treatment
Corn	17.7	42.2
Wheat	9.5	65.0
Potatoes	1.0	48.0
Soybeans	5.5	20.8
Cotton	14.5	51.1

The greatest benefit from insecticides, however, has been the protection from human diseases spread by insects. The control of malaria vectors in many countries was first made possible by applying DDT. It is estimated that approximately one-quarter of the world population is now protected from parasites as the result of control measures.

Information is fragmentary on the benefits of pesticides to the protection of stored products. Losses occur mainly due to insects, fungi, mites and rodents, which are all susceptible to pesticide control. All estimates agree that at least 10% of the harvested crops worldwide are lost during storage (Hayes, 1981).

The worldwide output of pesticides was valued in 1980 at $11.5 billion with 34% for insecticides, 19% for fungicides and 42% for herbicides. In the developed countries the relative amounts of insecticides and herbicides used has dramatically changed in the past two decades. In 1964 in the USA, only 34 500 tons (3.1×10^4 t) of herbicides were used together with 64 900 tons (5.9×10^4 t) of insecticides. 18 years later the use of herbicides has been increased to 196 400 tons (1.8×10^6 t) and that of insecticides decreased to 26 800 tons (2.4×10^4 t).

The herbicides atrazine and alachlor were each applied in greater amounts than all insecticides together. Today in the developed countries, herbicides are the main pesticides, while in Africa, Latin America and Asia insecticides are the most important products.

9.2 CLASSIFICATION OF PESTICIDES

9.2.1 Biological activity

Pesticides are classified according to their biological effect. The main groups with respect to the amount used are

— insecticides
— fungicides
— herbicides

followed by

— acaricides and aphicides
— nematicides
— molluscicides
— rodenticides.

Other compound groups in use are bactericides, pheromones, and plant growth regulators.

9.2.2 Chemical structure

Another means of classification of more importance to the food chemist in the pesticide residue laboratory is that according to chemical structure. There are an estimated 500 to 600 pesticides in use all over the world. Compilations of active compounds worldwide in use are given in the newest edition of The Pesticide Manual (1987) published by the British Crop Protection Council and by Perkow (1983/1988). Monographs of the chemistry and biological activity were published by Büchel (1977) and Wegler (1970/1982). A textbook covering aspects of application, analysis and regulation of pesticides appeared recently (Thier & Frehse, 1986). Here, only a short review of the main chemical classes can be provided.

9.2.2.1 Chlorinated pesticides (CPs) (Fig. 9.1)

These compounds played an important role in the history of chemical plant protection as well as human hygiene. These compounds combined high insecticidal effectiveness with low mammalian toxicity and provided long-lasting protection against a broad spectrum of insects owing to their chemical stability. The development and continuing improvement of trace analysis, however, resulted in the knowledge of their persistence in all sections of our environment. They are not only persistent but also lipophilic and therefore accumulate in plant and animal tissue. An increasing public concern about their ubiquitous occurrence resulted in their use being banned in most of the industrialized countries.

The most important compound of this group is DDT, which was introduced as an effective insecticide in 1939 in human hygiene and a few years later for agricultural use. Its name is an abbreviation of Dichloro-Diphenyl-Trichloroethane. Residues of DDT are always found together with its metabolites DDE and DDD. All three compounds were highly accumulated in the food chain and eventually in the human body. Other compounds of this chemical group are methoxychlor, dicofol, chlorobenzilate and chloropropylate; the latter three are effective acaricides. Another important CP is lindane, the γ-isomer of the mixture of hexachlorohexane isomers (HCH or BHC).

Cyclodiene insecticides are a group of compounds that have been used extensively in the USA and in many tropical countries. They were all produced with a Diels–Alder diene synthesis and the most important products had been named Dieldrin and Aldrin after the discoverers of this chemical reaction. The whole group is now banned worldwide but traces can be found in soil, water and fish owing to their persistence in the environment. Endosulfan is the only compound of this type which is still in use. It is not persistent and one of the few insecticides of low toxicity with respect to honeybees.

A last group with acaricidal activity consists of chlorinated phenyl moieties linked by a sulphur bridge: tetradifon, tetrasul, chlorfenson.

9.2.2.2 Organophosphorus pesticides (OPs) (Fig. 9.2)

This class of insecticides was developed after the second world war, they have a broad spectrum of effectiveness with additional nematicidal and acaricidal action. Thousands of compounds were tested and about a hundred introduced onto the market. Many of them are highly toxic against man and mammals but their persistence in the environment is only short. OPs may exhibit systemic properties or

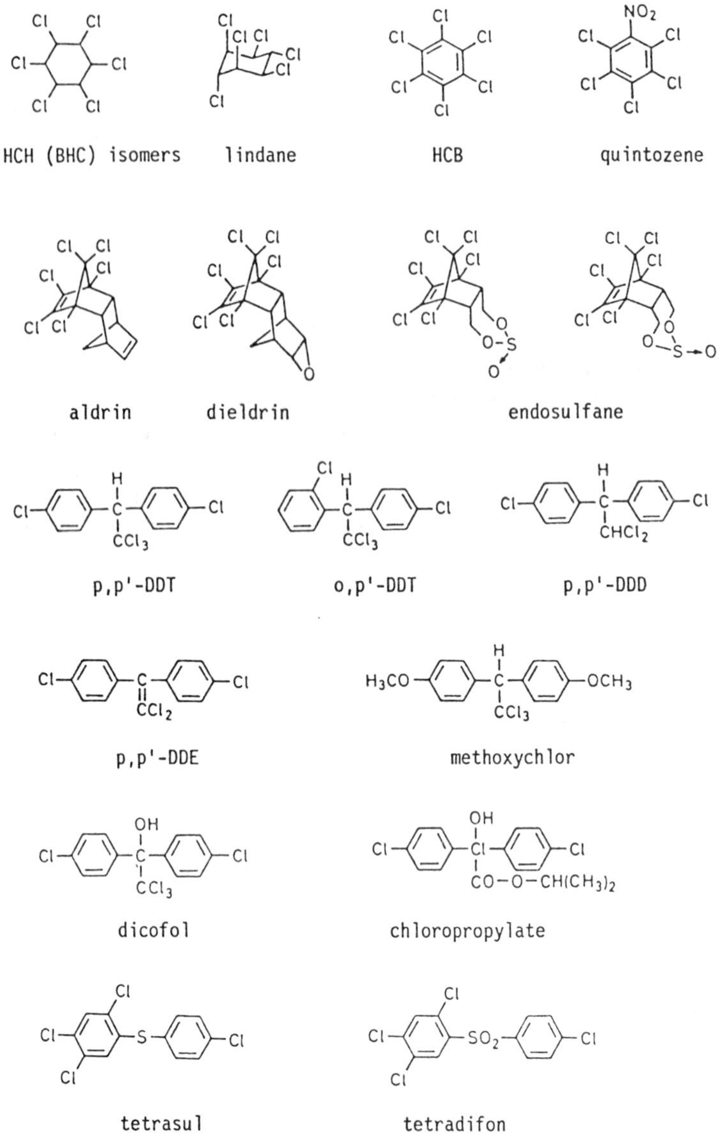

HCH (BHC) isomers lindane HCB quintozene

aldrin dieldrin endosulfane

p,p'-DDT o,p'-DDT p,p'-DDD

p,p'-DDE methoxychlor

dicofol chloropropylate

tetrasul tetradifon

Chlorinated Pesticides

Fig. 9.1

fungicidal or even herbicidal effectiveness. A few compounds were applied in animal production against endo- or ektoparasites. OPs are effective as inhibitors of acetylcholinesterase in the nervous system.

OPs can be classified in four groups:

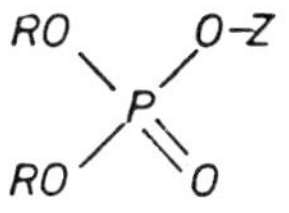

phosphate 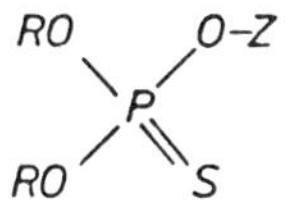 — thionophosphate 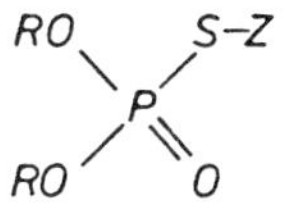 — thiolophosphate 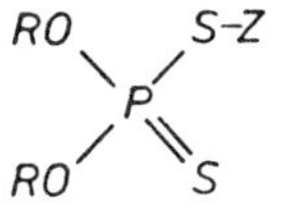 — dithiophosphate

The group R is mainly methyl or ethyl, the large size of the class is due to the variety of the moiety Z. According to the 'acyl rule' of Schrader, the inventor of this insecticidal class, Z must represent a good leaving group with electron-withdrawing properties.

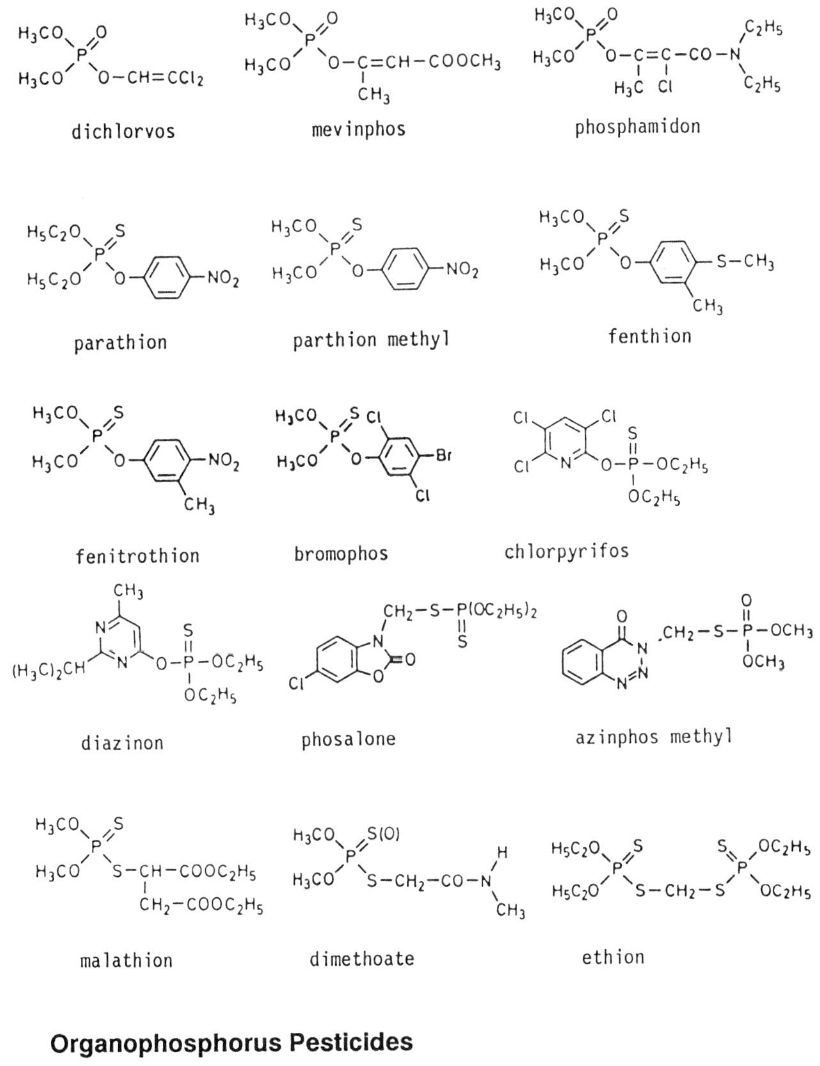

dichlorvos — mevinphos — phosphamidon

parathion — parthion methyl — fenthion

fenitrothion — bromophos — chlorpyrifos

diazinon — phosalone — azinphos methyl

malathion — dimethoate — ethion

Organophosphorus Pesticides

Fig. 9.2

Phosphates are more subject to hydrolysis and have a short life time in the environment. The more stable thiophosphates have to undergo a metabolic conversion into the corresponding oxo-compounds to result in an active inhibitor of acetylcholinesterase. The most prominent compound in this group is parathion, an insecticide with a very broad spectrum of activity but highly toxic to mammals and man. Other economically important compounds are malathion, the first OP developed with lower mammalian toxicity, azinphos, dimethoate, fenthion, chloropyrifos, dichlorvos and diazinon to mention only a few.

9.2.2.3 Carbamate pesticides (Fig. 9.3)

Carbamates are found to be active as insecticides, fungicides and herbicides. Insecticides are derived from a basic structure:

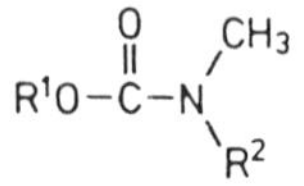

$$R^1O-\overset{\overset{\displaystyle O}{\|}}{C}-N\begin{matrix} CH_3 \\ R^2 \end{matrix}$$

with R^2 being H (*N*-methylcarbamates) or CH_3 (dimethylcarbamates) and R^1 a substituted aryl moiety. A different structure is found with the oxime carbamates, which have an oxime group at the position R^1.

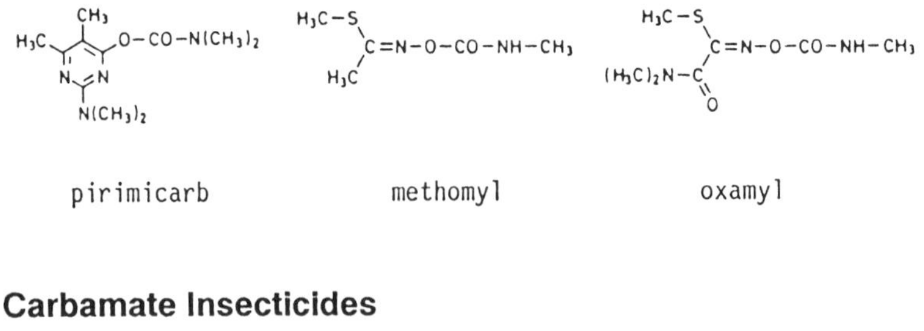

O−CO−NH−CH₃ carbaryl

$O-CO-NH-CH_3$; H_3C; $CH(CH_3)_2$ promecarb

$O-CO-NH-CH_3$; $O-CH(CH_3)_2$ propoxur

CH_3; H_3C; $O-CO-N(CH_3)_2$; N; N; $N(CH_3)_2$ pirimicarb

H_3C-S; $C=N-O-CO-NH-CH_3$; H_3C methomyl

H_3C-S; $C=N-O-CO-NH-CH_3$; $(H_3C)_2N-C$; O oxamyl

Carbamate Insecticides

Fig. 9.3

Carbamate insecticides act with the same biological mechanism as the OP in that they inhibit cholinesterases.

Carbamate herbicides can be divided into two structural groups:

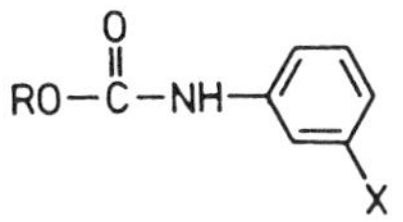

N-arylcarbamates 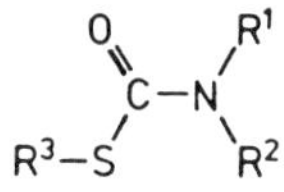 *S*-alkylthiocarbamates

They are active as soil herbicides, mainly against germinating broad-leaved and grassy weeds, by inhibiting cell division. Propham and chlorpropham are also used to inhibit the sprouting of potatoes. Dithiocarbamates are used as protective fungicides to control many fungal diseases of field crops, fruits, nuts and vegetables as well as downy mildews in grape vines. Sodium salts are soluble in water but metal salts with iron, manganese or zinc are applied as suspensions.

9.2.2.4 Phenol pesticides

Substituted phenols, their esters and ethers exhibit many biological activities. They are mainly used as herbicides but they have also fungicidal, insecticidal and acaricidal activity. DNOC (2-methyl-4,6-dinitrophenol) is now used as herbicide although it was introduced in 1892 by Bayer as the first synthetic insecticides. The nitrophenol ether nitrofen is a selective herbicide. Pentachlorophenol (PCP) was extensively used as a common biocide in the protection of wood (lumber).

9.2.2.5 Acidic and urea herbicides and derivatives (Fig. 9.4)

Many chlorinated benzoic acids (e.g. dicamba) are phytotoxic compounds interfering with growth regulation in plants. This is true in particular for the important group of phenoxy herbicides with 2.4D and MCPA [(4-chloro-2-methylphenoxy) acetic acid] being the most important active compounds. Urea herbicides (e.g. monuron, linuron), however, are potent inhibitors of photosynthesis. Chloroacetamides inhibit protein synthesis (e.g. alachlor). Propyzamide and chlorpropham are inhibitors of mitosis.

9.2.2.6 Dinitroaniline herbicides

2,6-Dinitroanilines (e.g. trifluralin) are active as soil herbicides against annual grasses and broad-leaved weeds in soybeans, cotton and field crops.

9.2.2.7 Triazine herbicides

This herbicide group, discovered in 1955, is based on 1,2,5,-triazine with two alkylaminogroups in positions 2 and 4:

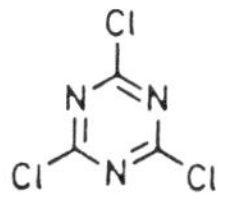

From this compound three structural chemical groups were derived differing in the substituent in position 6, which can be a chlorine, a methoxy or a methylthio group. The common names end in -zine, -tryn, or -ton, respectively. All compounds are inhibitors of photosynthesis. Atrazine and simazine are extensively used on corn because of its high tolerance to these herbicides.

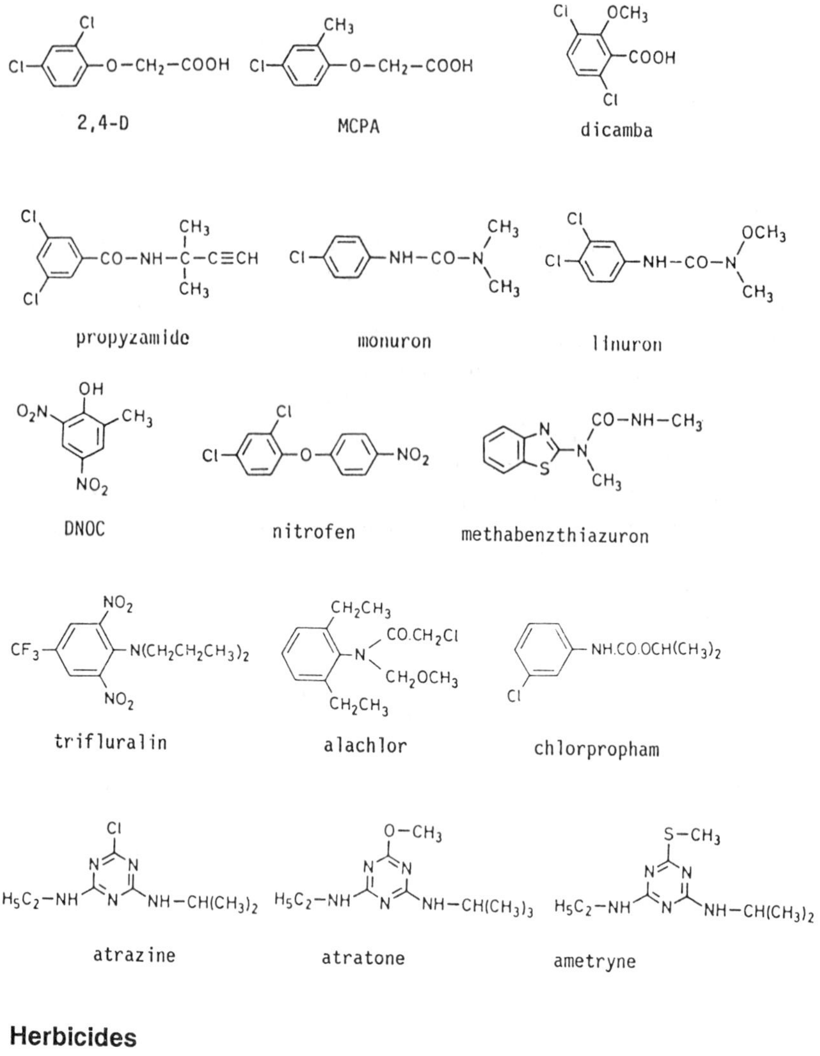

2,4-D MCPA dicamba

propyzamide monuron linuron

DNOC nitrofen methabenzthiazuron

trifluralin alachlor chlorpropham

atrazine atratone ametryne

Herbicides

Fig. 9.4

9.2.2.8 *Mercaptoimide fungicides* (Fig. 9.5)

Fungicides of this structural group (captan, folpet, captafol, dichlofluanid) show a wide range of protective activity against the most important fungal diseases of vegetables, fruit and vines, such as *Venturia* spp. on apples and pears, *Botrytis* spp. and downy mildews.

9.2.2.9 *Imidazole and triazole fungicides*

These are recent systemic fungicides (e.g. triadimefon) giving excellent control of powdery mildews on apples, cereals and grapes and of other fungal diseases.

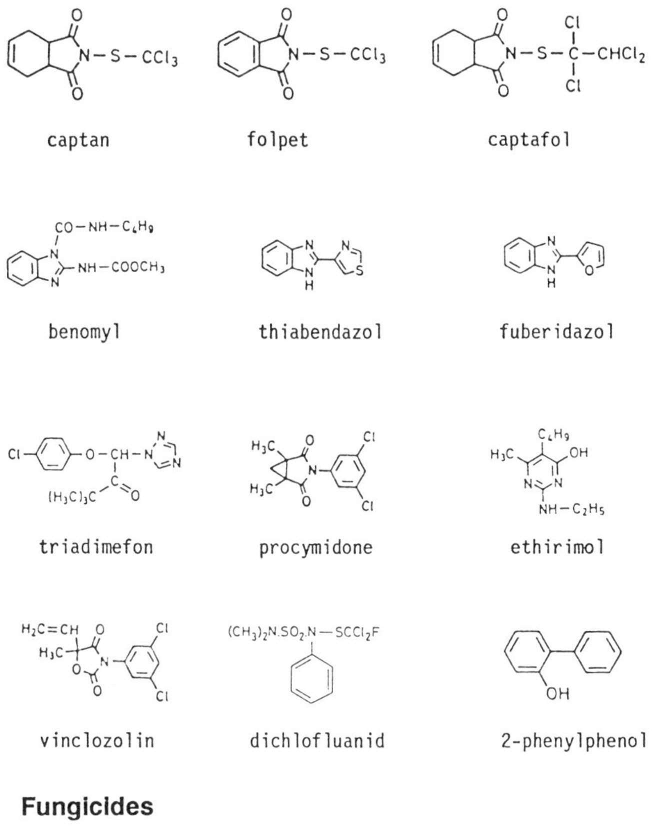

captan folpet captafol

benomyl thiabendazol fuberidazol

triadimefon procymidone ethirimol

vinclozolin dichlofluanid 2-phenylphenol

Fungicides

Fig. 9.5

Another group of systemic fungicides that are effective against a wide range of fungi affecting field crops and fruit are the benzimidazole derivatives benomyl and thiabendazol. Fuberidazol is used for seed treatment, thiabendazol is also employed for the post-harvest treatment of citrus (as also is 2-phenylphenol).

9.2.2.10 Pyrethrins and pyrethroids

Pyrethrins are natural insecticidal constituents present in extracts of *Pyrethrum* flowers that were used from the beginning of the nineteenth century to control insects in the house. Pyrethroids are synthetic insecticides of the same structural group but of higher stability.

9.3 AIM AND PRINCIPLES OF PESTICIDE RESIDUE ANALYSIS

The worldwide agricultural use of pesticides described can be expected to result in residues in food and drinking water. Regulatory systems have been introduced in most countries to ensure that pesticide residues do not constitute an unacceptable health risk. Tolerances are established for individual pesticides in specific commodities. Only an efficient monitoring of food samples from the market can protect the consumer from the potential hazards of pesticide application. In most of the developed countries about 300 pesticides currently have tolerances in food. In addition, there are pesticides for which approvals for domestic use have been cancelled, but which may contaminate food because of their persistence in the environment, and pesticides used only in countries foreign with respect to the consumer. Another problem arises from metabolites and alteration products that result from the use of pesticides.

9.3.1 Analytical methods

No single analytical method or combination of methods is applicable to all possible residue/commodity combinations. Analytical methods for pesticide residues generally require a procedure for extracting the residues, clean-up procedures to separate the residues from co-extracted matrix compounds, a technique to measure the residues and a technique to confirm their identity. Similarities in physical and chemical properties permit the application of methods that can be used to analyse a single sample for certain groups of pesticides. Such analytical methods are referred to as multiresidue methods; they must provide reliable identification and quantitation of a large number of compounds at very low concentrations (Thier & Zeumer, 1987; Moye, 1981; Ambrus *et al.*, 1981b; FDA Pesticide Analytical Manual, 1985; Luke & Masumoto, 1986).

9.3.2 Multiresidue methods applying gas chromatography

Gas chromatography (GC) with the selective electron-capture (ECD) and nitrogen–phosphorus (NPD) detectors allows the detection of contaminants at trace level concentrations in the lower p.p.b. range in the presence of a multitude of compounds extracted from the matrix to which these detectors do not respond. The number of compounds that is required to be monitored certainly now surpasses 500. Additionally, the input of pollutants into the environment has increased and so it is impossible to separate all these compounds in a single analysis, even with the application of high-performance capillary columns.

Reliability of multiresidue gas chromatographic analysis can be enhanced by different approaches:

9.3.2.1 Effluent splitting to selective detectors

One technique is effluent splitting after the capillary column to two selective detectors. Each peak is characterized by its retention time and the response factors in the different detectors. Confirmation is accomplished by applying the same technique to a second capillary column coated with a separation phase of different polarity (Goebel & Stan, 1983; Stan & Goebel, 1983).

9.3.2.2 Two-dimensional gas chromatography

A second approach is two-dimensional capillary gas chromatography with pneumatic switching between two columns of differing polarity (Stan & Mrowetz, 1983b).

Again, it is possible to increase the information content of the chromatographic data by applying effluent splitting after the first or after both columns to several detectors (Stan & Mrowetz, 1983; Stan, 1988a).

9.3.2.3 Silica-gel fractionation

A third approach to make the chromatogaphic data obtained more reliable is the application of a more extensive clean-up. One procedure widely used in Germany divides the sample into up to six fractions applying chromatography on a small silica-gel column (Specht & Tillkes, 1985). Many of the pesticides are separated from overlapping matrix compounds by means of this fractionation method. At the same time the classification of pesticides according to their partition coefficients in a system of water and organic solvents of increasing polarity is a valuable independent analytical parameter.

9.3.2.4 Application of small computers

With these methods data processing on-line by means of a mini computer may be used as a support in identifying suspected pesticide residues by checking the data against a database of calibrated compounds (Lipinski & Stan, 1988; Stan, 1988b).

9.3.2.5 Gas chromatography–mass spectrometry

The most sophisticated approach to obtain reliable data combines high-resolution GC with mass spectrometry (GC–MS). The mass spectrometer is without any doubt the most specific detector available in multiresidue analysis. The specificity is based on the fact that molecules when bombarded with electrons of particular energy under vacuum conditions fragment following strict rules. The resulting fragmentation pattern reflects the individual molecular structure in a mass spectrum that is often considered as the fingerprint of the substance. These mass spectra show such specific characteristics that it is possible to differentiate many ten thousands of compounds. It is of great importance that these mass spectra do not depend on the instrument used for measuring but only on the ionization conditions applied. Standarized ionization conditions can be easily reproduced. Therefore, it is possible to compile all the mass spectra recorded all over the world in libraries. Unknown compounds can be identified by comparing their spectra with such well-established library spectra (Hites, 1985; Cairns *et al.*, 1987; Damico, 1972; Safe & Hutzinger, 1973; Freudenthal & Gramberg, 1975; Stan *et al.*, 1977).

The formerly laborious task of comparing mass spectra is now performed with the help of computers and sophisticated software programs. Within a few seconds such a computer program performs a search of more than 40 000 documented spectra and draws up a list of a few mass spectra ranked according to their strongest resemblance to the one just recorded (McLafferty *et al.*, 1974; Pensyna *et al.*, 1976; In Ki Mun *et al.*, 1981). In this way it is possible to identify an unknown peak in a gas chromatogram without having the corresponding test substance available.

9.3.2.6 The advent of mass selective detectors

Until recent years the mass spectrometer appeared to be an instrument of great complexity that required a lot of maintenance and technical skill when using it as an analytical tool. Nevertheless, from the beginning of reliable environmental studies the mass spectrometer was recognized as the ultimate gas chromatographic detector for confirming results obtained by means of the very sensitive selective detectors.

Today, with the mass selective detector MSD, a small mass spectrometer is commercially available designed as a sensitive and most specific detector for high-resolution GC. The MSD offers the full capacity of any quadrupole mass spectrometer when coupled with a gas chromatograph. However, the MSD is as easy to maintain and to handle as any other gas chromatographic detector.

9.4 SAMPLE PREPARATION: CLEAN-UP

9.4.1 Principle of sample preparation

Food samples range in complexity from relatively simple matrices such as drinking water, wine and beverages to the more complex food matrices of plant and animal origin. The extent of clean-up required prior to the final determination by GC may be minimal but the more complex samples require procedures involving sophisticated physical and chemical separation techniques to eliminate undesired coextractives.

A typical screening analysis for a great number of pesticides and contaminants involves at least three basic clean-up steps:

(1) Extraction of residues from sample matrix
(2) Partitioning from the extraction solvent to another solvent
(3) Adsorption chromatography

In recent years gel permeation chromatography has played a role of growing importance in multiresidue methods due to its ability to eliminate the undesired non-volatile coextractants of higher molecular weight.

Sample extraction and clean-up in a screening method have to cover pesticides of varying polarity and they have to allow for fat and moisture content of the sample. Therefore, food samples are commonly classified into four categories:

(1) high moisture — low fat (fruits and vegetables)
(2) high moisture — high fat (meat, milk)
(3) low moisture — low fat (flour)
(4) low moisture — high fat (butter, oil, cocoa)

9.4.2 Sample extraction

Initial methods of residue analysis for fruits and vegetables started by homogenizing the food in high-speed blenders and adding a non-polar extracting solvent, usually hexane or benzene. Later acetonitrile was used as extracting solvent for organophosphorus and organochlorine residues, resulting in quantitative recoveries and cleaner sample extracts. This method was elaborated by Mills, Onley and Geither (1963) and is often referred to as the MOG method or Mills procedure. It was extensively used for residue analysis in plant materials over the past 20 years and adopted as official method in many countries. To samples with a lower moisture content water was added to produce a constant ratio between water and acetonitrile in the sample.

Similar extraction procedures have been developed by Luke *et al.* (1975) with acetone and Krause (1980) with methanol as extracting solvent. The reason for the use of water-miscible solvents is that they are able to break up the sample matrix and better dissolve the more polar pesticide residues. Acetone has the further advantage of its lower expense and toxicity and ease of removal due to its higher volatility. In this extraction step again the ratio of water to acetone must be kept constant to ensure a quantitative recovery of pesticides (Specht & Tillkes, 1980, 1985).

From samples with relatively high fat content the lipid material is extracted in the first step using one of the classical procedures, e.g. Soxhlet-extraction. Samples consisting of virtually 100% lipids (lard and oils) are dissolved in a non-polar solvent. Butter requires merely warming and then filtering of the separated fat. Tissues, milk, cheese are frequently ground with addition of sodium sulphate to disintegrate the sample and remove water through solvent extraction.

9.4.3 Solvent partitioning

The extraction has to ensure quantitative recovery; the next step in clean-up is to reduce the amount of coextractives and to remove the water. In the MOG procedure additional water was used to remove the water-miscible solvent. In the Luke procedure there is enough water present in the extraction step to recover more polar as well as non-polar residues and to remove parts of the acetone. Dichloromethane is the most popular solvent for partitioning against the aqueous acetone extract. Sodium chloride or sulphate is added to increase the transfer of pesticides to the dichloromethane phase by a salting-out effect.

Lipid samples are dissolved with petroleum ether or hexane and partitioned with acetonitrile. The bulk of the lipids remains in the petroleum ether phase while the acetonitrile layer contains the pesticides. From this solution the pesticides are partitioned again into hexane by adding an excess of water and sodium chloride to dissolve acetonitrile in the aqueous phase. Another procedure introduced by De Faubert Maunder *et al.* (1964) partitioned pesticides from hexane solutions of extracted fats into dimethyl formamide and then extracted the pesticides back into hexane after the addition of aqueous sodium sulphate solution.

9.4.4 Adsorption chromatography

In all the multiresidue methods the solvent partitioning step was followed by column chromatography using various adsorbents. Deactivated silica gel, silica gel, carbon, deactivated neutral or basic alumina, magnesium oxide and florisil (a synthetic magnesium silicate) are widely used. Florisil is widely used with fat-containing extracts, alumina for clean-up of fatty samples for the analysis of organochlorine pesticides and carbon clean-up to remove the pigments.

This clean-up step is designed to remove non-volatile high molecular weight coextractants that may cause the efficiency of the gas chromatographic system to deteriorate, and to eliminate compounds interfering with pesticide signals in the sensitive selective detectors, of which the ECD is the most vulnerable.

9.4.5 Sweep codistillation

Storherr and Watts (1965) described a sweep codistillation technique for clean-up of fat samples. The method can also be applied to the separation of organophosphorus and organochlorine pesticides from plant and animal samples as well as for separating triazine herbicides from soil. The concentrated extract is introduced into a continuous stream of nitrogen and transferred to a hot zone where the material is condensed on the large surface of glass wool. The more volatile pesticides are codistilled by means of a sweeping flow of volatile solvent, such as hexane, and trapped in a cooler (Storherr *et al.*, 1971; Pflugmacher & Ebing, 1973).

9.4.6 Gel-permeation chromatography (GPC)

In 1972 Stalling *et al.*, developed a manual gel-permeation chromatographic system for the clean-up of several pesticides and PCBs (polychlorinated biphenyls) in fish extracts that was later automated (Tindle & Stalling, 1972) and is now commercially available. A column of cross-linked polyvinyl resin (Bio-Beads S-X3) is used with an organic solvent as eluant to separate the pesticides by size from interfering substances of higher molar mass. The column can be loaded with about 1 g fat or oil dissolved in the eluting solvent. The method gives excellent clean-up with both plant material and animal samples (Specht & Tillkes, 1980) and can be applied to all kinds of environmental samples (Ambrus *et al.*, 1981b).

9.4.7 General clean-up for multiresidue analysis (DFG-Method S19)

A general clean-up procedure in a multiresidue method includes sample extraction according to moisture and fat content followed by solvent partitioning. The crude extract is then subjected to gel permeation to remove the non-volatile high-molecular weight coextractants, in particular lipids. The concentrated extract is then separated by adsorption chromatography on silica gel into several fractions that are used for gas chromatographic analysis.

A diagram of the clean-up procedure of DFG-Method S19 is shown in Fig. 9.6.

9.5 GAS CHROMATOGRAPHY

9.5.1 Introduction

The analysis of pesticide residues in food as well as in the environment is dependent on trace analytical methods employing gas chromatography. The invention of electron-capture detection (ECD) by Lovelock (1960) and Lovelock & Lipsky (1961) provided the first selective detector with extemely high sensitivity for halogenated compounds. In the decade of the late 1950s to 1960s chlorinated hydrocarbons represented the class of insecticides mostly used. In the middle of the 1960s the invention of the alkali flame ionization detector (AFID) and the flame photometric detector (FPD) allowed selective detection of organophosphates and nitrogen-containing compounds with high sensitivity (Karmen & Guiffrida, 1964; Brody & Chaney, 1966). The combination of the separation power of GC with the sensitive selective detectors made this technique the most important tool in pesticide residue analysis.

9.5.2 Packed columns

The stationary phase materials commonly used in pesticide analysis are relatively few, most of them are of the non-polar to medium-polar type. An important

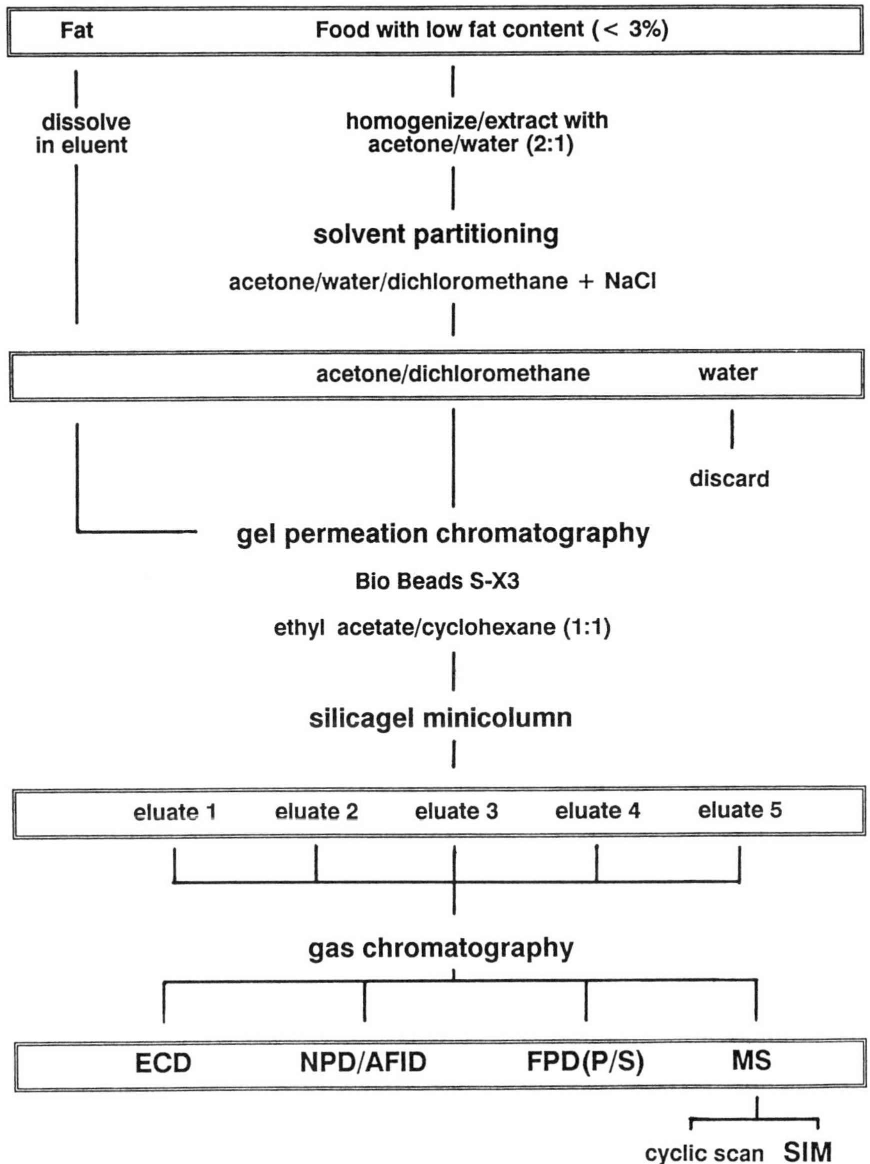

Fig. 9.6 — Clean-up procedure according to DFG Method S 19.

prerequisite for application in multiresidue analysis is thermostability. Therefore, most of the common phase materials are silicone polymers with differing substituents. A compilation of these liquid phases is given in Table 9.2.

Table 9.2 — Stationary phase materials commonly used in pesticide analysis

Chemical name	Phase name	Equivalent designation
Methyl silicone	OV-1	OV-101, SE-30, SP2100 DC-200, DC-11, SF-96
Methyl silicone, 5% phenyl	OV-73	SE-52
Methyl silicone, 5% phenyl, 1% vinyl	SE-54	
Methyl silicone, 10% phenyl	OV-3	
Methyl silicone, 20% phenyl	OV-7	
Methyl silicone, 35% phenyl	OV-11	
Methyl silicone, 50% phenyl	OV-17	SP-2250
Methyl silicone, 50% trifluropropyl	OV-210	QF-1, SP-2401
Methyl silicone, 25% cyanoethyl	OV-225	XE-60
Methyl silicone, 14% cyanopropylphenyl	OV-1701	

The liquid phases are spread in a thin layer on chemically inert support materials of narrow particle size range with a large surface area. Most of the support material in use is prepared from marine or terrestrial diatoms, as the Chromosorb, Gas-Chrom, Anakrom or Supelcoport series, to name a few from various manufacturers. Column packings are prepared by dissolving the calculated amount of liquid phase in an appropriate solvent and mixing with the support material. After evaporation of the solvent the packing is filled into a gas chromatographic column and conditioned by gradual heating above the maximum temperature that it is intended to use for residue analysis. A 'curing' at 40°C higher than the operating temperature is recommended. The tubes are predominantly made from borosilicate glass with an inner diameter of 2 mm or 4 mm and about 2 m length. Shorter columns are often used for specific compound analysis to reduce analysis time and the thermal burden to thermolabile pesticides.

The percentage of stationary phase in a column packing exerts a strong influence over column efficiency and retention time. Columns of low loading exhibit higher efficiency than those of high loading and produces shorter retention times.

In pesticide residue analysis mixtures of liquid phases have gained a broad acceptance; the packings are as easy to prepare as those of pure polymers. Such columns can be optimized with respect to selectivity in order to separate so-called critical pairs that cannot be resolved on a pure separation phase. A very popular mixed phase is 1.5% OV-17+1.95% OV-210 originally introduced by Thompson *et al.* (1969) as a mixture of OV-17 and QF-1. Other mixed phases are 4% SE-30+6%

OV-210 or 3% OV-61+7.5% QF−1+3% XE-60 (DFG-Method S 19) and 10% DC-200+1.5% QF-1 (DFG-Method S 16).

These column packings were introduced into the official analytical methods in several countries and there exist many listings of retention times. With packed columns, retention-time data are published relative to a reference compound. Aldrin has been most commonly used as a reference compound for all pesticides detected by ECD. Parathion has been widely used when analysing organophosphorus compounds by FPD or AFID. With the latter detector it can also be used as a reference compound for nitrogen-containing pesticides. Relative retention time (RRT) data are reported mostly at one or more given temperatures for the specified column. Thompson *et al.* (1975) published tables of RRT for 48 pesticides on six different columns with Aldrin and Parathion being the reference compounds. An example is given in Table 9.3.

The columns described have been mostly optimized with respect to a particular elution pattern for persistent chlorinated pesticides that were widely used in the 1950s and 1960s in the United States and many other countries. The non-polar methyl silicone phases showed an elution pattern as given in Fig. 9.7 (DC 200) with an acceptable separation of the HCH isomers but no separation of the critical pair *p,p'*-DDE and dieldrin. The trifluoropropylmethyl silicone phase OV-210 exhibits the elution pattern shown in Fig. 9.7 (QF-1) with a full separation of the HCH isomers and excellent separation of *p,p'*-DDE and dieldrin.

With a mixed stationary phase of methyl silicone with trifluoropropylmethyl silicone in a basic ratio 2:3 excellent separation of *p,p'*-DDE and dieldrin is observed with no separation of β-HCH and lindane and poor separation of *o,p'*-DDT and *p,p*-DDD, formerly also called TDE (Fig. 9.7; SE-30/QF1).

The polular mixed phase of 50% phenyl methyl silicone with trifluoropropyl methyl silicone in a 1.5:1.95 ratio (OV-17/OV-210) shows an elution pattern with the best separation of all common chlorinated pesticides (Fig. 9.7/OV-17/QF1).

Some pesticides are not as stable as one might expect from their persistence in the environment. DDT, to give an example, is subjected to on-column decomposition. Mostly a dirty injector tube or deposits from the biological matrix are the causes of this phenomenon. The replacement of the injector tube or pre-column and renewing of the front part of the packing will rejuvenate the column in most cases.

In 1973, Aue *et al.* (1973) demonstrated that Carbowax 20M could be chemically bonded to the solid support. These columns showed very little bleeding and high inertness after coating with other liquid phases. Therefore, they gained popularity with a number of pesticide laboratories and those packings became commercially available. Hall and Harris (1979) and Moseman (1978) reported on gas chromatography of intact carbamate pesticides and polar pesticide metabolites using Carbowax 20M deactivated columns. Recently, Suprock and Vinopal (1987) reported on the behaviour of 78 pesticides and metabolites on four different columns of this type. Relative retention times with respect to aldrin are presented for two columns in Table 9.4.

9.5.3 Capillary columns

Capillary columns have a number of advantages over packed columns. They offer higher resolution, more rapid analysis time, less decomposition and less adsorption

Table 9.3 — Relative retention times (RRTs) for pesticides on two packed columns. Reference compound: aldrin = 1.00. Reference compound for organophosphates: parathion[a]

Pesticide	4% SE-30/6% OV-210		1.5%OV-17/1.95% OV-210	
	(180°C)	(200°C)	(180°C)	(200°C)
Mevinphos	0.29	0.32	0.32	0.33
Tecnazene	0.36	0.40	0.36	0.40
2,4-D, methyl ester	0.40	0.44	0.45	0.47
Hexachlorobenzene	0.41	0.45	0.44	0.48
α-BHC	0.44	0.49	0.50	0.54
Sulfallate	0.46	0.50	0.55	0.56
2,4-D, isopropyl ester	0.55	0.55	0.56	0.56
Chlordene	0.56	0.60	0.56	0.60
Diazinon	0.59	0.58	0.66	0.64
Quintozene	0.61	0.64	0.66	0.68
Lindane	0.56	0.60	0.67	0.69
2,4,5-T, methyl ester	0.65	0.64	0.75	0.73
β-BHC	0.58	0.61	0.81	0.80
Heptachlor	0.81	0.83	0.82	0.82
2,4,5-T, isopropyl ester	0.87	0.83	0.91	0.85
Dimethoate	0.93	0.86	1.11	1.00
Aldrin	1.00	1.00	1.00	1.00
Frenchlorphos (Ronnel)	0.98	0.91	1.14	1.07
Parathion methyl	1.44	1.34	1.62	1.45
DCPA	1.57	1.44	1.72	1.52
Hepatachlorepoxide	1.50	1.43	1.65	1.54
Malathion	1.61	1.42	1.92	1.63
γ-Chlordane	1.61	1.52	1.85	1.69
o,p'-DDE	1.60	1.46	2.03	1.82
Parathion	1.98	1.76	2.16	1.84
α-Chlordane	1.78	1.67	2.05	1.86
Endosulfan I	1.93	1.79	2.11	1.95
p,p'-DDE	2.05	1.82	2.58	2.23
Dieldrin	2.33	2.12	2.67	2.40
*o,p'*DDD	2.22	1.98	3.11	2.65
Chlordecone	2.83	2.56	3.09	2.77
Endrin	2.64	2.42	3.29	2.93
o,p'-DDT	2.80	2.39	3.71	3.16
p,p'-DDD	2.98	2.55	4.26	3.48
Endosulfan II	3.04	2.72	4.17	3.59
p,p'-DDT	3.73	3.12	5.11	4.18
Ethion	3.76	3.05	5.49	4.28
Carbophenothion	3.78	3.16	5.76	4.56
Mirex	5.68	4.79	7.10	6.10
Methoxychlor	5.98	4.60	11.00	8.10
Tetradifon	10.30	7.80	14.90	10.90

Table 9.3 — (Continued)

Pesticide	4% SE-30/6% OV-210 (180°C)	4% SE-30/6% OV-210 (200°C)	1.5%OV-17/1.95% OV-210 (180°C)	1.5%OV-17/1.95% OV-210 (200°C)
TEPP	0.10	0.12	0.11	0.15
Dichlorvos	0.14	0.16	0.17	0.20
Mevinphos	0.30	0.32	0.32	0.37
Thionazin	0.36	0.37	0.45	0.46
Ethoprop	0.42	0.42	0.48	0.48
Phorate	0.44	0.44	0.54	0.53
Sulfotepp	0.46	0.48	0.56	0.53
Oxydemeton methyl	0.50	0.49	0.60	0.59
Diazinon	0.55	0.56	0.71	0.66
Naled	0.48	0.48	0.71	0.70
Disulfoton	0.65	0.65	0.82	0.77
Dioxathion	0.61	0.62	0.84	0.79
Diazoxon	0.75	0.70	0.86	0.79
Dichlorfenthion	0.81	0.79	0.99	0.90
Cyanox	1.03	0.97	1.14	1.05
Dimethoate	0.93	0.86	1.17	1.07
Fenchlorphos (Ronnel)	0.97	0.92	1.23	1.12
Fenchlorphosoxon (Ronnoxon)	1.11	1.04	1.27	1.16
Monocrotophos	1.19	1.07	1.60	1.23
DMPA (Zytron)	1.17	1.09	1.45	1.29
Chlorpyrifos	1.17	1.09	1.62	1.40
Parathion methyl	1.45	1.34	1.62	1.45
Paraoxon methyl	1.78	1.58	1.71	1.49
Malaoxon	1.35	1.23	1.97	1.60
Malathion	1.58	1.37	1.97	1.64
Bromophos	1.82	1.55	1.86	1.64
Fenthion	1.25	1.14	1.97	1.66
Fenitrothion	1.68	1.51	1.97	1.69
Phosphamidon	2.32	1.97	2.16	1.77
Schraden	1.64	1.58	2.51	1.82
Parathion	1.98	1.76	2.16	1.84
Dicapthon	1.92	1.72	2.29	1.91
Paraoxon	2.42	2.06	2.27	1.91
Bromophos ethyl	1.78	1.60	2.40	2.01
Amidithion	2.00	1.78	2.59	2.21
Crufomate	2.18	1.92	2.94	2.34
Phenthoate	1.82	1.58	2.87	2.36
Folex	2.30	1.99	3.18	2.54
DEF	2.30	1.97	3.18	2.56
Iodofenophos	2.02	1.81	3.35	2.76
Tetrachlorvinphos	2.65	2.27	3.61	2.91
Methidathion	2.42	2.09	3.69	3.05
Carbophenoxon	3.64	3.04	5.55	4.27
Ethion	3.58	2.97	5.90	4.47
Carbophenothion	3.64	3.04	6.20	4.82
Fensulfothion	7.43	5.84	9.55	7.25
Phenkapton	6.00	4.82	11.23	8.19
Famphur	8.73	8.89	12.20	8.87
EPN	8.73	6.86	13.31	9.60
Imidan	8.71	6.88	15.94	11.48
Azinphos methyl	13.84	10.42	21.82	15.27
Azinphos ethyl	13.78	10.40	28.08	18.95
Coumaphos	30.89	21.47	43.20	28.52

[a] RRTs were reported relative to parathion and recalculated to aldrin. Pesticides are ordered according to their RRT on 1.5% OV-17/1.95% OV-210. Isothermal analysis at 200°C.

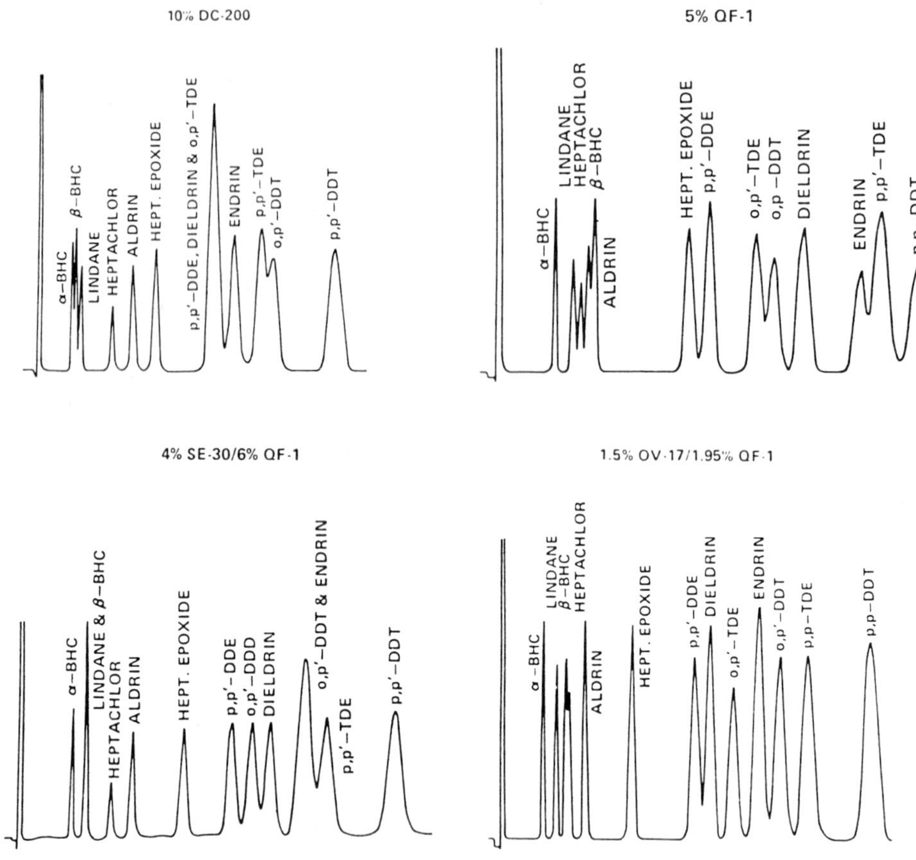

Fig. 9.7 — Elution pattern of 13 chlorinated pesticides using packed columns with single and mixed stationary phases. For phase names see Table 9.2. Source: Thompson & Watts (1981) (courtesy John Wiley & Sons, New York).

on the column. Many compounds that form critical pairs on packed columns can easily be separated on the more efficient capillary columns. Although these advantages are evident, many laboratories hesitated to introduce the new technique into routine residue analysis. The main reasons were the rapid contamination of the capillary column by non-volatile coextractants from the food matrix and the difficulties with the sampling techniques. The injection of a sample into a capillary column is far more prone to errors than that into a packed column.

In general the application of capillary columns requires more sophisticated equipment: special efficient sample introduction devices, an electronically controlled oven, sensitive detectors, high-speed recorders and integrators or chromatography software working at high sampling rate.

Table 9.4 — RRTs for pesticides on two ultra-bond columns. Reference compound: aldrin = 1.00; reference compound for organophosphates: parathion[a]

Pesticides	1% OV-210 (195°C)	0.5% OV-210/ 0.65% OV-17 (200°C)
Etridiazol (Ethazol)	0.24	0.20
Dichlobenil	0.30	0.23
Benfluralin	0.38	0.28
Trifluralin	0.38	0.29
Chloroneb	0.38	0.30
2,4-D, methyl ester	0.84	0.55
α-BMC	0.76	0.59
Diazinon	0.73	0.59
Chlordene	0.68	0.59
Fenoprop methyl ester	0.81	0.62
PCNB	—	0.63
Diazinon oxygen analogue	0.92	0.66
Heptachlor	0.92	0.85
2,4,5-T, methyl ester	1.35	0.87
Lindane	1.16	0.88
α-Chlordene	0.97	0.90
Dichlofenthion	1.14	0.90
Propazine	1.62	0.92
Aldrin	1.00	1.00
2,4-DB, methyl ester	1.46	1.00
Alachlor	1.43	1.07
Propyzamide	2.08	1.07
Atrazine	2.05	1.08
γ-Chlordene	1.38	1.19
Fenchlorphos (Ronnel)	1.59	1.20
Simazine	2.68	1.30
Chlorothalonil	1.73	1.32
β-BHC	2.35	1.32
Cyanophos	2.46	1.35
Chlorpyrifos	1.70	1.39
Oxychlordane	1.65	1.44
DMPA (Zytron)	2.32	1.52
Malathion oxygen analogue	2.73	1.58
DCPA	2.30	1.61
Hepatochlor epoxide	2.16	1.68
δ-BHC	3.22	1.73
Dimethoate	3.49	1.75
Monocrotophos	4.05	1.76
Malathion	2.73	1.78
trans-Nonachlor	2.05	1.83
trans-Chlordane	2.35	1.88
Triadimefon	3.22	1.93
cis-Chlordane	2.41	1.97
Methyl parathion	3.46	1.97
Isofenphos	2.86	2.02
Fenithrothion	3.59	2.07
Parathion	3.49	2.08
Endosulfan I	2.37	2.10
o,p'-DDE	2.73	2.10
Chlorfenvinphos	3.49	2.32
p,p'-DDE	3.54	2.62
Dieldrin	3.38	2.66
Oxadiazon	3.76	2.67

Table 9.4 — (Continued)

Pesticides	1% OV-210 (195°C)	0.5% OV-210/ 0.65% OV-17 (200°C)
Propanil	7.11	3.00
Endrin	3.55	3.14
o,p'-DDD	5.11	3.34
Captan	5.38	3.47
o,p'-DDT	4.78	3.63
Picloram, methyl ester	8.24	3.77
Ethion	6.62	4.51
Endosulfan II	7.00	4.73
p,p-DDD	8.49	4.90
Nitrogen	8.95	5.24
p,p-DDT	7.97	5.31
Carbophenothion	8.43	5.62
Bromacil	15.76	6.10
Endrin aldehyde	10.33	6.30
Mirex	8.08	7.02
Endosulfan sulphate	14.26	8.20
Phencapton	14.27	9.23
Methoxychlor	18.24	10.87

[a] RRT, were reported relative to parathion and recalculated with respect to aldrin. Pesticides are ordered according to their RRT on 0.5% OV-210/0.65% OV-17. Source: Suprock & Vinopal (1987).

9.5.3.1 Glass and fused-silica capillary columns

With the development of the glass drawing machine by Desty *et al.* (1960), glass became the material of choice for capillary columns. Although the glass columns were somewhat fragile, they replaced the earlier materials because of better inertness and ease of modification. Glass columns were drawn to length and dimensions of choice. The inner surface was deactivated and prepared for coating with the thin film of liquid phase according to the needs of the type of analysis. The introduction of flexible fused-silica columns by Dandeneau and Zerenner (1979) initiated a new era in open tubular column gas chromatography. Glass columns were successively replaced by the new columns exhibiting stronger mechanical properties and higher intrinsic inertness. The technology involved in drawing fused silica, however, is far more complicated than drawing glass. Therefore, it is generally not possible for analysts to fabricate their own column material and the columns have to be brought from manufacturers.

Prior to stationary-phase coating, the inner wall surface of the column must be treated in order to deactivate active surface sites and to enhance surface wettability. Generally, non-polar stationary phases are easily coated on deactivated wall surfaces, but with increasing polarity of the liquid phase it becomes more difficult to coat evenly. This is the reason why the production of columns with phases of good separation properties for pesticide analysis such as OV-17 turned out to be difficult. In comparing glass and fused silica, fewer stationary phases are available on fused-silica columns. The glass surface can be easily roughened, increasing both wettability and film stability for a wider range of stationary phases. Recently, the introduction of the *in situ* free radical cross-linking of coated liquid phases marked an advance in overcoming these problems with fused-silica columns. The cross-linked stationary phases are non-extractable and, therefore, well suited for on-column injections where large amounts of solvent recondense in the column (Lee *et al.*, 1984).

9.5.3.2 Early applications

Early applications of glass capillary columns by Sissons and Welti (1971) as well as by Schulte and Acker (1974a,b) and Schulte *et al.* (1976) demonstrated their tremendous separation power in the analysis of PCB isomers and their differentiation from chlorinated pesticides of the DDT group. From the chromatograms shown in Fig. 9.8 it is evident that it appears to be impossible to detect and determine quantitatively CP in the presence of PCBs in food samples using packed columns with ECD. This aim can be achieved, however, by applying a capillary column with its higher separation efficiency (Fig. 9.9).

Some impressive examples of this technique are the analyses of the polychlorinated dibenzodioxin (PCDD) isomers in soil by Buser (1976, 1977) and in marine fish by Ballschmiter and coworkers (Ballschmiter *et al.*, 1981; Zell & Ballschmiter, 1980).

Several groups described the analysis of organophosphorus pesticide residues in fruit and vegetables on glass capillary columns (Kriigsman & van de Kamp, 1976; Stan, 1977a,b; Hild *et al.*, 1978) and chlorophenoxy acid herbicides in flour (Gilsbach & Thier, 1982). The length of columns ranged from 20 to 60 m. Individual compounds in several classes were separated in capillary columns, including *N*-methylcarbamates (Wehner & Seiber, 1981), triazines (Roseboom & Herbold, 1980) and phenylurea herbicides (Deleu & Copin, 1980).

9.5.3.3 Application in multiresidue analysis

A first compilation of retention time data of about 200 pesticides and metabolites was published by Ripley and Braun (1983). They used a 15 m×0.25 mm fused-silica column coated with SE-30 (J & W), helium as carrier gas and temperature programming. The column was connected either to an ECD or to an NPD. The authors reported very good reproducibility of retention times for nearly all pesticides, whereas the reproducibility of peak area or peak height was very dependent on the individual compounds.

Stan and Goebel (1983b) described an automated gas chromatographic analysis of food samples by means of fused-silica capillary columns and data processing. In one conventional gas chromatograph equipped with one system for splitless injection and one for on-column injection and two selective nitrogen–phosphorus and electron capture detectors, two fused-silica columns were used. One column was 25 m long and coated with dimethylsilicone (SGE, BP-1) and the other 12 m long and coated with methylphenylsilicone (SGE, BP-10) (Goebel & Stan, 1983). Both coatings were 'bonded-phase' and the columns joined in an effluent splitter, which was connected to the two detectors (Fig. 9.10). The chromatograms of an analysis of a test mixture of OP obtained with the 25-m BP-1 column and parallel detection with NPD and ECD are presented in Fig. 9.11. The experimental conditions, chromatograms and the corresponding result tables are discussed in section 9.7.3.

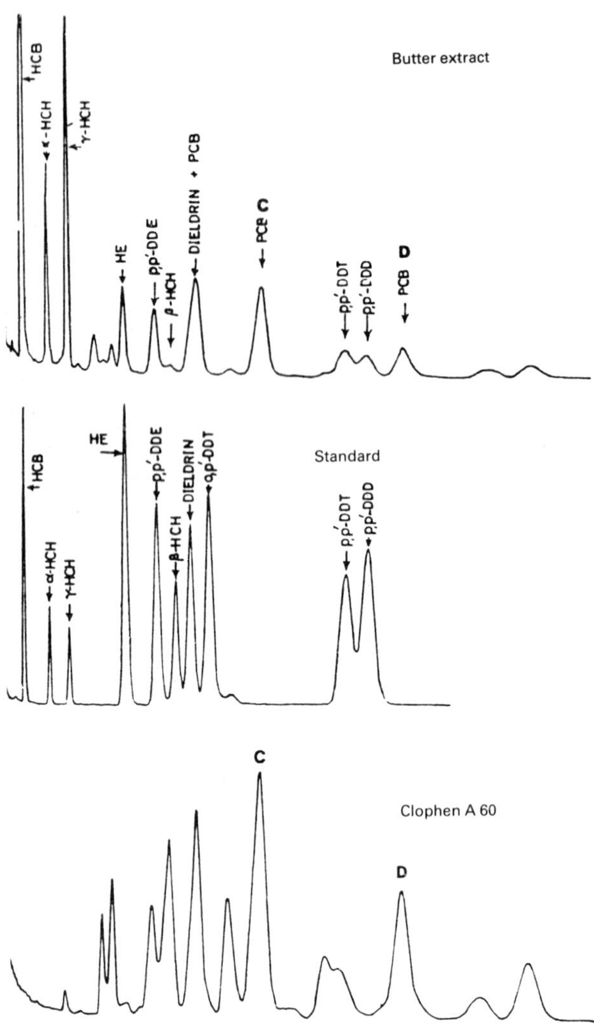

Fig. 9.8 — Residue analysis of chlorinated pesticides and PCBs in butter using a packed column: 3 m×2 mm 5% XE-60 on Chromosorb W AW CMS HP; isothermal at 190°C. Source: Schulte *et al.* (1976) (courtesy Wissenschaftliche Verlagsgesellschaft, Stuttgart).

9.5.3.4 Selection of capillary columns

Summarizing the reported applications of capillary columns in pesticide residue analysis from the beginning up to present, it can be stated that only a few stationary phases are in use. They are all silicones with varying substitution. A compilation of popular columns is given in Table 9.5.

Non-polar dimethyl polysilicones or those with 5% diphenyl substitution are best suited for screening analysis applying the more polar phases OV-1701 and OV-17 for confirmation. A new trend with the commercial suppliers may bring, however, some confusion in the future. As can be seen from the table, J & W introduced recently a

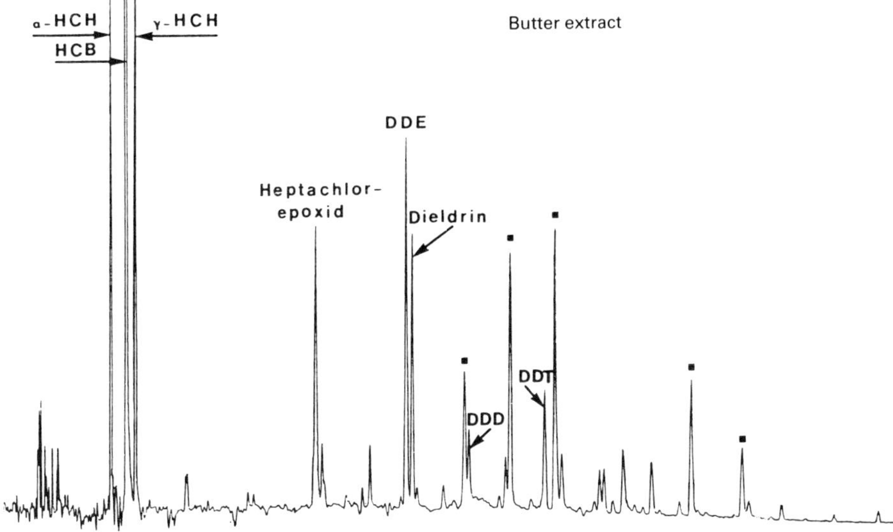

Fig. 9.9 — Residue analysis at chlorinated pesticides and PCBs in butter using a glass capilary column: 60 m SE-30; splitless injection: 180°–250°C. PCB peaks labelled with square. Source: Schulte *et al.* (1976) (courtesy Wissenschaftliche Verlagsgesellschaft, Stuttgart).

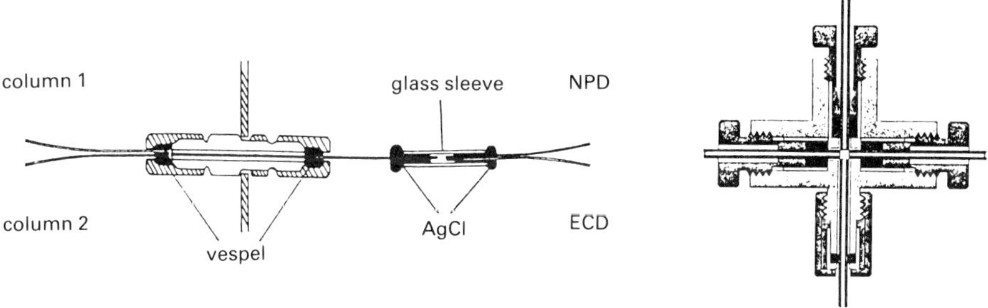

Fig. 9.10 — Effluent splitters for use with capillary columns. Left: SGE, middle, homemade; right: Gerstel.

special column for pesticide residue analysis called DB-608. The manufacturer advertises the column as tailor-made for executing the EPA method 608 for detecting a specified mixture of chlorinated pesticides. Unlike the good practice of declaring the type of stationary phase and the film thickness, the customer is only provided with the statement that the column is optimized for doing the job properly and to

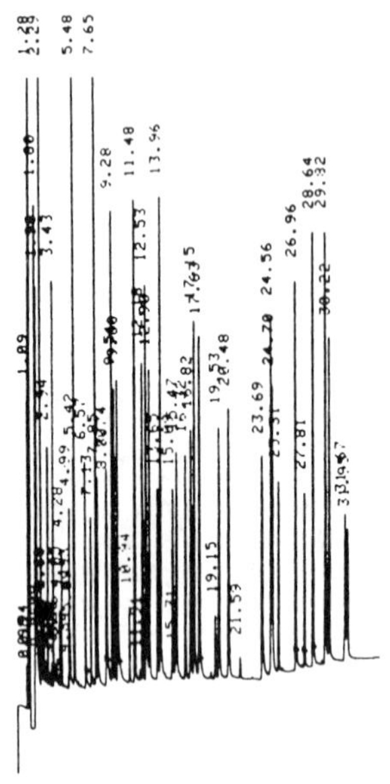

STANDARD MIXTURE I FOR ORGANOPHOSPHORUS PESTICIDES IN NPD
ISTD

RT	EXP RT	AREA		CAL	AMOUNT	NAME
2.29	2.28	4499.20		1	2.000	DIMEFOX
3.43	3.42	414.46		2	2.000	DICHLORVOS++
4.28	4.26	146.90		3	6.000	DIOXATHION+
4.99	4.97	295.07		4	2.000	MEVINPHOS++
5.42	5.45	322.32		5	2.000	PT
6.57	6.54	590.11		6	4.000	DEMEPHION
7.13	7.11	391.02		7	2.000	HEPTENOPHOS+
7.65	7.62	1836.17		8	2.000	THIONAZIN+
7.85	7.83	605.24		9	2.000	DEMETON-S-ME
8.70	8.69	440.66		10	2.000	DICROTOPHOS
9.28	9.26	1458.34		11	2.000	SULFOTEP
9.51	9.48	901.51		12	2.000	PHORATE++
9.80	9.77	1506.04		13	2.000	DIMETHOATE++
10.94	10.91	218.10		14	6.000	DIOXATHION
11.48	11.45	1723.80		15	2.000	FONOFOS+
12.18	12.17	1137.68		16	2.000	DIAZINON++
12.53	12.51	1748.96		17	4.000	FORMOTHION+
12.90	12.89	1159.01		18	2.000	ETRIMFOS+
13.65	13.65	778.18		19	4.000	PHOSPHAMI+
13.96	13.94	1934.59		20	2.000	PARATH-ME++
15.09	15.09	763.84		21	2.000	FENCHLORVOS+
15.47	15.47	1025.60		22	2.000	FENITROTH+
16.32	16.33	927.38		23	2.000	MALATHION+
16.82	16.82	1054.22		24	2.000	PARATHION+
17.15	17.16	1567.42		25	4.000	CHLORTHION+
17.63	17.63	1644.61	+	26	ISTD 1	NT
19.15	19.18	275.05		27	2.000	CHLORFENVIN+
19.53	19.55	1173.74		28	2.000	METHIDATH++
20.48	20.49	1429.64		29	2.000	BROMO-ET+
23.69	23.73	954.47		30	4.000	FENSULFO++
24.56	24.60	1236.36		31	2.000	ETHION++
24.70	24.74	1004.19		32	2.000	TRIAZOPHOS+
25.31	25.35	584.37		33	2.000	CARBOPHENO++
26.96	26.99	1334.33		34	4.000	PHOSMET+
27.81	27.85	531.51		35	2.000	PHENKAPT++
28.64	28.70	1532.84		36	4.000	PHOSALON+
29.82	29.99	1681.72		37	2.000	AZINPHOSETH+
30.22	30.30	1128.81		38	4.000	DIALIFOR++
31.67	31.76	708.64		39	4.000	COUMAPHOS++

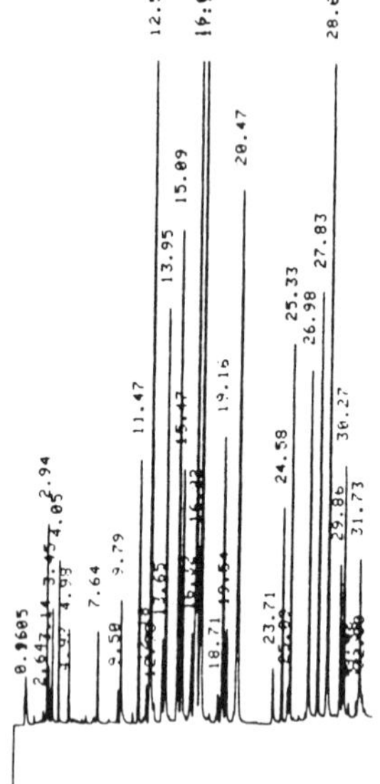

STANDARD MIXTURE I FOR ORGANOPHOSPHORUS PESTICIDES IN ECD
ISTD

RT	EXP RT	AREA		CAL	AMOUNT	NAME
3.45	3.44	26931.30		1	2.000	DICHLORVOS++
4.98	5.00	27155.30		2	2.000	MEVINPHOS++
7.64	7.66	40050.80		3	2.000	THIONAZIN+
9.50	9.52	15925.20		4	2.000	PHORATE++
9.79	9.82	88219.50		5	2.000	DIMETHOATE++
11.47	11.48	144786.00		6	2.000	FONOFOS+
12.18	12.20	23417.50		7	2.000	DIAZINON++
12.53	12.55	466195.00		8	4.000	FORMOTHION+
13.65	13.68	49002.90		9	4.000	PHOSPHAMI+
13.95	13.97	261011.00		10	2.000	PARATH-ME++
15.09	15.11	300473.00		11	2.000	FENCHLORVOS+
15.47	15.50	165759.00		12	2.000	FENITROTH+
16.32	16.35	59711.40		13	2.000	MALATHION+
16.67	16.67	444884.00	+	14	ISTD 1	ALDRIN
16.82	16.85	87279.30		15	2.000	PARATHION+
17.15	17.17	495662.00		16	4.000	CHLORTHION+
19.16	19.19	182273.00		17	2.000	CHLORFENVIN+
19.54	19.56	59946.90		18	2.000	METHIDATH++
20.47	20.49	389298.00		19	2.000	BROMO-ET+
23.71	23.74	39621.80		20	4.000	FENSULFO++
24.58	24.59	103229.00		21	2.000	ETHION++
25.33	25.33	158355.00		22	2.000	CARBOPHENO++
26.98	26.98	164983.00		23	4.000	PHOSMET+
27.83	27.83	178742.00		24	2.000	PHENKAPT++
28.67	28.67	332640.00		25	4.000	PHOSALON+
29.86	29.87	91452.10		26	2.000	AZINPHOSETH+
30.27	30.27	137449.00		27	4.000	DIALIFOR++
31.73	31.73	118603.00		28	4.000	COUMAPHOS++

Fig. 9.11 — Parallel recording of NPD and ECD signal analysing a test mixture of organophosphates. Source: Goebel & Stan (1983) (courtesy Elsevier, Amsterdam).

Table 9.5 — Popular capillary columns used in pesticide analysis

Manufacturer	100% Methyl	5% Phenyl	20% Phenyl	35% Phenyl	50% Phenyl	14% Cyanopropyl
Ohio valley	OV-1, OV-101	OV-73	OV-7	OV-11	OV-17	OV-1701
Alltech	RSL-100, RSL-150	RSL-200			RSL-300	RSL-1701
Chrompack	CP SIL 5CB	CP SIL 8CB				CP SIL 19CB
H.P.	HP-1, ULTRA-1	HP-2, ULTRA-2			HP-17	
J&W	DB-1	DB-5		DB-60B[a]	DB-17	DB-1701
Nordion	OV-1	SE-54			OV-17	OV-1701
Quadrex	007-1	007-2	007-7	007-11	007-17	007-1701
Restek	Rtx-1	Rtx-5	Rtx-20	Rtx-35	Rtx-17	Rtx-1701
SGE	BP-1	BP-5				
Supelco	SPB-1, SP-2100	SPB-5	SPB-20	SPB-35	SP-2250	
Other names	SE-30, SE-33	SE-54, SE-52				

[a] Exact phase composition not declared.

increase laboratory productivity. Another manufacturer, Supelco, followed with a corresponding column and called it SPB-608.

In a recent review of the state of the art of preparation and application of capillary columns, Blum (1988) made the same point of asking whether analytical procedures can be validated under the rules of good laboratory practice without exactly knowing the characteristics of the columns used.

9.5.3.5 Column length, inner diameter and film thickness

Fused-silica columns are supplied in various sizes. Established sizes are 0.2 mm, 0.25 mm and 0.32 mm inner diameter for the normal capillary column with a film thickness of 0.25 μm. These columns are offered by most manufacturers in standard lengths of 15 m, 30 m and 60 m or 15 m, 25 m and 50 m. Additionally, wide-bore capillary columns are offered with 0.53 mm inner diameter and lengths of 10 m, 15 m or 30 m. Recently, a new size is offered as a wide-bore capillary column 60 m$\times$0.75 mm with a film thickness of 1.0 μm. Several manufacturers also offer capillary columns with 0.18 mm inner diameter and narrow-bore capillary columns with 0.1 mm or 0.15 mm inner diameter.

The earlier applications of capillary columns for pesticide residue analysis were mostly executed on rather short columns of 10 m to 20 m length. Now the trend is going towards longer columns. In our laboratory, we apply a 50 m$\times$0.32 mm capillary column coated with methylsilicone (SE-54) at a film thickness of 0.25 μm for screening analysis. These columns retain good separation power in routine analysis for about one year. Confirmation analysis can be performed on shorter columns coated with OV 1701, OV 17 or OV 35.

The best combination was found in our laboratory to be a 50-m SE54 column with a 50-m OV17 column. In an investigation of about 270 pesticides and metabolites we have only observed six critical pairs that cannot be separated with one of these two columns (Lipinski & Stan, 1989).

9.6 SELECTIVE DETECTORS

9.6.1 Application to pesticide residue analysis

Pesticide residue analysis is based on the sensitivity and selectivity of gas chromatographic detectors. Pesticides residues are only present in food in trace amounts. In the course of the clean-up, the bulk of coextractants is decreased, but the final extract that is injected into the gas chromatographic column contains nevertheless many other organic substances from the food matrix. Fortunately, pesticide compounds in use possess heteroatoms that are not found in volatile biomolecules. Halogens are rare in biomolecules, nitrogen, sulphur and phosphorus are only found in amino acids, purines, pyrimidines and amines or sugar phosphates that are water-soluble and non-volatile.

The selectivity of a detector is expressed as the ratio of the detector sensitivity of two compounds. The selectivity factors are in the range of 100 000 for N and P against carbon in molecules without these heteroatoms. This means one phosphorus or nitrogen atom produces the same response as 100 000 carbon atoms. The relative response factors for halogenated compounds vary between 100 and 1 000 000 when compared with hydrocarbons. The sensitivity or response factor of a detector is defined as response per unit of substance, mostly in Coulomb/mole or Coulomb/g. A more convenient and practical measure is to report the minimum quantities of a few common pesticides that can be detected in a gas chromatographic analysis.

The four most popular selective detectors in pesticide residue analysis are the electron-capture (ECD), the flame photometric (FPD), the alkali flame ionization (AFID), and the thermionic nitrogen–phosphorus (NPD) detectors.

9.6.2 Flame photometric detector (FPD)

The most selective detector is the FPD which can be operated in a phosphorus- or sulphur-specific mode. The FPD is based on the element-specific chemiluminescence produced when compounds are burnt in a hydrogen-rich flame (see section 1.6.5).

The response of the FPD to phosphorus is linear, and for sulphur it should be a square function, but in practice, however, is dependent on the oxygen/hydrogen ratio and the jet design. The selectivity P/C is better than 100 000, that of S/C is better than 1000. The selectivity P/S is reported in the range 500 to 5000 and the opposite selectivity S/P is only 10 to 50 (Holland & Greenhalgh, 1981). Detection limits as low as 0.5 pg/s phosphorus and 50 pg/s sulphur have been reported (Patterson, 1978a).

9.6.3 Alkali-flame ionization detector (AFID) and thermionic detector (NPD)

In 1964, Karmen and Guiffrida showed that doping a conventional FID with a sodium salt enhanced its response to phosphorus or halogen-containing compounds. Two years later it was demonstrated that under different operating conditions this detector could selectively detect nitrogen-containing compounds. The detector in its present design is a modified FID in which a small electrically heated glass or ceramic bead, doped with rubidium, is positioned between the flame and the collector electrode (see section 1.6.3). Although the nature of the response in the detector is not well-understood, the following reactions may proceed in the gas phase. Organic molecules containing nitrogen or phosphorus are pyrolysed and form CN or PO_3

radicals, which then extract electrons from vaporized alkali metal atoms in an electronically excited state:

$$Rb^{*}+CN \rightarrow Rb^{+}+CN^{-}$$

$$Rb^{*}+PO_3 \rightarrow Rb^{+}+PO_3^{-}$$

The formed ions are collected at an electrode. The two operation modes flame-ionization (AFID) and flameless thermionic ionization (NPD or TD) differ in the sample decomposition and ionization process. It occurs in the flame (AFID) or on the hot surface of the alkali ceramic or glass bead (TD) (Patterson & Howe, 1978). In the TD a small hydrogen flow rate of 3 ml/min is normally required. By increasing the hydrogen flow-rate the detector's response to nitrogen is decreased. So, the NPD is usually sensitive for phosphorus but may lack sensitivity for nitrogen. The selectivity P/C is reported to be about 100 000. The most sensitive detection is reported as 50 fg/s nitrogen in azobenzene and 25 fg/s phosphorus in malathion (Patterson, 1978b).

9.6.4 Electron capture detector (ECD)

The electron-capture detector is, without any doubt, the most popular selective detector in use today. It responds to compounds capable of capturing thermal electrons. In the detector cell a high concentration of thermal ions is produced by the bombardment of the carrier gas with β particles from a radioactive foil, normally nickel-63 which has the advantage of a long half-life and also the detector can be operated at temperatures up to 400°C. This is important because the detector must be operated above the column oven temperature to avoid condensation of column effluents. The alternative radioactive source is a foil of titanium loaded with tritium. It is not suited for pesticide residue analysis because of a temperature limit of about 225°C and hydrogen should not be used with this detector foil because it exchanges for tritium with resulting decrease in the detector lifetime and emission of tritium into the atmosphere.

The β particles collide with the gas molecules (nitrogen or argon), ionising them according to the following reactions

$$N_2+\beta \rightarrow N_2^{+}+e^{-}+\beta'$$

$$Ar+\beta \rightarrow Ar^{+}+e^{-}+\beta'$$

$$[Ar+\beta \rightarrow Ar^{*}\ (11.6\ eV)+\beta' \quad]$$

$$[Ar^{*}+CH_4 \rightarrow Ar+CH_4+\text{energy} \quad]$$

The β particles β′ continue to collide with additional gas molecules and lose energy until their energy is reduced to the thermal level.

These thermal electrons e^{-} react with sample molecules undergoing associative electron capture

$$AB + e^- \rightarrow AB^-$$

or dissociative electron capture

$$AB + e^- \rightarrow A^{\cdot} + B^-$$

The formed negative ions recombine with positive carrier gas ions

$$N_2^+ + AB^- \text{ or } B^- \rightarrow N_2 + AB \text{ or } B$$

and deplete the concentration of ions and electrons.

These charged species are collected by applying a small voltage to produce a standing current. Compounds entering the cell and capturing thermal electrons reduce their concentration and decrease the standing current producing a chromatographic signal.

The effect of the carrier gas on the performance of the ECD must be carefully considered. With packed columns, either nitrogen or argon moderated with 5 or 10% methane must be used according to the detector specifications. Both possess high β-particle ionization cross-sections producing a high standing current. With argon, long-lived metastables, Ar*, are produced, which can directly ionize solute molecules. Therefore, methane must be added as a quenching gas. Nitrogen must be added as a carrier gas when the effluent is to be split to a second detector. Using capillary columns, the analyst is free to use helium or hydrogen as carrier gas with nitrogen or argon/methane used as make-up gas.

A number of operating modes for the ECD have been developed as well as the original d.c. constant voltage mode, namely the d.c. constant current mode, the constant frequency mode and the variable frequency mode. They differ in the way of measuring the electron and ion concentration in the detector cell.

The relationship between compound structure and ECD response has been extensively studied (Pellizzari, 1974; Devaux & Guichon, 1970). The probability for electron capture depends on several factors that are not readily calculable. Therefore, from many observations empirical rules have been formulated. A list of approximate relative molar responses for a variety of chemical classes expressed relative to chlorobenzene is given in Table 9.6.

9.6.5 Mass spectrometer

9.6.5.1 Coupling the gas chromatograph to the mass spectrometer

GC–MS is the most powerful available tool for the analysis of pesticide residues because of its inherent high selectivity and good sensitivity. It combines a high-performance separation method with a high-performance measuring technique. This holds true for packed column GC–MS but in particular for capillary GC–MS. The coupling of a gas chromatograph to a mass spectrometer originally caused problems because of the considerable pressure difference between the two. A packed column

Table 9.6 — Relative response factors for electron-capture detection

Chemical class	Response[a]
Hydrocarbons	0.01
Aliphatic ethers and esters	0.1
Aliphatic alcohols, carbonyls, amines, monochloro and -fluoro	1
Enols, monobromo, dichloro, hexafluoro	10
Trichloro, carbamates, triazines, organophosphates	100
Monoiodo, dibromo, trichloro-nitro, di- and trisulphides	1000
1,2-Diketone, conjugated esters, quinones	10000
Diiodo, tribromo, polychloro, polychloro-dinitro	10000

[a] With respect to chlorobenzene.

must be coupled via a separator interface. The Watson–Biemann glass frit separator (Watson & Biemann, 1964) and the Ryhage jet separator (Ryhage, 1964) were in widespread use (Stan, 1981). Recent development is characterized by mass spectrometers designed for coupling to gas chromatographs with the necessary high pumping rate and the use of capillary columns, in GC–MS work. This allows the introduction of the gas chromatographic effluent into a mass spectrometer in the simplest way, by the direct connection of the capillary column to the ion source with a vacuum-tight seal, as first described by Henneberg and Schomburg in 1964. This connection can be easily achieved today with a heated transfer line of deactivated fused-silica capillary. It guarantees the transfer of the total amount of substance and highest detection sensitivity.

Alternatively, an open-split coupling can be applied (Fig. 9.12), exhibiting the following advantageous properties (Henneberg *et al.*, 1975; Stan & Abraham, 1978):

(a) Atmospheric pressure at the end makes chromatograms directly comparable with separate GC analyses.
(b) Rapid and safe changing of columns is possible without losing the vacuum in the system.
(c) Versatility with respect to column types and flow rates.
(d) By introducing a large helium flow via a scavenger gas line it is possible to protect the ion source from undesired fractions, such as solvent peak and matrix compounds that may deteriorate sensitivity in the ion source (Stan & Abraham, 1978).

Capillary columns have often been criticized in connection with mass spectrometry for low sample capacity. Comparing the characteristic data for different column types, Eyem (1975) calculated that the minimum sample quantities that can be analysed with the MS are at least 100 times higher on packed columns than on capillary columns. Polychlorinated dibenzodioxins have been detected and quantified at the lower p.p.q. level ($1:10^{15}$) in cow's and human milk and food (Beck *et al.*, 1987). One femtogram is about as low as mass spectrometry can detect.

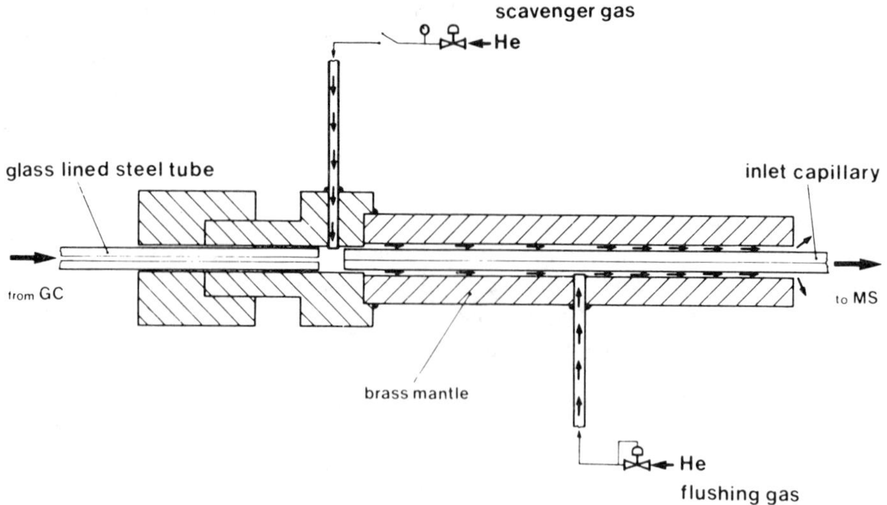

Fig. 9.12 — Open-split interface for GC–MS. Source: Stan & Abraham (1978) (courtesy American Chemical Society, Washington, DC).

9.6.5.2 Cyclic scan

The mass spectrometer as a gas chromatographic detector can be operated in two modes. With pesticide concentrations in the sample allowing the introduction of at least a few nanograms of the individual substances into the ion source, the mass spectrometer may be scanned over the entire mass range from m/z 50 to m/z 500. This cyclic scanning gives individual mass spectra from the substances in the effluent stream. The total ion current (TIC) is used as an universal detector signal. All spectra are stored by means of a data system on a magnetic disk. Background spectra can be subtracted from the sample spectra and the corrected spectra can be used for direct comparison with mass spectra in a library.

9.6.5.3 Selected-ion monitoring (SIM)

In most real-life samples the concentration level of pesticide residues is much lower, in the picogram range with respect to the injected sample. For such samples selected-ion monitoring (SIM) must be applied where the ion current is measured for a few selected masses. Adequate confirmation requires that at least three ions should be measured. The ions should have the same retention time and intensity ratio as a standard sample. The SIM method is best suited for confirmation of results obtained with other selective detectors. However, it also can be applied to screening for a group of pesticides with common fragments in their fragmentation pattern as found with isomers, such as hexachlorcyclohexanes, the DDT-group or organophosphorus insecticides (Stan, 1981). With modern instruments, SIM can be applied with time programming. Each group of selected ions is specified for a certain time window around the retention time of the compounds to be detected (Lipinski and Stan, 1989).

9.7 MULTIRESIDUE ANALYSIS WITH CAPILLARY COLUMNS

In the analysis of pesticide residues in foods using GC with selective detectors, a major problem is the large number of compounds to be detected. The efficiency of packed columns permits the separation of only a limited number of substances. The determination of a pesticide from its retention time on one column is definitely not sufficient. Therefore, analysis on several other columns with stationary phases of different polarity is necessary in order to confirm the identity of a compound detected on the chromatogram. This procedure is time-consuming and requires a whole set of gas chromatographs.

Capillary columns exhibit excellent efficiencies, which allow the separation of complex mixtures and the determination of the retention times of compounds with high accuracy and reproducibility. The high resolution facilitates the differentiation of substances belonging to the same structural class, such as OPs or CPs.

Although, a large number of papers, reviews and textbooks as well as several manuals on pesticide residue analysis exist, a full description of a multiresidue method as really applied in the laboratory is rare. One method was published by Stan and Goebel (1983a) and is reported here. The procedure can easily be adapted to the instrumentation and requirements of other laboratories.

9.7.1 Instrumentation and method

A gas chromatograph equipped with two injection ports for capillary columns and two selective detectors (ECD and NPD) is the basic instrument. The injection ports may be both designed for split/splitless injection or one of them for on-column injection. The capillary columns must be of different polarity, the one applied for screening analysis should be coated with methyl silicone (OV-1, SE-30) or with 5% phenyl substitution (SE-54) the other applied for confirmation analysis should be coated with OV-17 or OV-1701. For cross-reference see Table 9.5. Both columns are connected via a commercial four-way connector to both detectors or via a home-made effluent splitter (Fig. 9.10). The effluent splitter is constructed using fused silver chloride as a thermally stable and chemical inert cement. The ends of the two fused-silica capillaries are sealed into a glass sleeve and connected to the two detectors. In order to protect and strengthen the splitter, the connector is coated with a liquid silicone rubber that cures overnight. The other side is connected to a splitting device that allows column exchange without destroying the splitter. This end can also be constructed using silver chloride as cement (Goebel & Stan, 1983). In recent years, a variety of effluent splitters have been introduced into the market by the chromatography suppliers. The effluent stream from each column is split equally to both detectors and the detector signals are processed in a dual-channel integrator. The pesticides are detected either in one of the detectors or frequently in both of them simultaneously. The response ratio between both detectors together with retention time is a very important piece of information for the recognition of an individual pesticide.

The instrument used for several years in our laboratory was a gas chromatograph 5880 A from Hewlett Packard equipped with an autosampler HP 7671 A for 36 sample bottles. The dual-channel integrator of the system was equipped with two terminals and a cartridge tape device in which analytical methods could be stored in specified analysis files. The entire pesticide GC analysis procedure was stored on separate files, each containing all instrument settings for producing a chromatogram as well as two calibration tables, created in parallel for the various pesticide test mixtures from the detector signals of ECD and NPD.

After entering sample numbers and names via the alphanumeric keyboard, the automatic analysis was controlled by a BASIC program.

9.7.2 Internal standards and calibration

All samples were analysed applying the internal standard method using aldrin and the thiophosphinates of phenol and β-naphthol as internal standards (Stan and Goebel, 1983a). Aldrin was the main internal standard that was used for recalculation of slightly shifted retention times and the relative response values in the ECD, while the two thiophosphinates were used for calculating the relative response data in the NPD. The method was elaborated for the determination of 94 pesticides in food in a daily routine analysis. The gas chromatographic system was daily recalibrated with all pesticides. Therefore, it was necessary to prepare test mixtures containing the greatest possible number of pesticides that could be separated, including the internal standards. Good resolution of all peaks is necessary for the recognition and quantitative recalibration of all compounds. Critical pairs of pesticides have to be analysed in different calibration mixtures. Three calibration mixtures containing 35CPs and 39 or 20OPs, respectively, together with internal standards, were prepared. The temperature program was optimized to resolve as many critical pairs as possible (Stan and Steinbach, 1984, 1985).

9.7.3 Automated screening analysis

Automated screening analysis with data processing was carried out on a 25m×0.2mm fused-silica column BP1 (SGE) which was connected to the hot splitless injector. Helium was used as carrier gas and make-up for NPD (20ml/min); the ECD was purged with 25ml/min argon and 10% methane. The temperature of both detectors was 300°C and that of the splitless injector was 240°C. The sample volumes were 1 μl for both the autosampler and the manual injection. Splitless injection according to Grob and Grob (1969) into the 'cold' column at 100°C was carried out with the split valve closed for 30s. One minute after injection, the following temperature program was started: 30°C/min to 150°C; 2min at 150°C; 3°C/min to 205°C; 10°C/min to 240°C; 2°C/min to 260°C; 10min at 260°C; stop; cool to the initial temperature.

Usually, the autosampler was loaded in a series with toluene, the three calibration mixtures, toluene and then the sample vials. In Fig. 9.11 the parallel chromatograms recorded with NPD and ECD are shown as examples. The results table of the NPD channel demonstrates that 39 compounds — 37 pesticides and the two internal standards O-2-Naphthyl dimethyl thiophosphinate (NT) and O-Phenyl dimethyl thiophosphinate (PT) — were recognized. In the results table of the ECD, 27 pesticides and the internal standard ALDRIN are presented, demonstrating that many OPs respond to the ECD. The names in the results tables may be shown in an abbreviated form owing to the limited number of characters allowed. Most of the pesticide names are followed by a cross or an asterisk. The cross indicates response parallel in both detectors, the asterisk indicates that other OPs are eluted with a similar retention time forming a critical pair.

Evaluation of the screening analyses was performed automatically with a BASIC program (Goebel & Stan, 1983; Stan & Goebel, 1984a) and by 'manually' reviewing the parallel chromatograms from the integrator terminals by means of the calibration tables.

Losses of pesticides in the system by adsorption or decomposition were indicated as large differences in the last-created calibration tables when comparing the data to previous ones. This is the time to change the insert liner and, if necessary, to cut off a section of column.

9.7.4 Confirmatory analysis

Confirmatory analysis was performed on the same instrument during daytime. Test mixtures consisting of the pesticides found positive in the screening were analysed in parallel to the samples using the second fused-silica column mounted to the other injector that is for on-column injection. This was a $12\,m \times 0.2\,mm$ BP-10 column (SGE). The initial oven temperature was set to 90°C and the injection of $1\,\mu l$ was carried out manually using a 10-μl syringe with a fused-silica needle.

The procedure described allowed for the first time a complete pesticide residue analysis in 'real life' samples for about 100 pesticides on a single gas chromatograph using high performance capillary columns and applying simultaneous effluent splitting to two selective detectors with parallel recording of the chromatograms and calculation of relative response data.

9.8 CALIBRATION

9.8.1 Preparation of standard solutions

In a pesticide residue laboratory more than 300 (up to 500) test substances are necessary to conduct the testing and calibration of the methods applied. Qualitative and quantitative analysis requires standard solutions of all these substances and mixtures of them with concentrations that must always be accurate and reliable. This causes major problems in controlling these standard solutions when the whole staff of a laboratory works with the same standards. On the other hand it is impractical to have for each member of staff the whole set of standards for individual use. The principles of control of producing and using the standard solutions include

(1) durable and clear labelling of all bottles
(2) fixed concentration for all stock solutions
(3) preparation of diluted test substances and test mixtures in the various concentrations only from stock solutions
(4) control of all manipulations with the stock solutions, dilutions and text mixtures.

One problem that is difficult to overcome is the loss of solvent during long time storage because of untight stoppers. Therefore toluene might be used as solvent for stock solutions whenever posssible. We found bottles with screw caps and teflon coatings better suited than all types of ground stoppered flasks. However, no seal is always so tightened that losses are not observed. The only way to keep control of such possible solvent evaporation is an accurate book-keeping of the weights for all standard solutions and mixtures from production to last usage.

In our laboratory, stock solutions are prepared with a fixed concentration of 1 g/l and dissolved in toluene. Each bottle is coded with a glass brand number between 1 and 1000. Test solutions are prepared by transferring a certain quantity to a bottle containing toluene by means of a micropipette, which is washed with toluene into the same bottle. The quantity is controlled by weighing on a microbalance with a resolution of 0.01 mg. In this way a complete transfer of the measured part of the stock solution into the test mixture is confirmed. After all selected substances from stock solutions are transferred the mixture is made up to the calculated weight by adding toluene. Test mixtures are stored in our laboratory in bottles with numbers above 1000.

The whole book-keeping can be done by means of a computer directly connected to the microbalance and a computer program called BALANCE (Stan & Lipinski, 1987). The preparation of any test mixture with this system can be started only after entering a code word. Then the numbers of the bottles with the stock solutions needed can be obtained. After collecting the corresponding bottles from the refrigerator and warming to room temperature the preparation of calibration standards can be executed. The analysis is conducted in dialogue by the computer, which is in permanent control of each preparation step through incoming data from the balance. Any mistake made during the preparation will be recognized by the data system.

9.8.2 Calibration and control of the gas chromatographic system

Calibration of the GC system is executed with standard mixtures of pesticides as usual. Pesticides, however, vary very much in volatility, polarity and thermostability. Therefore, test mixtures representing these different chemical groups must always be analysed together with the samples in order to recognize any deterioration in chromatographic performance. Reproducibility of sample analysis depends on the matrix and on the injection technique. Hot splitless injection was found to be suited for screening analysis, although a discrimination of higher-boiling compounds was observed. Repeatability was found to be best when using cold on-column injection. This injection technique, however, is vulnerable to matrix compounds from food samples (Stan & Goebel, 1984b). An evaluation of hot splitless, cold splitless or programmed-temperature vaporization and cold on-column injection was performed for the analysis of OPs (Stan & Müller, 1988) and for the analysis of thermolabile carbamate insecticides (Müller & Stan, 1989).

Another important problem arises from the fact that the response to the individual selective detectors of several pesticides is observed to be not high enough to detect them in all foods with the appropriate sensitivity. Thus spiking of food samples with such critical pesticides and estimating their recovery is a necessity in the daily work.

9.9 SCREENING ANALYSIS

9.9.1 Principle and aim

The aim of a screening analysis in multiresidue analysis is to find out which of the food samples taken to the laboratory may be contaminated and which not. The analysis of any food starts with weighing a representative sample, followed by extraction and clean-up and injection of the cleaned extracts into a gas chromatograph equipped with a suitable column and selective detectors. The chromatograms with the corresponding listings of retention times and peak areas calculated by the integrator are the first visible results. The pesticides that may be present as contaminants are characterized by their retention data and their response factors in the selective detectors. Peaks found in the chromatograms arise from different sources: pesticide residues, environmental contaminants that are not pesticides, coextractants from the food matrix and trace contaminants in solvents and all chemicals used for analysis. Other sources may be packing material or cross-contamination. Cross-contamination is avoided by observing correct cleaning of the glassware. Trace contamination caused by chemicals and solvents is recognized in analysing blanks. They are of concern if they overlap with pesticides in the chromatograms.

9.9.2 Screening on packed columns

Screening can be executed with packed or capillary columns of various polarities. Ambrus *et al.* (1981a) described the gas chromatographic detection with packed columns using five basic packings for screening and confirmation. They described in detail the residue analysis and used Pyrex glass columns with 3 mm inner diameter and as short as 45 to 90 cm packed with 3% OV-22 or 3% OV-101 on Gas Chrom Q for general screening purposes. The OV-22 column operated isothermally at 180°C was the preferred column for screening the organophosphorus pesticides with an NPD, and the OV-101 column was used for chlorinated and other electron-capturing compounds. Highest sensitivity was achieved with 45 cm long columns applying temperature-programming. Seventeen organophosphorus pesticides were separated on such an OV-22 column within 20 min.

9.9.3 Screening on capillary columns

Screening methods applying fused-silica capillary columns were reported from our group (Stan & Goebel, 1983a; Goebel & Stan, 1983). The 25 m×0.2 mm columns were coated with bonded methylsilicone phase and purchased from two different manufacturers. They were well suited to screen for about 100 pesticides applying daily calibration of all pesticides (see section 9.7).

The growing number of pesticides included in our screening method and the laborious evaluation of chromatograms of critical foods such as leeks, cabbage, onions and others led us to change our general clean-up procedure and to apply longer columns. For general screening purposes we use fused silica columns of 50 m length, with 0.32 mm inner diameter coated with 0.25-μm SE-54 bonded phase.

9.9.4 Control of the gas chromatographic system

The reliability of a screening run depends on a careful control of the gas chromatographic system with respect to stability of retention time and response values of the selective detectors. A major cause for deterioration of the system is contamination of the injection port and the first part of the column by non-volatile deposits that may catalyse thermal degradation or give rise to adsorption. Thermal degradation and adsorption may reduce the response, produce tailing peaks or in the worst case lead to a complete loss of the residue. Therefore, test mixtures must be created from standard solutions containing stable internal standards to check for thermodegradation and adsorption. The best means of control is to analyse complex mixtures containing pesticides of all kinds of chemical structures. Two examples are given in Fig. 9.13 and 9.14. They show two test mixtures MIXLEG and MIXLER, each containing about 30 pesticides that are properly separated on the SE-54 column used for screening analysis as well as on the OV-17 column used for confirmatory analysis. An important step for increasing the validity of a screening analysis is fractionation within the clean-up procedure according to polarity. The application of a small silica gel column has proved to be very efficient. The pesticides are eluted with solvents of growing polarity and separated from each other as well as from coextractants from the food matrix. This way of enhancing the analytical validity, however, is time-consuming and occupies the gas chromatograph for at least 3 h for one food sample, but it reduces on the other hand the analyst's occupation with evaluating the chromatograms, and many confirmatory analyses appear unnecessary because of the

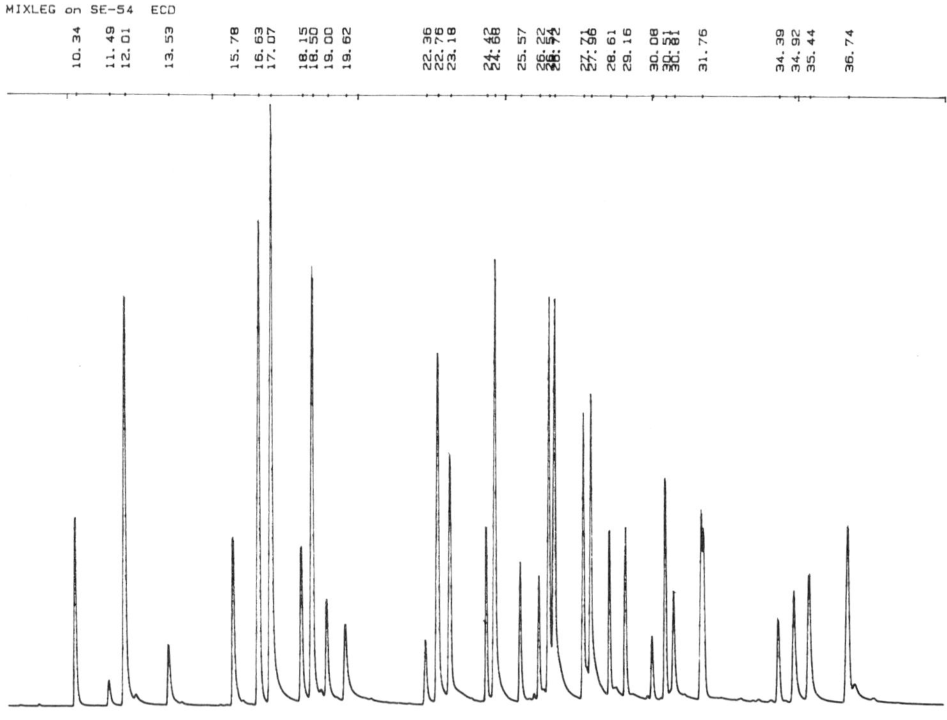

Fig. 9.13 — Chromatogram of the mixture MIXLEG containing pesticides eluting in the fraction 1–2 from the silica-gel minicolumn (DFG Method S 19) — ECD. Fused silica column 50 m×0.32 mm coated with SE-54 'bonded phase', 0.25 μm film thickness (Nordion) Hot splitless injection at 100°C; carrier gas: hydrogen. Temperature program: 100°C for 1 min; 30°C/min to 150°C, held for 2 min, 3°C/min to 205°C, 10°C/min to 260°C, held for 20 min. All pesticides are dissolved in toluene at 2.5 mg/ml. Injected volume: 1 μl.

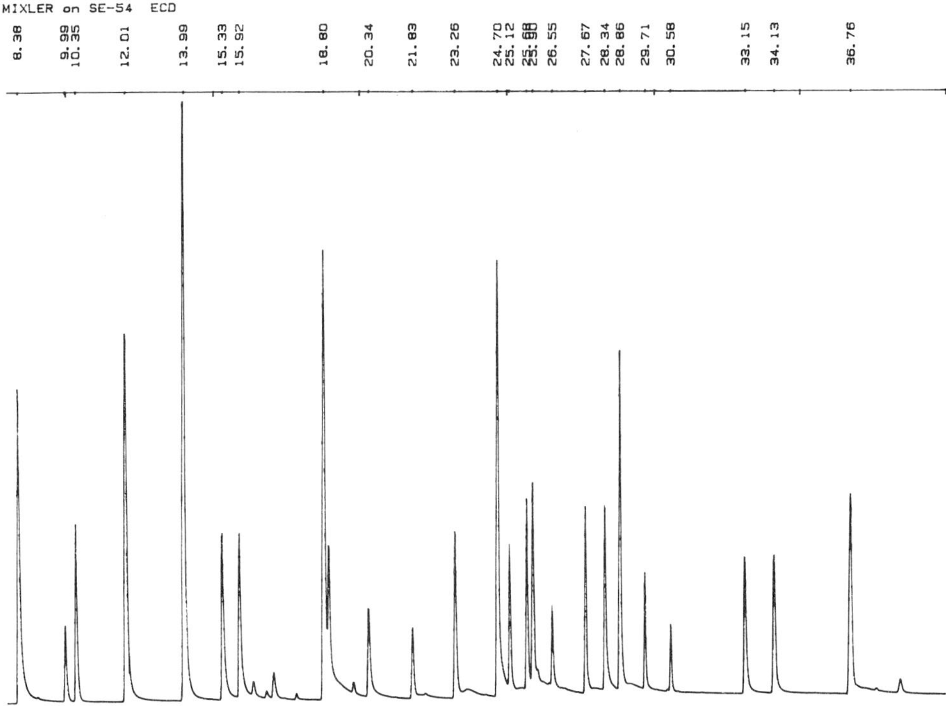

Fig. 9.14 — Chromatogram of the test mixture MIXLER containing pesticides eluting in the fraction 1–2 from the silica-gel minicolumn (DFG Method S 19) — ECD. Experimental conditions see Fig. 9.13.

clear screening result. The chromatograms of the two calibration mixtures MIXLEG and MIXLER that are presented in Figs. 9.13 and 9.14 show the three internal standards, aldrin, pentachlorobenzene and mirex and about 60 pesticides that are found in fraction 1–2 of silica-gel mini column separation. Although the majority of the pesticides in these mixtures are detected with ECD a few are only detected with NPD or FPD and cannot be found in the chromatograms shown. A compilation of all pesticides that are eluted in fraction 1–2 is presented in Table 9.7, with the pesticides being indicated that are included in the two calibration mixtures.

9.9.5 Recognition of matrix compounds

An estimated 100 different food species may be analysed in a pesticide residue laboratory. These foods contain a variety of coextractants that may be seen as peaks in the selective detectors when present in appropriate concentrations. It is good practice to keep them in a list for cross-checking or to file entire chromatrograms of food samples found to be non-contaminated. This can be performed conveniently with a chromatography data system connected to the integrator output (Stan &

Table 9.7 — Retention times of pesticides on a 50 m SE-54 capillary column. Pesticides eluting in fraction 1–2 from silicagel minicolumn. Detector response: ★ = positive, < = weak. Recovery in the fraction: #=25%; ##=50%; ###=75%; ####=90%; #####=100%

Pesticide	Silica-gel minicolumn							
	RT	1–2	3–4	5	ECD	NPD	FPD	MIX
Aldrin	24.70	####			★			LEG+R
Chlorthiamid	8.03	#		###	★	<		
Dichlobenil	8.38	####			★	★		LER
Chlormephos	9.99	####			★	★	★	LER
Nitrapyrin	10.34	####			★	★		LEG
Etridiazole	10.35	###			★	★		LER
Chloroneb	11.49	####			★			LEG
Pentachlorobenzene	12.00	####			★			LEG+R
Chlorfenprop methyl	13.53	####			★			LEG
Tecnazene	13.99	####			★	<		LER
Chlorpropham	15.25	#	###		★	★		
Ethalfluralin	15.33	####			★	★		LER
Trifluralin	15.78	####			★	★		LEG
Benfluralin	15.91	####			★	★		LER
Sulfotepp	16.07	###	#			★	★	LEG
Phorate	16.38	#			★	★	★	LER
α-HCH	16.63	####			★			LEG
Diallate	16.84	###	#		★			LER
HCB	17.07	####			★			LEG
Dicloran	17.25	####			★	<		
β-HCH	18.15	####			★			LEG
Lindane	18.50	####			★			LEG
Quintozene	18.80	####			★	<		LER
Profluralin	18.98	####			★	★		LER
Fonofos	19.01	###	#		★	★	★	LEG
Dichlone	19.62	###			★			LEG
Disulfoton	19.81	#		##	★	★	★	LER
γ-HCH	20.05	####			★			
Chlorothalonil	20.26	####			★	<		
Triallat	20.34	####			★	★		LER
Dichlofenthion	21.83	####			★	★	★	LER
Parathion methyl	22.36	###	#		★	★	★	LEG
Chloropyrifos methyl	22.38	####			★	★	★	
Vinclozolin	22.40	###	#		★	<		
Hepatchlor	22.76	####			★			LEG
Tridiphane	23.18	####			★			LEG
Fenchlorvos	23.26	#####			★	★	★	LER
Dinosebacetat	23.46	####			★	<		
Fenitrothion	24.11	###			★	★	★	
Dichlofluanid	24.42	##			★	★		LEG
Aldrin	24.70	####			★			LEG+R
Fenthion	25.04	##				★	★	
Chlorpyrifos	25.12	####			★	★	★	LER
Parathion	25.12	###		#	★	★	★	
Dicofol	25.21	####			★			
Chlorthal dimethyl	25.44	####			★			
Nitrothal	25.50	#		###	★	★		
Chlorthion	25.57	####			★	★	★	LEG
Fenson	25.58	####			★			

Table 9.7 — (Continued)

Pesticide	Silica-gel minicolumn							
	RT	1–2	3–4	5	ECD	NPD	FPD	MIX
Trichloronate	25.69	#####			★	★	★	LER
Bromophos	25.89	####			★	★	★	LER
Isopropalin	26.53	####			★	★		LEG
Heptachlorepoxide-c	26.53	####			★			LEG
Pendimethalin	26.54	#####			★			LER
Chlozolinate	26.67	####	#		★			
t-Hepatachlorepoxide	26.72	####			★			LEG
Tolylfluanid	26.96	##	##		★	★		
Phenthoate	27.02	#	###		★	★	★	
Folpet	27.13	##	#		★			
Quinomethionate	27.40	#	###		★	<		
Bromophos ethyl	27.67	#####			★	★	★	LER
o,p-DDE	27.70	####			★			LEG
Endosulfan-I	27.95	####			★			LEG
Chlorfenson	28.34	#####			★			LER
Iodofenphos	28.57	####			★	★	★	
Prothiophos	28.62	####			★	★	★	LEG
DDE	28.86	####			★			LER
Dieldrin	29.16	####			★			LEG
Binapacryl	29.71	####			★			LER
Nitrogen	29.76	#####			★			
o,p-DDD	29.76	####			★			
Chlorthiophos-I	29.96	##			★	★	★	
Perthane	30.00	####			★			
Endosulfan-II	30.07	####			★			LEG
DDD	30.51	####			★			LEG
Ethion	30.58	####			★	★	★	LER
o,p-DDT	30.81	####			★			LEG
Tetrasul	31.26	####			★			
Fenazaflor	31.30	###			★			
Endosulgansulfat	31.76	####			★			LEG
Carbophenothion	31.94	###			★	★	★	
Cyanofenphos	32.15	##	###		★	★	★	
DDT	32.52	####			★			
Nitralin	33 14	#	####		★	<		LER
EPN	34.14	####			★	★	★	LER
Methoxychlor	34.38	####			★			LEG
Bifenox	34.92	##	##		★			LEG
Tetradifon	35.44	####			★			LEG
Phenkapton	35.89	####			★	★	★	
Mirex	36.75	####			★			LEG+R
Dialifos	38.49	##	##		★	★	★	LER
Permethrin	44.17	####			★			

Lipinski, 1985). In Fig. 9.15 two electron-capture chromatograms from strawberries from Italy are compared, one of them representing a non-contaminated sample, the other containing two additional peaks (see arrows in Fig. 9.15). One of these peaks was confirmed as a contamination with vinclozolin (crossed arrow); the other's nature could not be identified.

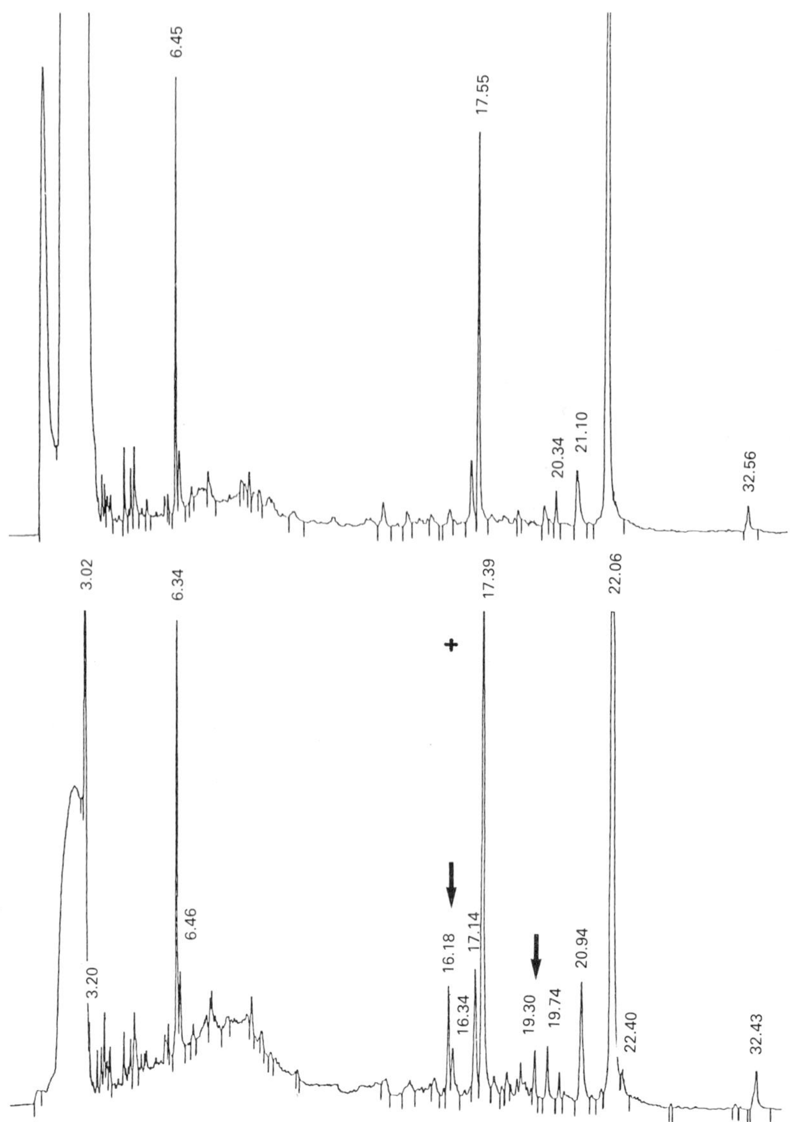

Fig. 9.15 — Electron-capture chromatograms of strawberries. Top: reference sample: bottom, actual sample with activated cursor (cross) and two peaks (arrows) from pesticides (?), one of them (cross) was confirmed of being vinclozolin, the other was unknown. Peak at 17.55 min. is aldrin. Source: Stan & Lipinski (1985) (courtesy Elsevier, Amsterdam).

9.10 CONFIRMATORY ANALYSIS

9.10.1 Confirmation by applying columns of different polarity

A tentative identification of a pesticide made by the screening analysis must be confirmed or rejected by alternative procedures. The first method most commonly used for confirming the identity of a pesticide peak is the simple comparison of the chromatogram of the food sample with that of the same sample spiked with the pesticide standard in the same concentration. If the gas chromatographic peak appears not to be identical in shape and of doubled area the confirmation procedures must be continued. On the other hand, if two peaks appear the screening result must be rejected. The validity of this type of confirmation depends very much on the resolution that can be achieved with the gas chromatographic column used for screening. This fact advocates strongly the use of long capillary columns. Confirmatory analysis is performed applying the same procedure with a column of different polarity. An identical retention time of test compound and unknown peak on two capillary columns of differing polarities constitutes strong evidence that the compounds are identical. When using shorter or packed columns retention times should be identical at least on three different columns. The analyst has to check the retention time tables of the selected columns to ensure that the different columns separate compounds forming critical pairs on the screening column.

Recently, we finished a compilation of retention data of about 270 pesticides, including a few metabolites recorded on three different capillary columns. Only a few critical pairs were found on a 50-m SE-54 column used for screening that cannot be separated on a 50-m OV-17 column used for confirmatory analysis. When defining a critical pair as being two pesticides eluting in a retention time window of 0.1 min only six of such critical pairs were observed. Three of them are separated in the clean-up procedure S19 and one can easily be distinguished by means of selective response to ECD (pentachlorophenol) and NPD (prometon). Only dicofol and its metabolite 4,4′-dichlorobenzophenon as well as allethrine and *cis*-heptachlor epoxide cannot be confirmed by using this combination of capillary columns (Lipinski & Stan, 1989).

An important step in the confirmatory analysis already referred to is the consequent application of effluent splitting to two selective detectors because many pesticides respond to more than one detector. The ratio of the detector signals contributes much to the confirmation of the identity of two compounds.

9.10.2 Confirmation using derivatization

An independent method for the identification of pesticide residues is the use of chemical derivatization. An excellent review is given by Cochrane (1975). A few instructive examples to illustrate the technique are given in Fig. 9.16. The general approach with compounds containing ester- or amide-bonds is hydrolysis followed by derviatization of the arising phenol, alcohol or amine moiety. All compounds containing free hydroxyl or amino groups can be derivatized in the same way. The trend is towards the use of perfluorinated reagents that form trifluoroacetyl, hepatafluorobutyryl or pentafluoroaryl derivatives, resulting in excellent sensitivity for electron-capture detection. Organophosphates containing the thiono group are oxidized to the corresponding oxons by using neutralized hypochlorite (Singh & Lapointe, 1974).

The most convincing confirmatory procedure, however, is the identification of a gas chromatographic peak by means of its mass spectrum (see Fig. 9.22).

9.11 TWO-DIMENSIONAL GAS CHROMATOGRAPHY

Another approach for increasing information about the identity of a possible pesticide peak is the application of two-dimensional GC. This technique was developed as early as 1968 by Deans (1968), who introduced pneumatic switching of two columns. For a long time its practical application was restricted to monitoring

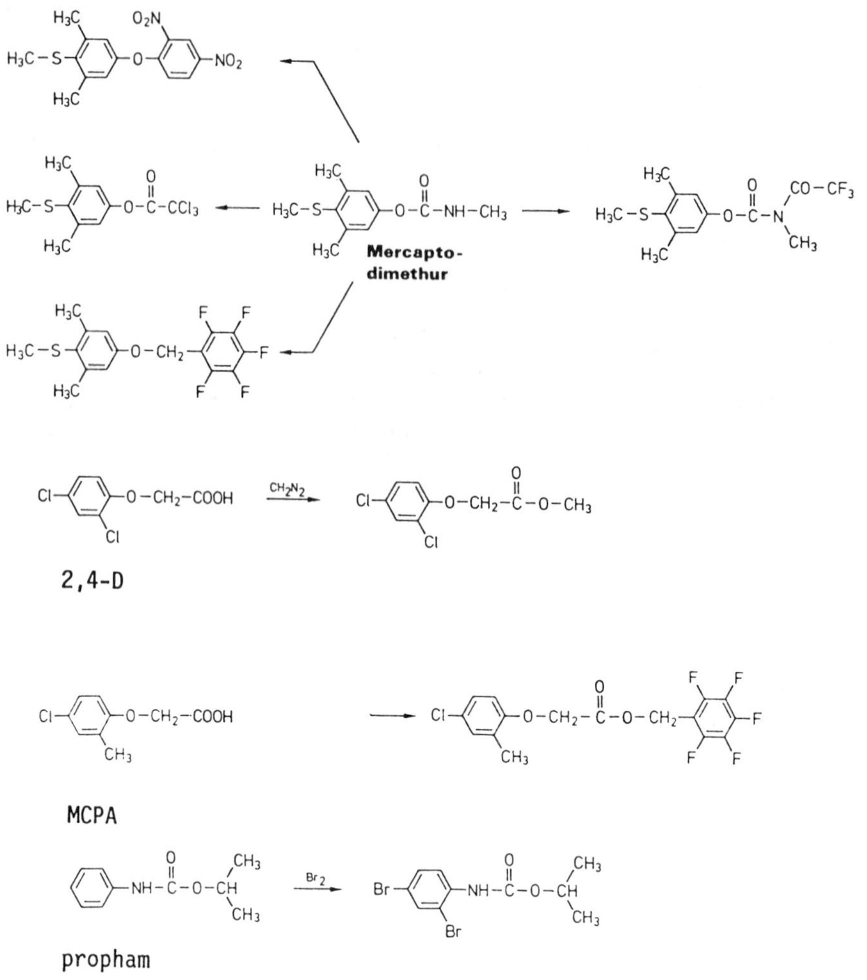

Mercapto-dimethur

2,4-D

MCPA

propham

Fig. 9.16 — Confirmation by chemical reactions producing derivatives suited for gas chromatrography.

process streams in industry. Only a few groups of chromatographers have used two-dimensional GC for the determination of complex mixtures (Schomburg *et al.*, 1975). A review of the technique and its application was given by Bertsch (1978). The first application to pesticide residue aanlysis was reported by Stan and Mrowetz (1983a,b).

9.11.1 Basic instrumentation and method

An instrument commercially available is the gas chromatograph Sichromat 2 (Siemens, Berlin, F.R.G.). It consists of two separately heated oven units and a special device for executing the pneumatic switching that permits column switching with immediate detection of the selected GC fraction. This technique, called 'live switching', is depicted in Fig. 9.17.

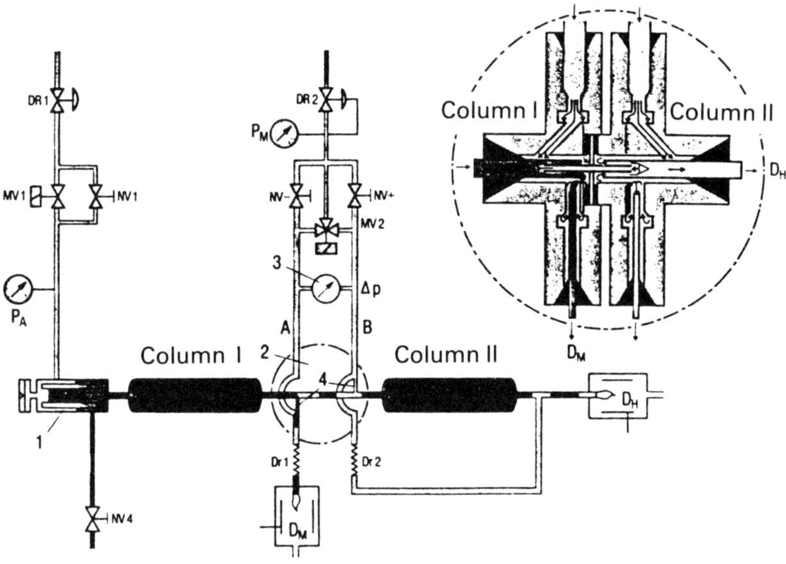

Fig. 9.17 — Two-dimensional gas chromatography applying 'live'-switching (courtesy Siemens, Berlin).

The original design is equipped with two detectors of the same type, mostly FIDs or ECDs. They serve different tasks, which can easily be understood when looking at the operation modes of the system. Pneumatic switching works with slight pressure differences between both ends of the T-piece that are generated by means of two make-up gas lines (A, B in Fig. 9.17). The gas flows through these make-up gas lines are adjusted with the two needle valves, NV^- and NV^+, and the pressure difference is indicated by the manometer. Any gas flow from one end to the other must pass through the tiny platinum–iridium capillary mounted in the centre and inserted loosely into the two capillary columns. The direction of flow inside the platinum–iridium capillary connecting the first and second column can be changed at will by simply opening and closing the external solenoid valve, MV2. By using a higher pressure in line B than in line A the effluent from column 1 is directed to the monitor detector (D_M) resulting in a conventional chromatogram. As long as the pressure at A exceeds that at B, the effluent from column 1 enters column 2 and is monitored with the main detector, D_M, a process usually called 'heart cut'. A third mode of operation is 'back-flush', which reverses the flow direction in the first column and can be performed by closing MV1.

9.11.2 Modified equipment for use in pesticide residue analysis

Application of the technique described to pesticide multiresidue analysis requires several modifications of the system. These are demonstrated in Fig. 9.18. The main

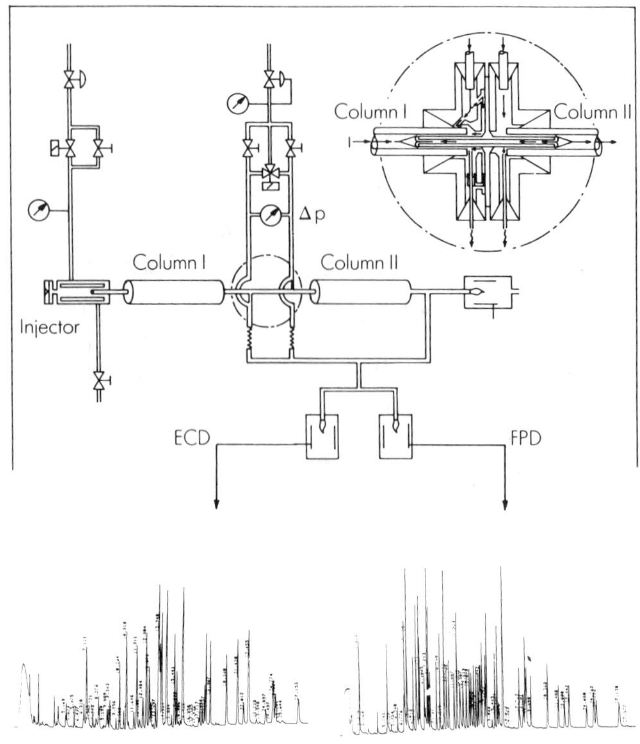

Fig. 9.18 — Two-dimensional gas chromatography with effluent splitting to two selective detectors.

differences from a conventional system are effluent splitting to two selective detectors and directing both the effluents from the first and the second column to the same parallel detectors. This arrangement results in a significant reduction of cost for the equipment. Only two detectors and two integrator channels are required to monitor and process data from two columns. While with the conventional system a peak recorded in one of the detectors indicates a compound eluted from the corresponding column, chromatograms with the modified system can be interpreted only when knowing the operating parameters at the switching device. After cutting a selected fraction from the first column into the second, back-flush of the first column is required in order to prevent interference of peaks eluted from the second with more retarded compounds from the first column. The situation is best demonstrated with a test mixture of pesticides. In Fig. 9.19 the separation of 51 CPs on a SP-2100 fused silica capillary column (17 m×0.2 mm i.d., Hewlett Packard) applying splitless injection into the 'cold' column at 100°C and temperature programming is demonstrated.

Obviously, there are a number of critical pairs within this mixture that cannot be resolved properly on this short capillary column. Integration results in 36 pesticide

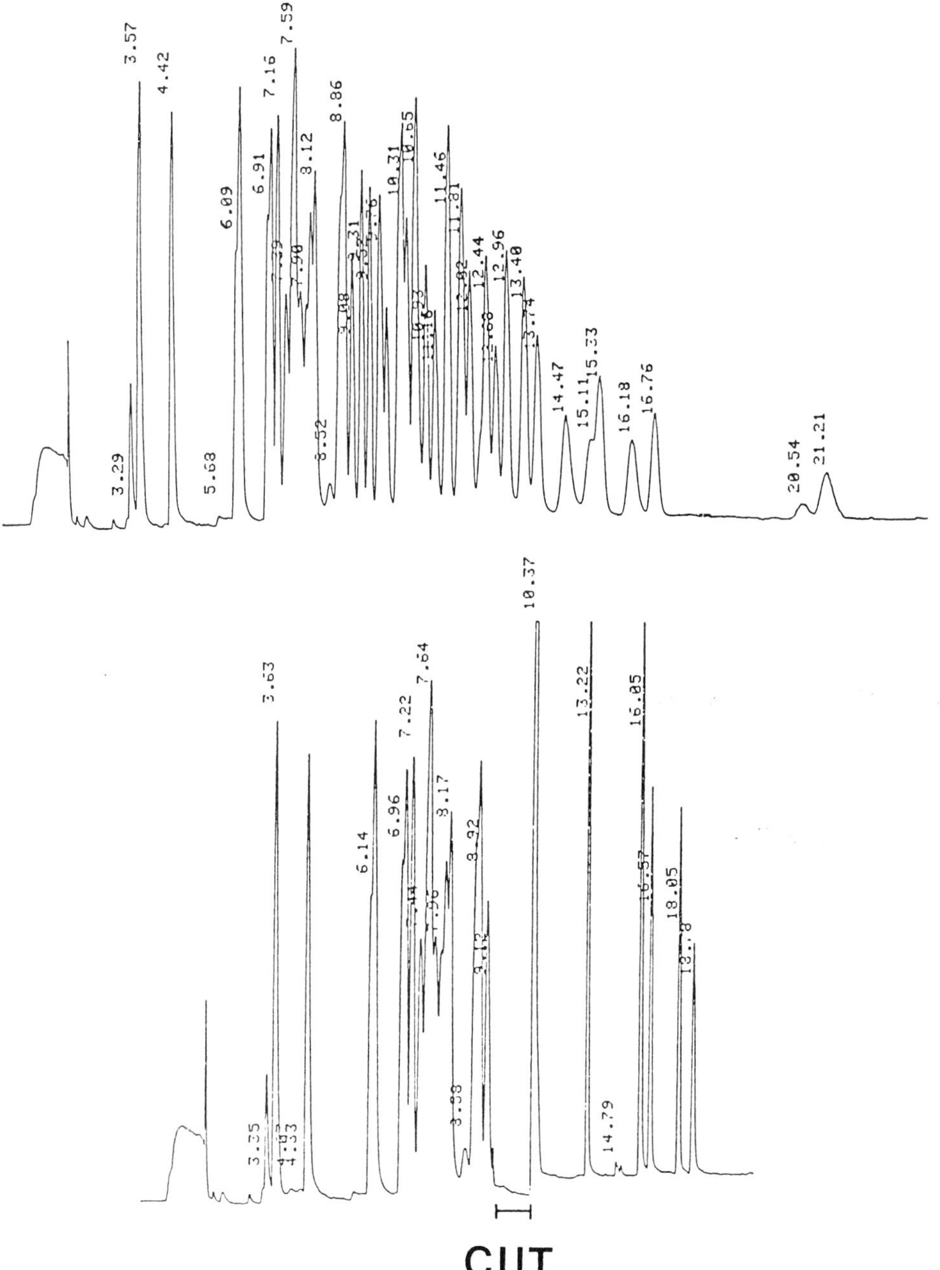

Fig. 9.19 — Demonstration of cut with a mixture of 51 chlorinated pesticides using ECD. Top: 51 CP result in 36 peaks; bottom: cut between 9.20 and 10.20 min; back-flush at 10.30 min. First column: 17 m×0.2 mm fused-silica SP 2100 (Hewlett-Packard); second column 20 m×0.32 glass OV-225 (homemade) Hot splitness injection, carrier gas: helium. Temperature program starting at 100°C. Result: aldrin at 13.22, chlorthal-dimethyl at 16.05, nitrothal-isopropyl at 16.57, dichlorfluanid at 18.05, and triadimefon at 18.79 min. Source: Stan & Mrowetz (1983a) (courtesy Elsevier, Amsterdam).

peaks (Fig. 9.19, top). With heart cut in the retention time window between 9.20 and 10.20 min followed by back-flush, a chromatogram resulted as shown in Fig. 9.19 (lower part). Within the selected time of 9.19 min five pesticides were transferred to the second column and separated properly. The huge peak following immediately the end of the cut results from eluting peaks that cannot be back-flushed because of a certain time delay in reversing the flow direction. The other five peaks were identified as aldrin used as internal standard (13.22 min), chlorthal-dimethyl (16.05), nitrothal-isopropyl (16.57), dichlofluanid (18.05) and tridimefon (18.79). How close these five pesticides were eluted can be learned from Table 9.8 where the relevant part of a list of relative retention times after the first and after both columns is extracted.

Table 9.8 — RRTs of chlorinated pesticides in two-dimensional gas chromatography

No.	Pesticide	Test ng/μl	First column RRT	Second column RRT
18.	Vinclozolin	1	0.922	1.188
19.	Heptachlor	1	0.931	0.948
20.	Dinoseb acetate	2	0.951	n.d.
21.	Dichlofluanid	2	0.975	1.369[a]
22.	Aldrin	1	1.000	1.000
23.	Chlorthal-dimethyl	1	1.025	1.217[a]
24.	Triadimefon	1	1.022	1.438[a]
25.	Nitrothal-isopropyl	3	1.041	1.257[a]
26.	Captan	2	1.077	1.940
27.	Hepatochlor epoxide	1	1.085	1.283
28.	Folpet	2	1.093	1.811
29.	Procymidone	5	1.113	1.873
30.	Chinomethionat	1	1.113	1.502
31.	Chlorbenside	1	1.114	1.435
32.	Endosulfane	1	1.146	1.368

[a]Pesticides transferred from the first column to the second column with CUT in Fig. 9.19.
n.d. = not detectable.

9.11.3 Application to a real-life sample

The application of the two-dimensional system described for confirmatory analysis is best demonstrated with a real-life sample. Zucchini were selected because they produce chromatograms with only a few matrix peaks in the ECD (Fig. 9.20a) and only one in the FPD eluting immediately after the internal standard NT with the retention time 5.21 min (Fig. 9.20b). Aldrin is recorded as the largest peak at 9.54 min on the first column; two other peaks are recognized as possible pesticides: bromophos ethyl at 10.84 and chlorfenson at 11.16 min (Fig. 9.20a). The FPD

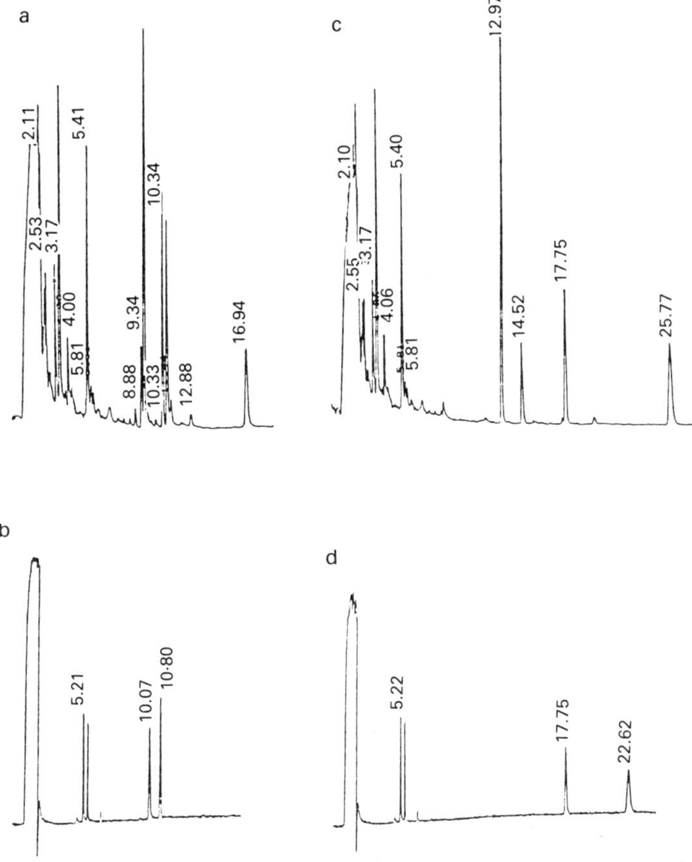

Fig. 9.20 — Identification of pesticide residues in a zucchini sample with two-dimensional GC. a: ECD signal — first column; b: FPD signal — first column; c: ECD signal — first and second column; d: FPD signal — first and second column; CUT between 8.50 and 11.70 min. Backflush at 11.75 min. Internal standards: aldrin — 9.55 min/12.97 min; NT — 5.27 min/5.22 min; PT — 10.07 min/22.62 min. Source: Stan & Mrowetz (1983a) (courtesy Elsevier, Amsterdam).

chromatogram (Fig. 9.20b) of the first column shows a peak at 10.80 min supporting the presence of bromophos ethyl in the sample. Performing the confirmatory run with the most interesting part between 8.50 and 11.70 min by transferring it to the second column results in a clear picture: after the second column, there are two peaks found in the ECD report that coincide with the two pesticides expected and that of the internal standard aldrin. The peak at 25.77 min coincides with chlorfenson and that at 17.75 min with bromophos ethyl. Bromophos ethyl is confirmed by a parallel signal in the FPD chromatogram at exactly the same time. The peak at 22.62 min shows the internal standard PT, which eluted from the first column at 10.07 min (Fig. 9.20d,b). This example illustrates both the merits of two-dimensional GC and of parallel recording with detectors of different selectivity.

9.12 DERIVATIZATION

Multiresidue methods applying a general clean-up (DFG-Method S19), or with small modifications as shown in Fig. 9.6, followed by GC determination using selective detectors are mainly suited for the detection of CPs and OPs, many fungicides and herbicides, in particular triazine compounds.

Other chemical groups of pesticides widely used as herbicides and insecticides cannot be analysed applying the multiresidue methods without employing derivatization procedures. These pesticide groups are not amenable to GC because of being free acids or thermolabile structures such as phenyl ureas and *N*-methyl carbamates. Derivatization is performed to produce thermostable volatile compounds that respond sensitively to selective detectors. Acidic herbicides are generally formulated as salts or esters which rapidly hydrolyse in soil, water and plants to the acidic form. Therefore, extraction of residues from aqueous solutions or food homogenates must be performed after acidifying the aqueous phase. The most popular derivatization procedure is forming methyl esters either by reacting with diazomethane or applying a mixture of methanol and concentrated sulphuric acid as reagent. A few compounds that are widely used in agriculture contain only one chlorine atom and are therefore not sensitively detected with the ECD. Derivatization is a means of introducing electron-capturing groups. Pentafluorobenzyl bromide is a derivatizing agent that is growing in popularity, yielding derivatives with high sensitivity in electron-capture detection. This and other reactions that are applied in order to enhance detection sensitivity with the ECD have been already presented in Fig. 9.16 as examples of chemical confirmation reactions. They include esterification of free acids with and without detector sensitivity enhancement, the bromination of aryl groups, the formation of phenyl and benzyl ethers and the acylation of the active hydrogen in *N*-methyl carbamates introducing electron-capturing groups. Phenyl urea herbicides are mostly hydrolysed with alkali to yield anilines that are acylated to anilides or brominated. In a special reaction sequence, called Sandmeyer reaction, the amino group is first transformed to a diazo group, which is finally replaced by an iodine atom.

9.13 GAS CHROMATOGRAPHY — MASS SPECTROMETRY (GC–MS)

9.13.1 GC–MS using cyclic scan

As mentioned already in 9.6.5, with higher levels of contamination pesticide residues can be detected applying cyclic scanning, as demonstrated in Fig. 9.21 with a sample of grapes. The total ion current chromatogram (TIC) contains thousands of mass spectra that can be successfully inspected and compared to those stored in a library. In the extract of grapes shown in Fig. 9.21, the peak at retention time 24.42 min resembles in chromatographic properties that of vinclozolin, a widely used fungicide. After background subtraction the mass spectrum is used as the basis for a library search with a special pesticide library (Stan & Lipinski, 1989). The search routine recognizes a good correspondence to the spectrum of vinclozolin, as shown in Fig. 9.22. A library search against the NBS-Library containing 4000 spectra confirmed this result. The quantitative determination resulted in 3.2 p.p.m. vinclozolin, which is within the maximum tolerance of 5 p.p.m. for vinclozolin in grapes in the F.R.G.

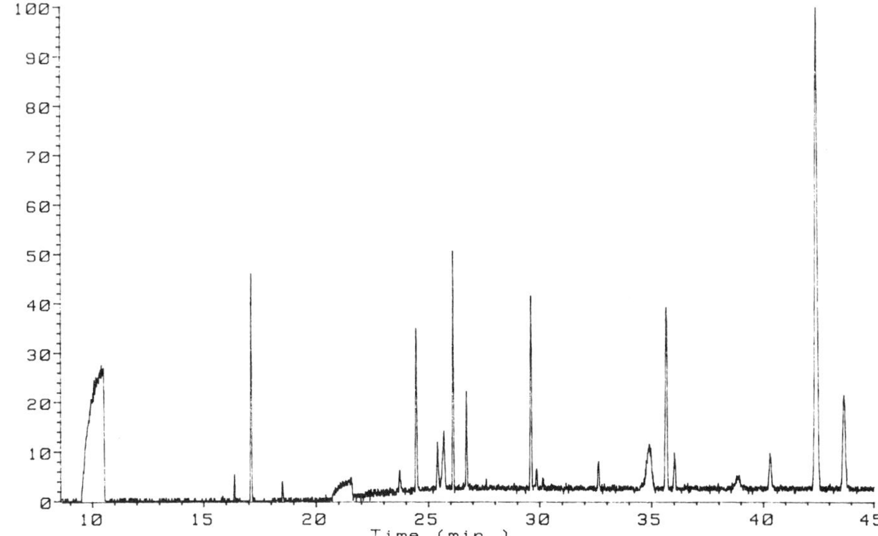

Fig. 9.21 — Total ion current (TIC) chromatogram of a GC–MS analysis of a grape sample contaminated with vinclozolin (RT= 24.42 min.).

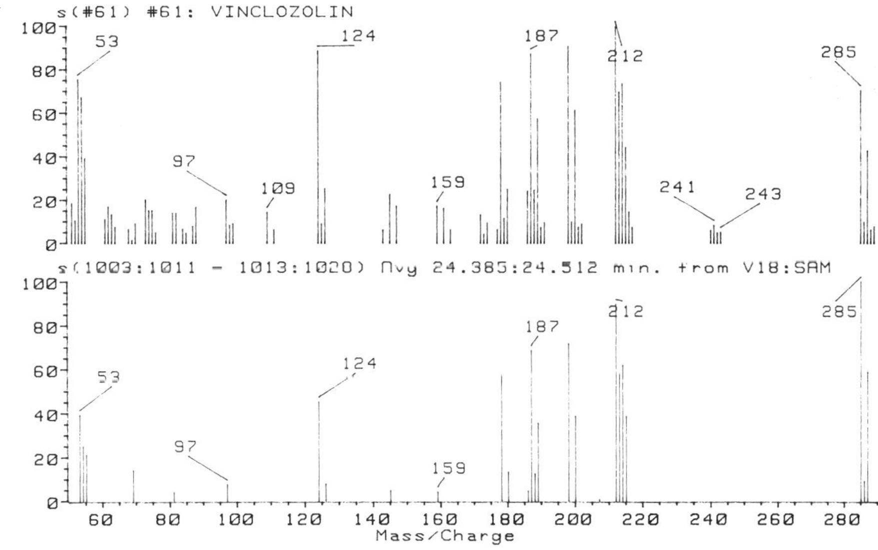

Fig. 9.22 — Result of a library search with the peak at RT = 24.42 min of the TIC shown in Fig. 9.21. Top: mass spectrum of vinclozolin from the library; bottom: corrected mass spectrum of peak at 24.42 min obtained by averaging and background subtraction. This peak was recognized as best fit with a PBM library search.

9.13.2 GC–MS using selected-ion monitoring

In a second example, ground water assessed for use as drinking water was first analysed with capillary GC and selective detectors, resulting in a tentative detection of about 0.1 p.p.m. atrazine. A standard solution containing 20 ng atrazine per microlitre was chromatographed, resulting in a chromatogram with one peak at 21.76 min. in the TIC. The chromatogram and the mass spectrum are shown in Fig. 9.23.

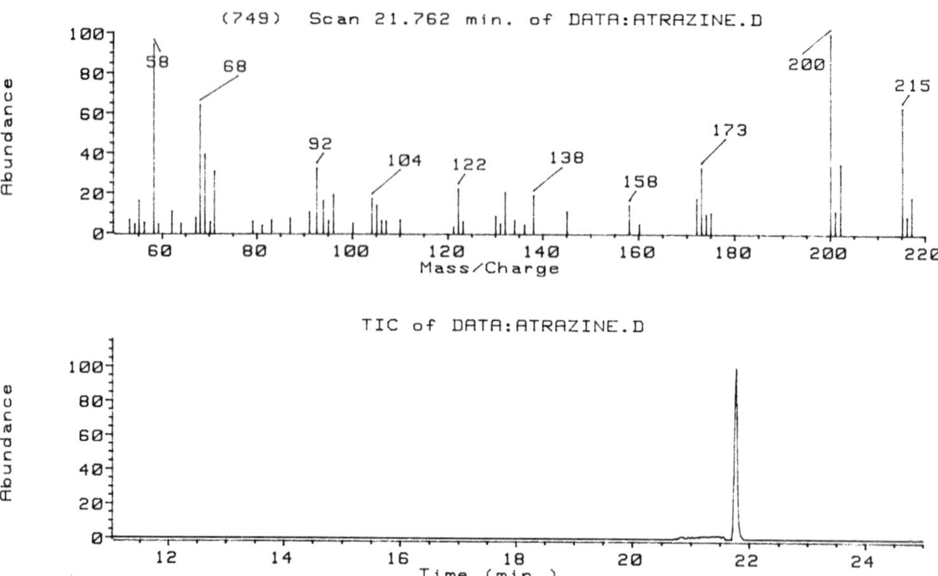

Fig. 9.23 — GC–MS of Atrazine. Top: mass spectrum of atrazine obtained from the peak at 21.76 min in the TIC chromatogram; bottom: TIC chromatogram of 20 ng atrazine recorded with cyclic scan.

Confirmatory analysis was performed using selected-ion monitoring with the ions *m/z* 200, 202, 215 and 217 that are characteristic for atrazine. The resulting mass chromatograms are given in Fig. 9.24 demonstrating all the same shape and maximum of peaks at the expected retention time. There is no other part of the chromatograms within the retention time window relevant for atrazine where the four different ion traces exhibit the same course with respect to forming a gas chromatographic peak. Quantification of the trace amount of atrazine resulted in 0.07 p.p.b., which is close to the maximum tolerance for drinking water of 0.1 p.p.b. recently introduced by the EEC. The amount injected into the GC–MSD system was 2.5 μl containing 180 pg atrazine. This example demonstrates the challenge to the analyst that arises from the new regulations for ground and drinking water.

9.14 APPLICATION OF PERSONAL COMPUTERS

The great number of pesticides in use makes the application of a data system a valuable tool in book-keeping of test substances and in evaluation of chromato-

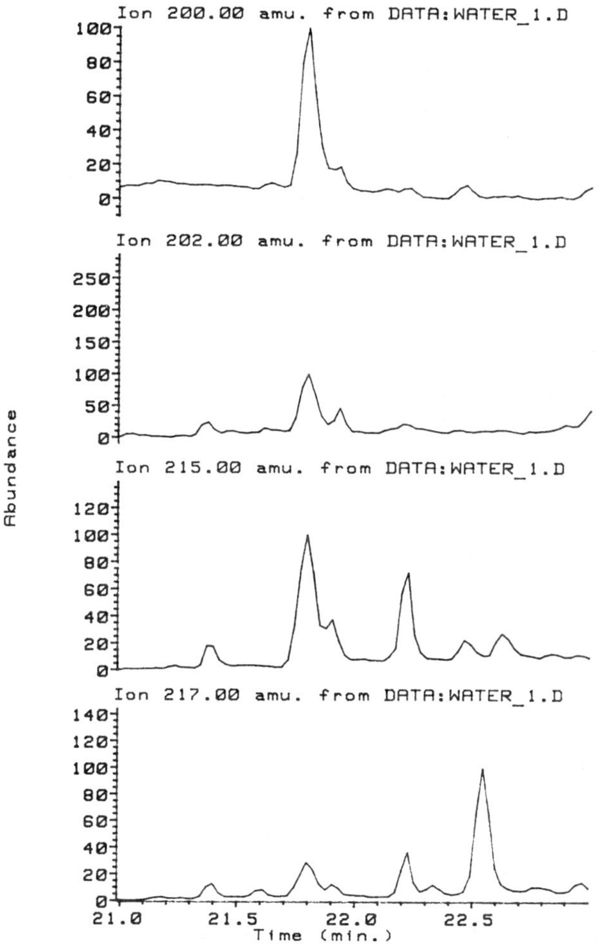

Fig. 9.24 — SIM analysis of a ground water sample confirming a contamination with atrazine (0.07 p.p.b.). For selected masses and retention time see Fig. 9.23.

grams. In a residue laboratory several hundred test substances and standard solutions, as well as test mixtures of varying concentration, are in stock and permanent use. Their handling has already been described. Computers are also valuable in archiving the literature and data about the individual compounds, their analytical data and occurrence in food commodities and the environment.

A major contribution can be made to chromatogram evaluation by applying appropriate software.

9.14.1 Computer aided pesticide analysis (CAPA)

A special computer program for automated pesticide screening analysis has been published recently by Lipinski and Stan (1988). This program CAPA (Computer

Aided Pesticide Analysis) is written in Turbo-Pascal and works under MS-DOS with all IBM-compatible personal computers equipped with 640-kilobyte RAM and a hard disk. CAPA consists of three major programs: EDITOR, INTERPRET and AUTOINTERPRET. EDITOR is used for creating and upgrading a relational database that contains all chromatographic data about the calibrated pesticides and documented matrix compounds from food samples. This database forms the foundation for the use of the other two programs. INTERPRET is designed for interactive evaluation of gas chromatograms obtained by standardized analytical procedures. A special feature is the parallel display of chromatograms and integrated data on two monitors. On both monitors the screens are divided into three windows. On the graphical screen the upper two windows are usually applied for displaying two chromatograms that were recorded simultaneously with two selected detectors from the same sample, while the bottom window is used for a corresponding matrix chromatogram. On the other screen at the same time, parts of the report table of the actual sample are displayed in the upper window. In the other two windows, the corresponding parts of the pesticide catalogue and the matrix peak table are displayed. After having called for an actual sample the program loads automatically the corresponding matrix chromatogram and the main calibration table. Evaluation of chromatograms is performed by parallel scrolling through the three tables with the sample table as leader. This happens in the following way: the cursor is set to any peak in the sample window and immediately the middle window presents the corresponding part of the main catalogue. This window contains four pesticides resembling in retention time the marked peak with the 'best fitting' pesticide highlighted. At the same time the peak corresponding closest in the matrix table is also highlighted. The computer calculates automatically from the area value of the marked peak in the sample window the concentration for the four most corresponding pesticides. The result is estimated on the assumption that if the unknown peak would be the corresponding pesticide it would have the concentration as displayed. The same scrolling can be executed by moving the cursor on the monitor displaying the chromatograms. When the operator considers a peak during screening as positive he transfers the result to a report file that is printed after ending the session. There are many other features of the program CAPA that cannot be described here. AUTOINTERPRET is the automated version of INTERPRET. After evaluating the screening analysis a report is prepared about the possible contaminants in the food sample.

9.14.2 Integrators programmable with BASIC

A recent review of the application of computers for the evaluation of gas chromatographic data in pesticide residue analysis was given by Stan (1988b). In laboratories without direct access to a large data system, retention times and areas of peaks are generally calculated by means of integrators directly connected to the detector signal outputs. These integrators can be used with external or internal standardization for calibrating the specified method. The great number of compounds detected in multiresidue analysis, however, cannot be resolved totally in one gas chromatographic run. Therefore, calibration tables must be compiled by analysing a series of test mixtures and may be designed individually according to the analyst's preferences. Several newer integrators offer the possibility of processing the chromatographic data by means of user-made programs written in a special BASIC. The HP–3993 A BASIC includes many integrator commands that allow direct access to integrator results from the BASIC program. This capability makes programming very effective and easy to learn (Stan, 1989).

9.14.3 Application of spreadsheet software

Calculation programs or spreadsheets have become widespread in the world of business in recent years. They are utilized in the preparation of calculation tables, preparing business reports and graphics, for modelling, and as small database managers. A number of spreadsheets have become commercially available for computers running under MS-DOS. LOTUS 1-2-3 (Lotus Development Corp.) and MULTIPLAN (Microsoft, Corp.) are the market leaders in the United States and in Europe, respectively. Other programs are QUATTRO (Borland) and, not to forget the first spreadsheet program of all, VisiCalc (VisiCorp). This last was the program that demonstrated the potential of this kind of software. Spreadsheets now are also becoming increasingly popular in analytical laboratories (Ouchi, 1987).

The various products follow the same fundamental concepts: an electronic worksheet that produces a large grid of cells ordered in rows and columns. These cells can be considered as boxes that can be filled with information in the form of text, numbers, or formulas. The program remembers the relationship between the cells on the worksheet and is capable of automatically performing calculations such as means and standard deviations from multiple measurements of recalculation of calibration data. A few suggestions are given on how to apply spreadsheets to the problems in a pesticide laboratory.

9.14.4 Spreadsheets used as database

A spreadsheet can be used to create a database for supervising pesticide standards or for compiling useful information about pesticides in a condensed format. In our laboratory about 600 compounds are compiled in a spreadsheet. Each row contains the information on one compound. In the first column we write the pesticide's name; in the other columns are abbreviations indicating biological activity and application in agriculture, the type of chemical references about detection procedures, MS etc. In Fig. 9.25 an extract from a reference table with respect to the detectiom limits reported in water analysis is given as an example of such a database. A feature that makes the spreadsheet very attractive is the SORT function. It enables sorting of columns containing numbers or letters in increasing or decreasing order. To give an example, it is easy to collect all pesticides of which mass spectra obtained with negative chemical ionization are documented or which are included in a specified multiresidue method for drinking water. Such sorted lists can be copied and printed or stored with another file name. Once created, a database can be modified by first copying and then deleting, adding or changing information.

9.14.5 Multiplan aided pesticide analysis (MAPA)

Spreadsheets can be understood as a kind of computer language that can be used to create special application programs. They have all the mathematical functions of a high-level language such as FORTRAN, PASCAL or BASIC and many commands that can be combined with keystrokes by so-called macros.

-1	1	2	3	4	5	6	7	8	9	10
1	Analysis of Pesticide Residues in Water by Gas Chromatography (ppb)									
2	------------------------------------									
3	Pesticide	References in review of J.Wilkens (7/88)								
4		27	6	12	13	14	15	23	24	26
5	------------------------------------									
6	Aldrin	0,050			0,02					
7	Ametryn	0,2		0,025						0,1
8	Atraton			0,025						
9	Atrazin	0,1		0,025	0,01	0,015	0,1	0,01	0,001	0,1
10	Azinphos-Methyl	0,3								
11	Aziprotryn	0,2								
12	Barban						0,1			
13	Benefin	0,3								
14	Benthiocarb	0,3								
15	Benzoylprop-Ethyl						0,025			
16	Bromophos	0,15								
17	Captafol	2								
18	Captan	2								

Fig. 9.25 — Using a spreadsheet as a database. Extract of a pesticide list sorted alphabetically.

Macros were originally designed to automate spreadsheet creation by allowing users to store keystrokes and then play them back with a single command. They have evolved into a programming language with the capability to perform loops, subroutines, jumps and conditional statements.

This language was used to create MAPA (Multiplan Aided Pesticide Analysis), a program for the evaluation of GC runs in pesticide residue analysis. It is particularly useful with screening analyses that contain a number of peaks produced from the food matrix as well as by pesticides or other environmental contaminants (Stan, 1988b). This application program is designed for use off-line. The analyst takes the chromatograms from the integrator to the personal computer and works in the same way as he is used to at his laboratory desk. Instead of searching retention times in lists and tables, he follows the menu-driven computer program. Inspection of the chromatograms is performed by scrolling through the catalogues displayed on the monitor, and each peak that coincides with the retention time of a pesticide is marked with the cursor. After having typed the area of the suspect peak into the corresponding cell, the estimated concentration of the pesticide is shown. This estimate can be copied to a results table, which is printed at the end of a session as part of the book-keeping in the laboratory. One example is given in Figs. 9.26 and 9.27. MAPA comes to the monitor with the results table of the last sample (Fig. 9.26). In the headlines the gas chromatographic method is indicated and in rows 4 and 5 the macro commands available. With the first keystroke 'ALT' st, the new evaluation procedure is started by setting the new date and demanding the input of the new sample number in cell r5c10 (old value: 88732), the actual retention time (old value 24.97) and the area of the internal standard A (old value 1904). With the next macro 'ALT' in, the result table is cleared and the screen split into a fixed header and a scrolling table. In Fig. 9.27, the screen shows the cursor at row 75 in column 10 because there appears a peak at 22.68 min in the chromatogram that may represent a chlorpyrifos-methyl residue of about 173 p.p.b. This estimate was calculated with the calibration value hidden in a cell outside the screen from the value of the peak area entered in cell r75c10. With another macro the whole row 75 is copied to the results table and the screening can be continued. After having evaluated all screening analyses the analyst returns with the printed reports to the laboratory and starts the necessary confirmatory analyses. The worksheet in Fig. 9.27 contains further information to support the analyst: column 2 — absolute retention times recalculated with that of the internal standard: columns 3 to 5 — occurrence in the clean-up fraction of the silica-gel mini column; columns 7 to 9 — detector response. Other data are not visible because of being in cells of column 12 and following. The application program MAPA can be copied to create a worksheet for other gas chromatographic conditions.

9.15 CONCLUSIONS

Screening analysis using capillary columns of about 50 m length in combination with effluent splitting to two selective detectors — ECD and NPD — provides the necessary body of information for the decision of accepting a food sample as not being contaminated with one of about 270 pesticides that can be analysed directly with GC. Confirmatory analysis can be performed on a second capillary column installed in the same instrument when applying the described effluent splitters.

Two-dimensional GC results in more reliable information when analysing samples containing many coextractants responding to the selective detectors, because overlapping in the chromatograms of the second column is minimized by cutting small fractions from the first column.

Personal computers with spreadsheets or special software are valuable tools for the support of the analyst's daily routine work at an attractive cost/benefit ratio.

Unambiguous identification of any pesticide residue in food or environmental samples can best be provided by means of coupling a high performance capillary column to a mass spectrometer or a mass-selective detector.

REFERENCES

Ambrus, A., Visi, E., Zaker, F., Hargitai, E., Szabo, L. and Papa, A. (1981a). General method for determination of pesticides in samples of plant origin, soil, and water. III. Gas chromatographic analysis and confirmation, *J. Assoc. Off. Anal. Chem.*, **64**, 749–768.

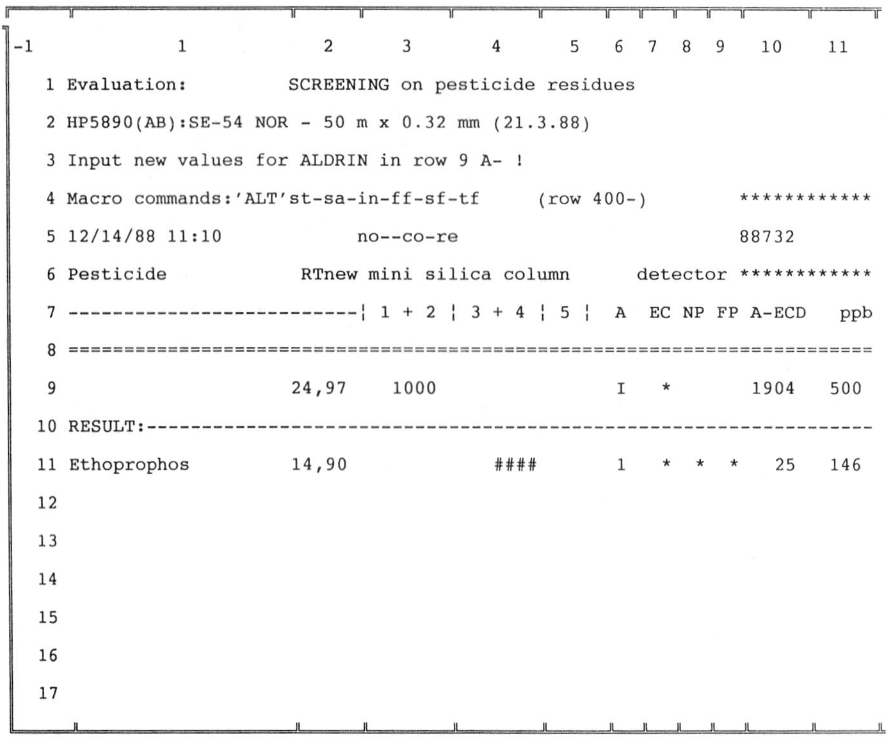

```
-1              1          2     3      4     5   6  7  8  9  10   11
 1 Evaluation:        SCREENING on pesticide residues
 2 HP5890(AB):SE-54 NOR - 50 m x 0.32 mm (21.3.88)
 3 Input new values for ALDRIN in row 9 A- !
 4 Macro commands:'ALT'st-sa-in-ff-sf-tf    (row 400-)        ************
 5 12/14/88 11:10           no--co-re                         88732
 6 Pesticide              RTnew mini silica column    detector ************
 7 -------------------------¦ 1 + 2 ¦ 3 + 4 ¦ 5 ¦  A  EC NP FP A-ECD   ppb
 8 ========================================================================
 9                        24,97   1000                I   *        1904   500
10 RESULT:-----------------------------------------------------------------
11 Ethoprophos            14,90            ####       1   *  *  *   25   146
12
13
14
15
16
17
```

Fig. 9.26 — MAPA: result table of sample 88 732 indicating a possible residue of ethoprophos with an estimate of 146 p.p.b.

Ambrus, A., Lantos, J., Visi, E., Csatlós, I., Sávári, L. (1981b). General method of determination of pesticide residues in samples of plant origin, soil, and water. I. Extraction and cleanup, *J. Assoc. Off. Anal. Chem.*, **64**, 733–742.

Aue, W. A., Hastings, C. R. and Kapila, S. (1973). On the unexpected behaviour of a common gas chromatographic phase, *J. Chromatogr.*, **77**, 299–307.

Ballschmiter, K., Buchert, H., Bihler, S. and Zell, M. (1981). Baseline studies of global pollution — IV. The pattern of pollution by organo-chlorine compounds in the North Atlantic as accumulated by fish, *Fresenius Z. Anal. Chem.*, **306**, 323–339.

Beck, H., Eckhart, K., Kellert, M., Mathar, W., Rühl, Ch.-S. and Wittkowski, R. (1987) Levels of PCDFs and PCDDs in samples of human origin and food in the Federal Republic of Germany, *Chemosphere*, **16**, 1977–1982.

Bertsch, W. (1978). Methods in high resolution gas chromatography. Two-dimensional techniques. *J. High. Resolut. Chromatogr. Chromatogr. Commun*, **1**, 85, 187, 289.

Blum, W. (1988). Säulen für die Gaschromatographie, *Nachr. Chem. Tech. Lab.*, **36**, 166–172.

Brody, S. S. and Chaney, J. E. (1966). Flame photometric detector — the appli-

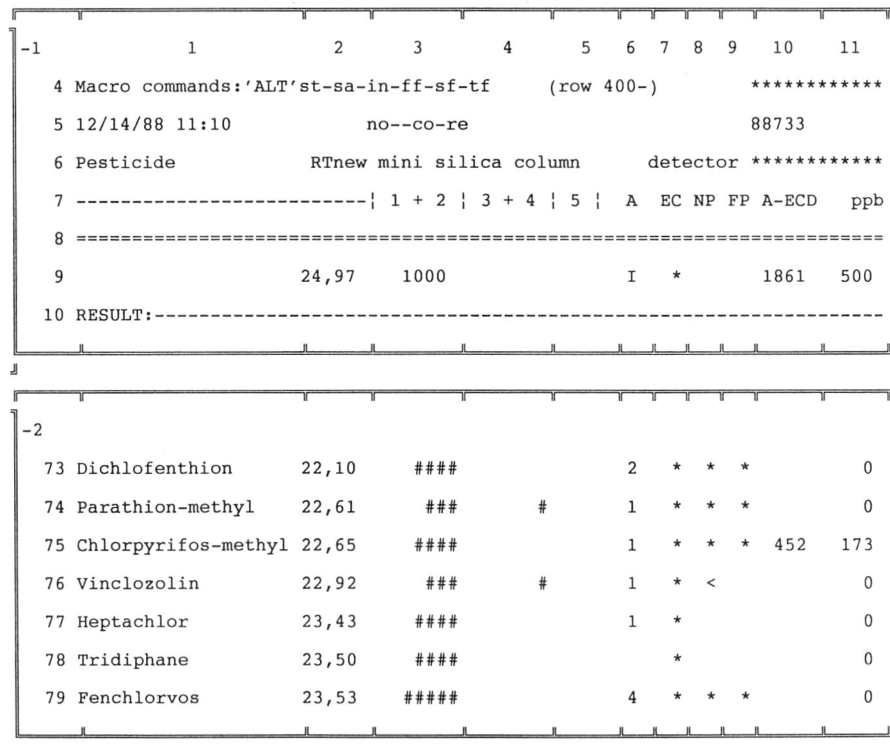

-1	1	2	3	4	5	6	7	8	9	10	11
4	Macro commands:'ALT'st-sa-in-ff-sf-tf				(row 400-)					************	
5	12/14/88 11:10		no--co-re							88733	
6	Pesticide	RTnew mini silica column				detector				************	
7	--------------------------¦		1 + 2 ¦	3 + 4 ¦	5 ¦	A	EC	NP	FP	A-ECD	ppb
8	==										
9		24,97	1000			I	*			1861	500
10	RESULT:--										

-2	1	2	3	4	5	6	7	8	9	10	11
73	Dichlofenthion	22,10	####			2	*	*	*		0
74	Parathion-methyl	22,61	###	#		1	*	*	*		0
75	Chlorpyrifos-methyl	22,65	####			1	*	*	*	452	173
76	Vinclozolin	22,92	###	#		1	*	<			0
77	Heptachlor	23,43	####			1	*				0
78	Tridiphane	23,50	####				*				0
79	Fenchlorvos	23,53	#####			4	*	*	*		0

Fig. 9.27 — MAPA worksheet during examination of sample 88 733 indicating a possible residue of 173 p.p.b. chlorpyrifos methyl.

cation of a specific detector for phosphorus and for sulfur compounds, sensitive to subnanogram quantities, *J. Gas Chromatogr.*, **4**, 42–46.

Büchel, K. H. (1977). *Pflanzenschutz and Schädlingsbekämfung*, Georg Thieme, Stuttgart.

Buser, H. R. (1976). High-resolution gas chromatography of polychlorinated dibenzo-*p*-dioxins and dibenzofurans, *Anal. Chem.*, **4B**, 1553–1557.

Buser, H. R. (1977). Determination of 2,3,7,8-tetrachlorodibenzo-*p*-dioxin in environmental samples by high-resolution gas chromatography and low resolution mass spectrometry, *Anal. Chem.*, **49**, 918–922.

Cairns, T., Siegmund, E. G. and Jacobson, R. A. (Ed.) (1987). *Mass Spectral Data Compilation of Pesticides and Industrial Chemicals.* Food and Drug Administration, Los Angeles, CA.

Cochrane, W. P. (1975). Confirmation of insecticide and herbicide residues by chemical derivatization, *J. Chromatogr. Sci.*, **13**, 246–253.

Damico, J. N. (1972). Pesticides. In *Bichemical Applications of Mass Spectrometry*, Waller, R. G. (ed)., Wiley, New York, p. 623.

Dandeneau, R. and Zerenner, E. H. (1979). An investigation of glasses for capillary chromatography, *J. High. Resolut. Chromatogr. Chromatogr. Commun.*, **2**, 351–356.

Deans, D. R. (1968). A new technique for heart cutting in gas chromatography, *Chromatographie*, **1**, 18–22.

De Faubert Maunder, M. J., Egan, H., Godly, E. W., Hammond, E. W., Roburn, J., Thomson, J. (1964). Clean-up of animal fats and dairy products for the analysis of chlorinated pesticide residues, *Analyst*, 89, 168–174.

Deleu, R. and Copin, A. (1980). Separation of pesticides by capillary gas chromatography, *J. High Resolut. Chromatogr. Chromatogr. Commun.*, **3**, 299–300.

Desty, D. H., Haresnape, J. N. and Whyman, B. H. F. (1960). Construction of long lengths of coiled glass capillary, *Anal. Chem.*, **32**, 302–304.

Devaux, P. and Guichon, G. (1970). Determination of the optimum operating conditions of the electron capture detector, *J. Chromatogr. Sci.*, **8**, 502–508.

DFG-Method S16, *see* Thier and Zeumer (1987).

DFG-Method S19, *see* Thier and Zeumer (1987).

Eyem, J. (1975). The role of wall-coated capillary columns in GC–MS techniques, *Chromatographia*, **8**, 456–462.

FDA Pesticide Analytical Manual (1985). Food and Drug Administration, Washington, D.C., vol. 1.

Freudenthal, J. and Gramberg, L. G. (1975). *Catalogue of Mass Spectra of Pesticides*, National Institute of Public Health, Bilthoven.

Gilsbach, W. and Thier, H.-P. (1982). Beiträge zur Rückstandsanalyse von Chlorphenoxycarbonsäure-Herbiciden in Weizenmehl, *Z. Lebensm.-Unters.-Forsch.*, **175**, 327–332.

Goebel, H. and Stan, H.-J. (1983). Automated gas chromatographic analysis of pesticide residues in food samples by means of fused-silica capillary columns and data processing, *J. Chromatogr.*, **279**, 523–532.

Grob, K. and Grob, G. (1969). Splitless injection on capillary columns, Part I, The basic technique, steroid analysis as an example, *J. Chromatogr. Sci.*, **7**, 584–586.

Hall, R. C. and Harris, D. E. (1979). Direct gas chromatographic determination of carbamate pesticides using carbowax 20M-modified supports and the electrolytic conductivity detector, *J. Chromatogr.*, **169**, 245–259.

Hayes, W. J., Jr., (1981). *Toxicology of Pesticides*, Williams & Wilkins, Baltimore, MD.

Henneberg, D. and Schomburg, G. (1964). Die kombinierte Anwendung von Gaschromatographie und Massenspektrometrie, *Fresenius Z. Anal. Chem.*, **211**, 55–61.

Henneberg, D., Henrichs, U. and Schomburg, G. (1975). Open split connection of glass capillary columns to mass spectrometers, *Chromatographia*, **8**, 449–462.

Hild, J., Schulte, E. and Thier, H.-P. (1978). Trennung von Organophosphor-Pestiziden und ihren Metaboliten auf Glaskapillarsäulen, *Chromatographia*, **11**, 397–399.

Hites, R. A. (1985). *CRC Handbook of Mass Spectra of Environmental Contaminants.* CRC Press, Boca Raton, FL.

Holland, P. T. and Greenhalgh, R. (1981). Selection of gas chromatographic detectors for pesticide residue analysis. In *Analysis of Pesticide Residues. Chemical Analysis, Vol. 58.* Moye, H. A. (Ed), John Wiley & Sons, New York, p. 51.

In Ki, Mun, Bartholomew, D. R., Stauffer, D. B. and McLafferty, F. W. (1981).

Weighted file ordering for fast matching of mass spectra against a comprehensive data base, *Anal. Chem.*, **53**, 1938–1939.

Karman, A. and Giuffrida, L. (1964). Enhancement of the response of the hydrogen flame ionization detector to compounds containing halogens and phosphorus, *Nature*, **201**, 1204–1205.

Krause, R. T. (1980). Multiresidue method for determining *N*-methylcarbamate insecticides in crops, using high performance liquid chromatography, *J. Assoc. Off. Anal. Chem.*, **63**, 1114–1124.

Krijgsman, W., and van de Kamp, C. G. (1976). Analysis of organophosphorus pesticides by capillary gas chromatography with flame photometric detection, *J. Chromatogr.*, **117**, 201–205.

Lee, M. L., Yang, F. J. and Bartle, K. D. (1984). *Open Tubular Column Gas Chromatography*, John Wiley & Sons, New York.

Lipinski, J. and Stan. H.-J. (1988). CAPA — Computer Aided Pesticide Analysis computer program for the automated evaluation of chromatographic data for residue analysis of foods. *J. Chromatogr.*, **441**, 213–225.

Lipinski, J. and Stan. H.-J. (1989) Compilation of retention data for 270 pesticides on 3 different capillary columns. *Poster presented at the 10th International Symposium on Capillary Chromatography*, Riva del Garda, Italy, May, 22–25, 1989.

Lovelock, J. E. (1960). An ionization detector for permanent gases, *Nature*, **187**, 49–50.

Lovelock, J. E. and Lipsky, S. R. (1961). Electron affinity spectroscopy — a new method for the identification of functional groups in chemical compounds separated by gas chromatography, *J. Am. Chem. Soc.*, **82**, 431–433.

Luke, M. A. and Masumoto, H. T. (1986). Pesticide residue analysis of foods. In, *Analytical Methods for Pesticides and Plant Growth Regulators*, G. Zweig and J. Sherma (eds.), Academic Press, Orlando, FL, vol. XV, P. 161.

Luke, M. A., Froberg, J. E., Masumoto, H. T. (1975). Extraction and cleanup of organochlorine, organophosphate, organonitrogen, and hydrocarbon pesticides in produce for determination by gas–liquid chromatography, *J. Assoc. Off. Anal. Chem.*, **58**, 1020–1026.

McLafferty, F. W., Hertel, R. H. and Villwock, R. D. (1974). Computer identification of mass spectra, VI. Probability based matching of mass spectra; rapid identification of specific compounds in mixtures, *Org. Mass Spectrom.*, **9**, 690–702.

Mills, P. A., Onley, J. H. and Geither, R. A. (1963). Rapid method for chlorinated pesticide residues in nonfatty foods, *J. Assoc. Off. Anal. Chem.* **46**, 186–191.

Moseman, R. (1978). Rapid procedure for preparation of support-bonded carbowax 20M gas chromatographic column packing, *J. Chromatogr.*, **166**, 397–402.

Moye, H. A. (ed.) (1981) *Analysis of Pesticide Residues. Chemical Analysis, Vol. 58*, John Wiley & Sons, New York.

Müller, H.-M. and Stan. H.-J. (1989). Pesticide residue analysis in food with capillary gas chromatography. *Poster presented at the 10th International Symposium on Capillary Chromatography* in Riva del Garda, Italy, May, 22–25, 1989.

Ouchi, G. I. (1987). *Personal Computers for Scientists*, American Chemical Society, Washington, D.C.

Patterson, P. L. (1978a). Comparison of quenching effects in single- and dual-flame photometric detectors, *Anal. Chem.*, **50**, 345–348.

Patterson, P. L. (1978b). Selective responses of a flameless thermionic detector, *J. Chromatogr.*, **167**, 381–397.

Patterson, P. L. and Howe, R. L. (1978). Thermionic nitrogen-phosphorus detection with an alkali-ceramic bead, *J. Chromatogr. Sci.*, **16**, 275–280.

Pellizzari, E. D. (1974). Electron capture detection in gas chromatography, *J. Chromatogr.*, **98**, 323–361.

Pensyna, G. M., Venkataraghaven, R., Dayringer, H. E. and McLafferty, F. W. (1976). Probability based matching system using a large collection of reference mass spectra. *Anal. Chem.*, **48**, 1362–1368.

Perkow, W. (1983/1988). *Wirksubstanzen der Pflanzenschutz- und Schädlingsbekämpfungsmittel*, Paul Parey, Berlin.

Pflugmacher, J. and Ebing, W. (1973). Reinigung Pestizidrückstände enthaltender Rohextrakte mit einer automatisch arbeitenden Apparatur nach dem Prinzip der kombinierten Spül- und Codestillation (Sweep Co-Distillation), *Fresenius Z. Anal. Chem.*, **263**, 120–127.

Ripley, B. D. and Braun, H. E. (1983). Pesticide residues — retention time data for organochlorine, organophosphorus, and organonitrogen pesticides on SE-30 capillary column and application of capillary gas chromatography to pesticide residue analysis, *J. Assoc. Off. Anal. Chem.*, **66**, 1084–1095.

Roseboom, H. and Herbold, H. A. (1980). Determination of triazine herbicides in various crops by capillary gas chromatography with thermionic detection, *J. Chromatogr.*, **202**, 431–438.

Ryhage, R. (1964). Use of a mass spectrometer as a detector and analyser for effluents emerging from high temperature gas liquid chromatography columns, *Anal. Chem.*, **36**, 759–764.

Safe, S. and Hutzinger, O. (1973). *Mass Spectrometry of Pesticides and Pollutants*, CRC Press, Cleveland.

Schomburg, G., Husmann, H. and Weeke, F. (1975). Aspects of double-column gas chromatography with glass capillaries involving intermediate trapping, *J. Chromatogr.*, **112**, 205–217.

Schulte, E. and Acker, L. (1974a). Gas-chromatographie mit Glascapillaren bei Temperaturen bis zu 320°C und ihre Anwendung zur Trennung von Polychlorbiphenylen, *Fresenius Z. Anal. Chem.*, **268**, 260–267.

Schulte, E. and Acker, L. (1974b). Identifizierung und Metabolisierbarkeit von polychlorierten Biphenylen, *Naturwissenschaften*, **61**, 79–80.

Schulte, E., Thier, H.-P. and Acker, L. (1976). Rückstandsanalytik polychlorierter Biphenyle in Lebensmitteln tierischer Herkunft: Erfahrungen und Vorschläge zur Vereinheitlichung, *Dtsch. Lebensm. Rundsch.*, **72**, 229–232.

Singh, J. and Lapointe, M. R. (1974). Confirmation of six organothiophosphorus pesticides by chemical derivatization at nanogram levels, *J. Assoc. Off. Anal. Chem.*, **57**, 1285.

Sissons, D. and Welti, D. (1971). Structural identification of polychlorinated biphenyls in commercial mixtures by gas–liquid chromatography, nuclear magnetic resonance and mass spectrometry, *J. Chromatogr.*, **60**, 15–32.

Specht, W. and Tillkes, M. (1980). Gaschromatographische Bestimmung von

Rückständen an Pflanzenbehandlungsmitteln nach Clean-up über Gel-Chromatographie und Mini-Kiesegel-Säulen-Chromatographie, 3. Mitteilung, *Fresenius Z. Anal. Chem.*, **301**, 300–307.

Specht, W. and Tillkes, M. (1985). Gaschromatographische Bestimmung von Rückständen an Pflanzenbehandlungsmitteln nach Clean-up über Gel-Chromatographie und Mini-Kieselgel-Säulen-Chromatographie, 5. Mitteilung, *Fresenius Z. Anal. Chem.*, **322**, 443–445.

Stalling, D. L., Tindle, R. C. and Johnson, J. L. (1972). Cleanup of pesticide and polychlorinated biphenyl residues in fish extracts by gel permeation chromatography, *J. Assoc. Off. Anal. Chem.*, **55**, 32–38.

Stan, H.-J. (1977a). Nachweis von Phosphorpestizidrückständen in Lebensmitteln durch Kapillargaschromatographie/Massenspektrometrie, *Chromatographia*, **10**, 233–239.

Stan, H.-J. (1977b). Nachweis von Organophosphorsäureester-Rückständen in Lebensmitteln im ppb-Bereich durch Kapillargaschromatographie/Massenspektrometrie-Kopplung, *Z. Lebensm.-Unters.-Forsch.*, **164**, 153–159.

Stan, H.-J. (1981). Combined gas chromatography-mass spectromery. In *Pesticide Analysis*, Das, K. G. (ed.), Marcel Dekker, New York.

Stan, H.-J. (1989). Application of computers for evaluation of gas chromatographic data. In *Analytical Methods for Pesticides and Plant Growth Regulators, Vol. XVII*, Sherma, J. (ed.), Academic Press, Orlando, Fl., p. 167–215.

Stan, H.-J. (1988) Automatisierte Rückstandsanalyse von Pflanzenschutzmitteln mit Hilfe der zweidimensionalen Kapillargaschromatographie, *Lebensmittelchem. Gerichtl. Chem.*, **42**, 31–36.

Stan, H.-J. and Abraham, B. (1978). All-glass open-split interface for gas chromatography–mass spectrometry, *Anal. Chem.*, **50**, 2161–2164.

Stan, H.-J., and Goebel, (1983a). Automated capillary gas chromatographic analysis of pesticide residue in food, *J. Chromatogr.*, **268**, 55–69.

Stan, H.-J. and Goebel, H. (1983b). Automated gas chromatographic analysis of pesticide residues in food samples by means of fused-silica capillary columns and data processing, *J. Chromatogr.*, **279**, 523–532.

Stan, H.-J. and Goebel, H. (1984a) Program in BASIC to combine DATA from two different selective detectors and its applications for screening of pesticides in residue analysis. *J. Automatic Chemistry*, **6**, 14–20.

Stan, H.-J. and Goebel, H. (1984b) Evaluation of automated splitless and manual on-columns injection techniques using capillary gas chromatography for pesticide residue analysis. *J. Chromatogr.*, **314**, 413–420.

Stan, H.-J. and Lipinski, J. (1985). Microcomputer programming in basic for the evaluation of capillary GC in the analysis of pesticides, *J. Chromatogr.*, **349**, 49–53.

Stan, H.-J. and Lipinski, J. (1987). BALANCE — a computer program for the handling and supervision of analytical standards, *Intelligent Instruments & Computer*, **5**, 103–104.

Stan, H.-J. and Lipinski, J. (1989). Mass spectral library for pesticides. *Poster presented at the 10th International Symposium on Capillary Chromatography* in Riva del Garda, Italy, May, 22–25.

Stan, H.-J. and Mrowetz, D. (1983a). Residue analysis of pesticides in food by two-dimensional gas chromatography with capillary columns and parallel detection with flame photometric and electron-capture detection, *J. Chromatogr.*, **279**, 173–187.

Stan, H.-J. and Mrowetz, D. (1983b). Residue analysis of organophosphorus pesticides in food with two-dimensional gas chromatography using capillary columns and flame photometric detection, *J. High. Resolut. Chromatogr. Chromatogr. Commun.* **6**, 255–263.

Stan, H.-J. and Müller, H.-M. (1988). Evaluation of automated and manual hot-splitless (PTV) and on-column injection technique using capillary gas chromatography for the analysis of organophosphorus pesticides. *J. High Resolut. Chromatogr. Commun.*, **11**, 140–143.

Stan, H.-J. and Steinbach, B. (1984). Automated development of optimum temperature programmes for gas chromatographic separation of complex mixtures on capillary columns, *J. Chromatogr.*, **290**, 311–319.

Stan. H.-J. and Steinbach, B. (1985). BASIC program for development of optimum temperature programs for gas chromatographic separation, *Intelligent Instruments & Computers*, **3**, 3–13.

Stan, H.-J., Abraham, B., Jung, J., Kellert, M. and Steinland, K. (1977). Nachweis von Organophosphorinsecticiden durch Gaschromatographie/Massenspektrometrie, *Fresenius Z. Anal. Chem.*, **287**, 271–285.

Storherr, R. W. and Watts, R. R. (1965). A sweep co-distillation cleanup method for organophosphate pesticide, *J. Assoc. Off. Anal. Chem.*, **48**, 1154–1160.

Storherr, R. W. and Ott, P. and Watts, R. R. (1971) A sweep co-distillation cleanup method for organophosphate pesticides, I. recoveries from fortified crops. *J. Assoc. Off. Anal. Chem.*, **54**, 513–516.

Suprock, J. F. and Vinopal, J. H. (1987). Behaviour of 78 pesticides and pesticide metabolites on four different ultra-bond gas chromatographic columns. *J. Assoc. Off. Anal. Chem.*, **70**, 1014–1017.

The Pesticide Manual (1987). Worthing, Ch. R. (ed.), The British Crop Protection Council, Thornton Heath, UK.

Thier, H.-P. and Frehse, H. (1986). *Rückstandsanalytik von Pflanzenschutzmitteln*, Georg Thieme, Stuttgart.

Thier, H.-P. and Zeumer, H. (1987). (Ed.). *Manual of Pesticide Residue Analysis.* DFG, Dt. Forschungsgemeinschaft, Pesticides Comm., VCH, Weinheim.

Thompson, J. F. and Watts, R. R. (1981). Gas-chromatographic columns in pesticide analysis. In *Analysis of Pesticide Residues. Chemical Analysis, Vol. 58*, Moye, H. A. (ed.), John Wiley & Sons, New York, p. 1–50.

Thompson, J. F., Mann, J. B., Apodaca, A. O. and Kantor, E. (1975). Relative retention ratios of ninety-five pesticides and metabolites in nine gas–liquid chromatographic columns over a temperature range of 170 to 204°C in two detection modes, *J. Assoc. Off. Anal. Chem.*, **58**, 1037–1050.

Thompson, J. F., Walker, A. C. and Moseman, R. F. (1969). Evaluation of eight gas chromatographic columns for chlorinated pesticides, *J. Assoc. Off. Anal. Chem.*, **52**, 1263–1277.

Trindle, R. C. and Stalling, D. L. (1972). Apparatus for automated gel permeation, cleanup for pesticide residue analysis, *Anal. Chem.*, **44**, 1768–1773.

Ware, G. W. (1983). Pesticides: chemical tools. In *Pesticides: Theory and Application*, H. W. Freeman, New York, p. 3–25.

Watson, J. T. and Biemann, K. (1964). High resolution mass spectra of compounds emerging from a gas chromatograph, *Anal. Chem.*, **36**, 1135–1137.

Wehner, T. A. and Seiber, J. N. (1981). Analysis of *N*-methylcarbamate insecticides and related compounds by capillary gas chromatography, *J. High Resolut. Chromatogr. Chromatogr. Commun.*, **4**, 348–350.

Zell, M. and Ballschmiter, K. (1980). Baseline studies of global pollution — III. Trace analysis of polychlorinated biphenyls (PCB) by ECD glass capillary gas chromatography in environmental samples of different trophic levels, *Fresenius Z. Anal. Chem.*, **304**, 337–349.

Zweig, G. and Sherma, J. (1963/1988) (ed.) *Analytical Methods for Pesticides and Plant Growth Regulators*, vol. I–XVII, Academic Press, Orlando, FL.

Wegler, R. (1970/1982). *Chemie der Pflanzenschutz- und Schädlingsbekämpfungsmittel*, Springer, Berlin.

10

Food additives

Nicholas P. Boley

10.1 INTRODUCTION

Gas–liquid chromatography (GLC) has been used widely over the past 20 years in the quantitative analysis of many classes of food additives. These have included preservatives, antioxidants, emulsifiers and artificial flavourings, as well as a number of additives in wine. In recent years, however, other techniques have gained more favour in the determination of food additives, particularly high-performance liquid chromatography (HPLC). This is probably due to the fact that additives are often, by their very nature, non-volatile compounds; artificial colours are a good example of this. Nevertheless, gas chromatographic techniques for the analysis of additives remain important. There are laboratories which do not always have access to the equipment to use other techniques, and the use of complementary techniques to verify results, or overcome problems arising from using other methods, is often necessary.

This partial shift away from the use of gas chromatography (GC) is reflected in the relative paucity of published methods using capillary columns. In food additive analysis, packed-column GC still holds sway, although many of these applications are capable of being successfully adapted, or even improved, by use of a corresponding capillary column. Some examples of this that have been successfully carried out in the author's own laboratory are cited later in this chapter.

10.2 PRESERVATIVES

10.2.1 Introduction

A number of preservatives are permitted for food use in the United Kingdom (Preservatives in Food Regulations, 1989). Those that may be determined by gas chromatography are listed, together with their 'E' numbers, in Table 10.1.

These preservatives are used at levels in foods ranging up to 10 000 mg/kg, depending on the preservative and the particular foodstuff.

Table 10.1 — Preservatives determined by GLC

Benzoic acid (and benzoates)	E210
Methyl-4-hydroxybenzoate	E218
Ethyl-4-hydroxybenzoate	E214
n-Propyl-4-hydroxybenzoate	E216
Propionic acid	E280
Sorbic acid (and sorbates)	E200
Biphenyl	E230
2-Hydroxybiphenyl (*o*-phenylphenol)	E231
Sulphur dioxide	E220
Nitrates	E251,252
Nitrites	E249,250
2-(Thiazol-4-yl)benzimidazole (Thiabendazole)	E233

10.2.2 Benzoates, sorbates and propionic acid

This class of preservatives, used in the form of free acids, salts, or esters, are largely non-volatile, polar compounds. A number of approaches have been made to determine them by GC employing steam distillation or solvent extraction followed by gas chromatography of the free acids or esters on polar phase columns, or they have been derivatized for chromatographing on apolar columns.

Propionic and sorbic acids in bakery products, and benzoic acid in margarine (Graveland, 1972) were determined following direct extraction of the samples with diethyl ether containing 3% (v/v) orthophosphoric acid. Valeric (pentanoic) acid was added as an internal standard. After centrifugation of the extract, an aliquot of the ether layer was examined by gas chromatography. A 2 m×2 mm i.d. glass column packed with Carbowax 20M–terephthalic acid on 60–80 mesh Chromosorb W was used, the oven temperature being programmed from 100°C to 210°C at 5°C/min, with a carrier-gas flow of 65 ml/min nitrogen. Propionic, sorbic and benzoic acids were separated under these conditions in 25 min. Levels of these preservatives down to 5 ng can be detected, and the method is considered to be applicable to a wide range of food samples.

A similar analytical scheme was suggested (Bertrand and Sarre, 1978) for the determination of benzoic and sorbic acids, as well as their sodium and potassium salts, in a range of soft drinks. The samples were acidified with sulphuric acid and an internal standard (undecanoic acid) added. The acid suppresses the ionization of the preservatives, and they could then be extracted as the free acids with diethyl ether. The extracted acids were chromatographed on a different polar phase column, 4 m×2 mm i.d. glass packed with 5% diethyleneglycol succinate (DEGS) and 1% orthophosphoric acid on 80–100 mesh Gas Chrom Q. This column is very similar in polarity to that previously described. An isothermal oven temperature of 170°C was used, with a carrier-gas flow of 20 ml/min nitrogen. Under these conditions, benzoic acid, the last eluting component, elutes after about 30 min. A similar gas chromatographic method has been described (Ro, 1972) for the separation of benzoic, sorbic and dehydroacetic acids, as well as butyl-*p*-hydroxybenzoate in meat products, dairy products and oriental foods, such as soy sauce and bean paste. A column temperature of 200°C was employed in this method, and acetanilide used as the internal standard.

The use of steam distillation to extract sorbates, benzoates and *p*-hydroxybenzoates from food matrices was also suggested (Isshiki *et al.*, 1980) followed by gas chromatography on a polar-phase column. This procedure was used on a very wide range of food and drink samples. The preservatives were steam-distilled from a matrix containing sodium chloride and tartaric acid into a mixture of dichloromethane and water (75:25). The preservatives were extracted into the organic layer, the volume of which was reduced using a Kuderna–Danish apparatus, for gas chromatographic analysis. Fluorene was added to the dichloromethane as an internal standard. The two alternative stationary phases that were employed were 10% FFAP, and 5% DEGS plus 1% orthophosphoric acid (q.v.), both on 60–80 mesh Chromosorb W AW DMCS. The oven temperature was programmed from 140°C to 210°C at 3°C/min to obtain optimum separation. The FFAP column does not resolve isopropyl *p*-hydroxybenzoate, which is not permitted for food use, and ethyl *p*-hydroxybenzoate. The DEGS/H_3PO_4 does not resolve the *n*-propyl ester from ethyl *p*-hydroxybenzoate. Recoveries of between 75 and 95% were obtained for these preservatives, using this procedure on the wide range of samples investigated, although in a few cases some interferences prevented quantification of particular preservatives. This could be overcome by use of a polar capillary column, such as OV-351 or FFAP, to resolve the interferences from the preservative peaks. The procedure described has an advantage in that it can quantitatively determine other preservatives, such as biphenyl and 2-hydroxybiphenyl, as well as the antioxidants butylated hydroxyanisole (BHA) and butylated hydroxytoluene (BHT).

The relative merits of steam distillation versus solvent extraction for the determination of sorbic, benzoic and dehydroacetic acids, as well as butyl-*p*-hydroxybenzoate have also been evaluated (Ro, 1979). It was concluded that steam distillation was the preferred technique. GC was carried out on two columns, both stainless steel (2 m×3.5 mm i.d.). One was packed with 2% DEGS+orthophosphoric acid, and the other with NPGS (neopentyl glycol succinate)+orthophosphoric acid, both on Chromosorb W 60–80 mesh, at an oven temperature of 150°C to 200°C.

The permitted *p*-hydroxybenzoate esters have also been determined in wheat germ oil (Iwaida *et al.*, 1977). The samples were blended with saturated sodium chloride, sulphuric acid and ethyl acetate. The ethyl acetate layer was taken and extracted with 0.4 M potassium hydroxide in 50% aqueous methanol. This extract was acidified and the preservatives finally extracted into diethyl ether. Gas chromatography was carried out using a 1.5 m×3 mm i.d. column packed with an apolar phase, 3% SE-30, on 60–80 mesh Chromosorb W. The apolar column can be used in this analysis as esters are being determined; these may be easily separated on such a column, whereas the free acids (sorbic and benzoic) will tail badly on an apolar column, leading to poor resolution and quantification.

Benzoic, sorbic and *p*-hydroxybenzoic acids have been determined in soft drinks (Neale & Ridlington, 1978). Samples were mixed with a sodium chloride/hydrochloric acid solution, and extracted into chloroform. Phenylacetic acid or *p*-toluic acid were used as an internal standard; phenylacetic acid was the preferred internal standard, but *p*-toluic acid was used if vanillin or ethyl vanillin were present in the sample, as these interfere chromatographically with phenylacetic acid under the conditions used. These conditions involved the use of a 1.5 m×4 mm i.d. glass column packed with 1% DEGS-PS on 80–100 mesh Chromosorb W-HP, operated isothermally at 185°C. Recoveries of between 96 and 102% were obtained using this procedure. (See Fig. 10.1.)

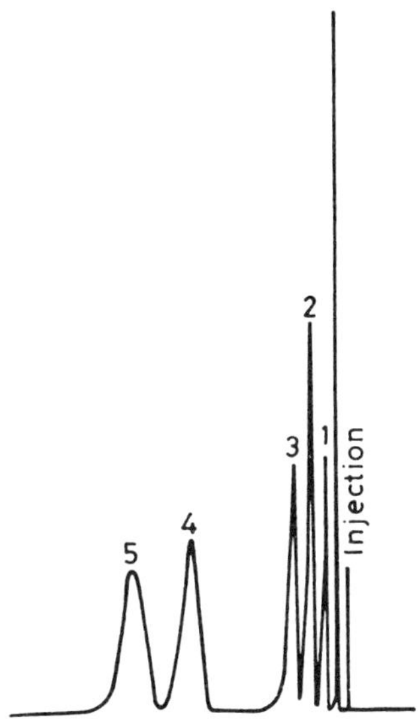

Fig. 10.1 — Gas chromatogram of sorbic, benzoic and *p*-hydroxbenzoic acids on a DEGS–PS column. 1, Sorbic acid; 2, benzoic acid; 3, phenylacetic acid; 4, methyl *p*-hydroxybenzoate; 5, *n*-propyl *p*-hydroxybenzoate. (Phenylacetic acid is the internal standard.) (Reproduced from Neal & Ridlington, 1978 with permission of *J. Assoc. Public Analysts*.)

Derivatization procedures that have been employed prior to gas chromatographic determination of these preservatives include ester formation (methyl, butyl) or silylation.

The formation of the butyl ester of sorbic acid extracted from dried prunes (Stafford & Black, 1978) was achieved using boron trifluoride/butan-1-ol reagent. The sorbic acid was extracted into dichloromethane, back-extracted into 0.5 M sodium bicarbonate, and finally re-extracted into dichloromethane following acidification, in order to reduce possible interferences. A 6′ (1.85-m) stainless steel column packed with 10% AT-1000 on 80–100 mesh Chromosorb W-HP, was used at 135°C, with decanoic acid, similarly derivatized, as the internal standard.

The formation of the methyl esters of benzoic and sorbic acids (Fogden *et al.*, 1974) has been used for a range of foods and drinks. A number of extraction techniques, based on direct solvent extraction or distillation, were investigated for use prior to reaction with boron trifluoride/methanol reagent. After cooling and the addition of water, the methyl esters were extracted with diethyl ether for gas chromatographic analysis on a 5′×¼″ i.d. (1.52 m×6 mm i.d.) glass column packed with 10% Apiezon L on Phasesep N AW DMCS 85–100 mesh. An isothermal oven temperature of 150°C was used, with a carrier-gas flow of 46 ml/min of nitrogen. A very apolar phase such as Apiezon L can be used in these applications because the less polar methyl esters have been formed.

The determination of sorbic and benzoic acids, and the methyl and propyl esters of *p*-hydroxybenzoic acid as their trimethylsilyl (TMS) ethers (Gosselé, 1971) has been described, on samples such as marmalade, mustard, fish and margarine. The samples were acidified with sulphuric acid and extracted into diethyl ether. The ether extract was back-extracted with sodium hydroxide solution, neutralized, and finally extracted into chloroform, methyl gallate being added as an internal standard. The volume of the chloroform solution was reduced to approximately 1 ml, and *N,O*-bis(trimethylsilyl)acetamide (BSA) was added in order to form the trimethylsilyl ethers. These were chromatographed on a 3-m column packed with 3% SE-30 on 100–120 mesh Aeropak, with the oven temperature being programmed from 90°C to 290°C at 8°C/min. Under these conditions the TMS ethers of the preservatives and the internal standard were eluted within 16 min.

Similarly, the TMS ethers of *p*-hydroxybenzoate esters have been determined by gas chromatography (Daenens & Laruelle, 1973). Samples were initially defatted by silica-gel column chromatography, prior to extraction of the preservatives and the formation of the TMS ethers using BSA. The chromatographic conditions used in this example were a 1.2 m×2 mm i.d. glass column packed with 3% OV-17 on 80–100 mesh Gas Chrom Q, at an isothermal oven temperature of 115°C.

Trimethylsilyl ethers of benzoic and sorbic acids have also been determined following reaction with *N*-methyl,*N*-trimethylsilyltrifluoroacetamide (MSTFA) in foods (Larsson & Fuchs, 1974; Larsson, 1983). A range of foods was investigated, including shrimp salad, cheese and jams. Caproic and phenylacetic acids were used as internal standards: phenylacetic acid was used for benzoic acid, and caproic acid for sorbic acid. Following solvent extraction with dichloromethane and derivatization, the TMS ethers were chromatographed on an OV-1 column (3% on Varaport 30) using a temperature programme from 80°C to 210°C at a rate of 8°C/min. Under these conditions, all four acids were chromatographed in 6 min (see Fig. 10.2). An alternative column used OV-225 as stationary phase (3% on Chromosorb W AW-DMCS, 100–120 mesh).

10.2.3 Biphenyl, 2-hydroxybiphenyl and thiabendazole

Biphenyl, 2-hydroxybiphenyl and thiabendazole are used almost exclusively to prevent fungal growth on fruits, usually citrus fruits, after harvesting. Different countries have set varying limits as to the residual levels of these preservatives which may carry through into the food chain.

Because of the common mode of use of these three preservatives, it is desirable to be able to screen for all three using a common scheme of analysis. This has not often been achieved by GC; neither has it by HPLC. Thiabendazole is, perhaps, less commonly used as a fruit preservative, biphenyl and 2-hydroxybiphenyl being more common. The structural similarity between these two compounds is obvious (see Fig. 10.3), but 2-hydroxybiphenyl is considerably more polar than the non-polar biphenyl.

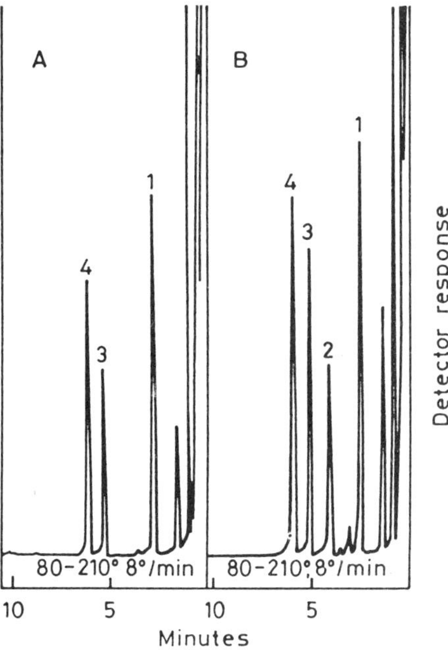

Fig. 10.2 — Gas chromatogram of TMS ethers of preservatives obtained from canned herring (A) and shrimp salad (B) on a 3% OV-1 column. Caproic acid, internal standard (1); sorbic acid (2); benzoic acid (3) and phenylacetic acid, internal standard (4). (Reproduced from Larsson, 1983 with permission of *JAOAC*.)

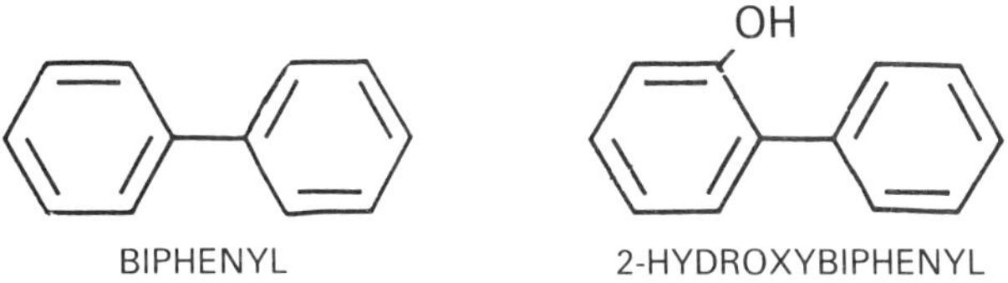

Fig. 10.3 — Structure of fruit preservatives.

Most workers have suggested methods for the determination of biphenyl and 2-hydroxybiphenyl based on an initial steam-distillation step from the fruit sample, although some methods based on direct solvent extraction have also been suggested. A common approach has been the use of a modified Clevenger trap for steam distillation (Westöö & Anderson, 1975). The sample was acidified with sulphuric acid and then steam-distilled into cyclohexane. The first distillate contained biphenyl, which was determined gas chromatographically on a column packed with 5%

Carbowax 20M on 60–80 mesh Chromosorb W-AW DMCS at an oven temperature of 160°C. In order to determine the 2-hydroxybiphenyl, two distillations needed to be carried out. The combined portions of both distillates were extracted with 1 M sodium hydroxide; this extract was acidified before being re-extracted into cyclohexane. This procedure removes compounds naturally occurring in the fruit which may interfere chromatographically. An aliquot from this extract was determined gas chromatographically using a column packed with a 1:1 mixture of 10% DC-200 and 15% QF-1, both on 80–100 mesh Chromosorb W-AW DMCS HP, using an oven temperature of 180°C. Quantification was achieved for both preservatives by comparison with standard solutions in cyclohexane injected under the same conditions.

A development of this procedure (Lord *et al.*, 1978) uses only one column for the gas chromatographic determination of both preservatives. The samples are again steam-distilled by use of the modified Clevenger apparatus, but after distillation heptadecane is added as an internal standard prior to gas chromatographic examination. The column used was 2 m×4 mm i.d. glass packed with 3% OV-17 on Gas Chrom Q. The oven temperature was programmed from 130°C to 160°C at 10°C/min, after a 10 min hold at the lower temperature. (See Fig. 10.4.)

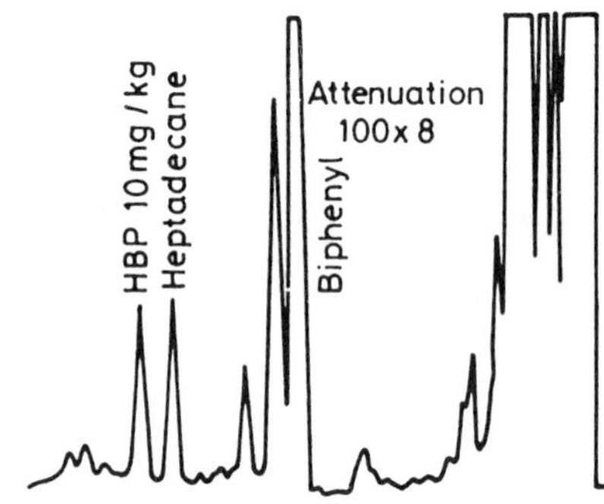

Fig. 10.4 — Gas chromatogram of biphenyl and 2-hydroxybiphenyl (HBP) on OV-17. (Reproduced from Lord *et al.*, 1978 with permission of *J. Assoc. Public Analysts*.)

A method that involves a straight steam-distillation has also been described (Tanaka *et al.*, 1978). Biphenyl and 2-hydroxybiphenyl were steam-distilled into a series of four receiver traps. The first three of these contained sodium hydroxide solution to trap the 2-hydroxybiphenyl; the fourth contained water for the collection of biphenyl. The biphenyl was extracted from this trap with heptane, and fluorene added as an internal standard, prior to injection onto a 1.5 m×3 mm i.d. glass column packed with 5% polyethylene glycol-20M on 80–100 mesh Chromosorb W-AW DMCS (an equivalent phase to Carbowax 20M). The oven temperature was 175°C, with a carrier-gas flow of 26 ml/min nitrogen. The procedure employed for the determination of the 2-hydroxybiphenyl was somewhat more involved. The contents of the three alkaline traps were combined, neutralized and treated with pentafluorobenzoyl chloride in order to form the pentafluorobenzoyl ester. This was extracted with heptane containing dichloro-2,2-bis(4-chlorophenyl)ethylene as an internal standard. This was injected onto a 2 m×3 mm i.d. column packed with 5% SE-30 on 80–100 mesh Gas Chrom Q. The oven temperature employed was 200°C, the carrier-gas flow 75 ml/min nitrogen, and detection achieved using an electron capture detector with a nickel-63 source. This allowed for levels of 2-hydroxybiphenyl down to 0.13 mg/kg to be detected.

A different approach to the determination of biphenyl involves looking at the biphenyl vapour in impregnated cartons used for the transportaion of citrus fruits (Davis & Munroe, 1977). Biphenyl was either extracted from these cartons with ethanol, or the biphenyl vapour itself withdrawn into a glass gas-tight syringe surrounded by a heating coil maintained at 80°C to prevent the biphenyl vapour from condensing on the sides of the syringe. The biphenyl was determined chromatographically using a 3 m×6 mm i.d. glass column packed with a stationary phase of 15% SE-30, with an oven temperature of 175°C and a carrier-gas flow of 90 ml/min nitrogen.

There have been successful attempts to determine simultaneously biphenyl, 2-hydroxybiphenyl and thiabendazole (Isshiki *et al.*, 1980a) in citrus fruits. They homogenized the samples, and added ammonium acetate to bring the pH to between 7 and 10 before extracting into ethyl acetate. Separation of the three preservatives was achieved on a column packed with 10% FFAP on 60–80 -mesh Chromosorb W-AW DMCs (0.5 m×3 mm i.d.), temperature-programmed from 110°C to 250°C at 8°C/min. A limit of detection of 1 mg/kg for each preservative was obtained.

Thiabendazole has been successfully determined following derivatization, as the methyl ester (Tanaka & Fujimoto, 1976), or as the pentafluorobenzoyl ester (Nose *et al.*, 1977). In the former procedure, the fruit sample was extracted with a mixture of ethyl acetate and sodium chloride–sodium acetate buffer. Following washing of the organic layer with sodium hydroxide solution and then water, the thiabendazole was extracted into acid solution, and back-extracted into ethyl acetate after adjustment of the pH to alkaline. Derivatization was carried out by reaction with dimethylformamide–dimethylacetal (DMF–DMA) in acetonitrile. The resulting methyl ester was determined on a 1.5 m×2 mm i.d. glass column packed with 10% DC-200 on 80–100 mesh Gas Chrom Q, at 240°C, with a carrier gas flow of 40 ml/min nitrogen. An injection volume of 2 μl was used, quantification being achieved by reference to a calibration curve prepared by injecting 2-μl aliquots of standard solutions of thiabendazole methyl ester under the same conditions. Recoveries of between 90 and 95% were obtained for fruit samples spiked at levels of 0.5 mg/kg.

The determination of thiabendazole as the pentafluorobenzoyl ester follows a similar extraction procedure as that for the methyl ester. The dried extract, after evaporation of the ethyl acetate, was derivatized by reaction at 120°C with pentafluorobenzoyl chloride in benzene, in a sealed tube. After derivatization was complete, decachlorobiphenyl was added as an internal standard. A 1.5 m×3 mm i.d. glass column packed with 5% OV-101 on 80–100 mesh Gas Chrom G(HP) was used, and the eluting compounds detected using an electron-capture detector, with a nickel-63 source. A column temperature of 230°C and a carrier-gas flow of 40 ml/min nitrogen were employed. This method proved to be extremely sensitive; concentrations down to 0.01 mg/kg could be detected, which makes the procedure attractive despite the fact that fairly noxious reagents are used in the derivatization procedure.

10.2.4 Other preservatives

Other preservatives have been determined by GC, but these are largely more recent developments. Sulphur dioxide, used as a preservative in many foods, and particularly in beers, wines and fruit juices, is very volatile, and therefore lends itself to determination by headspace gas chromatographic methods, using, for example, a conductivity detector. Such an approach has been described (Barnett & Davis, 1983) for application mainly to packaged foods. Samples were placed in flexible-film pouches with two silicone septa attached on different corners. Following replacement of air in the pouch by nitrogen, the samples were equilibrated, and then an aliquot of headspace withdrawn by means of a gas-tight syringe, which had been preheated to 90°C. A 1.2 m×2 mm i.d. FEP Teflon column packed with Chromosorb 108 was used, and the detector used was a Hall electrolytic conductivity detector operated in non-catalytic mode. Sulphur dioxide was separated from both carbon dioxide and water under these conditions. Work currently being undertaken in the author's laboratory is seeking to evaluate the use of a porous-layer open tubular (PLOT) column (10 m×0.32 mm i.d.) coated with Poraplot Q for the separation of sulphur dioxide evolved from samples equilibrated and injected by means of an automatic headspace sampler. Detection would be achieved using a Finnegan-MAT ion-trap detector operated in selected-ion monitoring (SIM) mode.

An unusual development has been in the determination of nitrates and nitrites by gas chromatography (Wu & Saschenbrecker, 1978) in meat and meat products. In this procedure the samples were defatted by carbon tetrachloride extraction, and deproteinated using Carrez I and II solutions. Nitrates were determined following conversion to nitrobenzene by nitration of benzene in the presence of 80% sulphuric acid. After neutralization with sodium carbonate, the organic layer was collected for gas chromatographic examination. Nitrite was similarly converted to nitrobenzene following oxidation of nitrite to nitrate with potassium permanganate. The nitrobenzene was determined using a 6′×1/8″ (1.83 m×3 mm) glass column packed with a mixture of 4% SE-30 and 6% QF-1 on 60–80 mesh Chromosorb W, and 2-chloronaphthalene as the internal standard. These compounds were detected with an electron capture detector (see Fig. 10.5). An oven temperature of 125°C was used, with a carrier-gas flow of 65–70 ml/min nitrogen, and a make-up gas flow of 20 ml/min nitrogen. The nitrite concentration is determined from the difference between the total nitrate/nitrite content, and the nitrate alone, which is determined separately. Nanogram levels of nitrobenzene can be determined using this procedure.

10.3 ANTIOXIDANTS

10.3.1 Introduction

Antioxidants are added to foods to prevent oxidation, which would result in the formation of off-flavours and a reduced shelf-life. Many antioxidants are added to oils and fats, or to foods containing high fat levels, to prevent autoxidation, which leads to rancidity. Table 10.2 shows those antioxidants, both permitted for food use, and non-permitted, which have been determined by GC.

As with preservatives, there are a number of techniques, both chromatographic

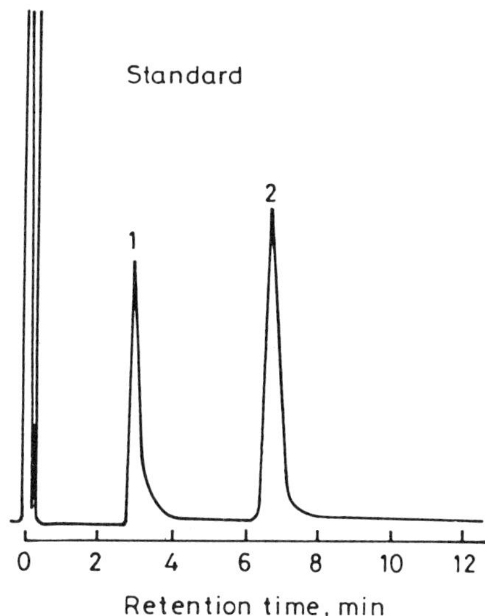

Fig. 10.5 — Gas chromatogram of a beef liver sample containing 5 ng nitrobenzene (1) derived from nitrate, and 4 ng 2-chloronaphthalene (2) (internal standard) (4% SE 30+6% QF-1 on chromosorb W). (Reproduced from Wu & Saschenbrecker, 1977 with permission of *JAOAC*.)

Table 10.2 — Antioxidants determined by GC

Butylated hydroxyanisole	(BHA)
Butylated hydroxytoluene	(BHT)
Gallic acid (including various gallate esters)	
tert-Butylhydroquinone	(TBHQ)[a]
Tocopherols	
Ionox-100	
Ethoxyquin	
2,4,5-trihydroxybutyrophenone	(THBP)[a]
Nordihydroguaiaretic acid	(NDGA)

[a]Not permitted for food use in the United Kingdom.

and classical, which have been used for the determination of antioxidants. Both GC and HPLC are used extensively, although there have been fewer developments in gas chromatographic methods in recent years.

10.3.2 BHA, BHT, TBHQ, Gallates, Ionox-100, THBP, NDGA

These antioxidants are those which are, or have been, commonly used in many foods, and particularly in oils and fats. There have been several methods published which determine quantitatively all, or some, of them simultaneously.

An early gas chromatographic procedure for the determination of BHA and BHT in breakfast cereals (Takahashi, 1970) involved the homogenized sample being placed in a sintered glass chromatography column. The antioxidants were eluted from this column using carbon disulphide, an aliquot of this eluate being injected onto the gas chromatograph. Two columns were used in order to confirm the identity of BHA and BHT — Apiezon L, and QF-1 — both phases being coated onto 80–100 mesh Gas Chrom Q. It was necessary to use two different phases, as a very large number of compounds were co-extracted with carbon disulphide, which led to many possible sources of chromatographic interference, which is an obvious drawback to this procedure.

Two procedures have been described that cover the determination of BHA, BHT, THBP, TBHQ, NDGA, Ionox-100 and propyl gallate (Kline *et al.*, 1978). These procedures apply to the determination of antioxidants in both oils and fats, and in dried foods. The first procedure determines BHA, BHT, Ionox-100 and TBHQ. Oils or fats were melted if necessary, and diluted as appropriate with ethyl acetate. Solid samples were extracted by shaking with ethyl acetate and filtering. The antioxidants were chromatographed using a 4′(1.22 m)×4 mm i.d. glass column packed with 3% OV-17 on 80–100 mesh Gas Chrom Q. A pre-column was placed in the injector of the gas chromatograph, packed with silanized glass wool, in order to absorb any non-volatile components of the oil, which otherwise would cause considerable interference in later chromatographic analyses. Care does need to be exercized when injecting several oil samples that the pre-column is changed as necessary, to prevent breakthrough of the oil onto the analytical column when it has become saturated. The oven temperature was programmed from 85°C to 175°C, and helium used as the carrier gas at a flow rate of 40 ml/min. Under these conditions all four antioxidants were eluted in about 14 min, and recoveries of between 90 and 100% were obtained; quantification was achieved by comparison with standard solutions of the antioxidants injected under similar conditions. This is similar to a procedure for determining BHA, BHT and TBHQ (Mancini *et al.*, 1977) in which a column of 10% OV-17 on Anakrom ABS 80–90 mesh was used at a temperature programmed from 160°C to 260°C at 10°C/min (see Fig. 10.6).

An alternative to direct injection techniques for the determination of BHA and/or BHT in vegetable oils has been described (Jedrych and Karlowski, 1979) in which the antioxidants are extracted from the oil by means of a modified Clevenger trap steam-distillation apparatus. BHA and BHT were collected in *n*-butanol and determined by injection onto a column (1.5 m×4 mm i.d.) packed with 10% OV-1 on 80–100 mesh Gas Chrom Q. An oven temperature of 160°C was used, with argon as carrier gas at 40 ml/min. A limit of detection of 0.5 mg/kg was obtained.

The second procedure described by Kline *et al.* (1978) involves the determination of TBHQ, propyl gallate, NDGA and THBP, as their trimethylsilyl ethers. Solid samples were extracted with ethyl acetate as previously described, and oil samples dissolved in carbon tetrachloride, prior to the antioxidants being extracted into 70% aqueous ethanol. After being taken to dryness, the TMS ethers were formed by reaction with a reagent containing pyridine, hexamethyldisilazane (HMDS) and trimethylchlorosilane (TMCS) in the proportion of 9:3:1. The TMS ethers were chromatographed on a 10′ (3.05 m)×2 mm i.d. glass column packed with 3% OV-225 on 80–100 mesh Gas Chrom Q, temperature-programmed from 100°C to 250°C.

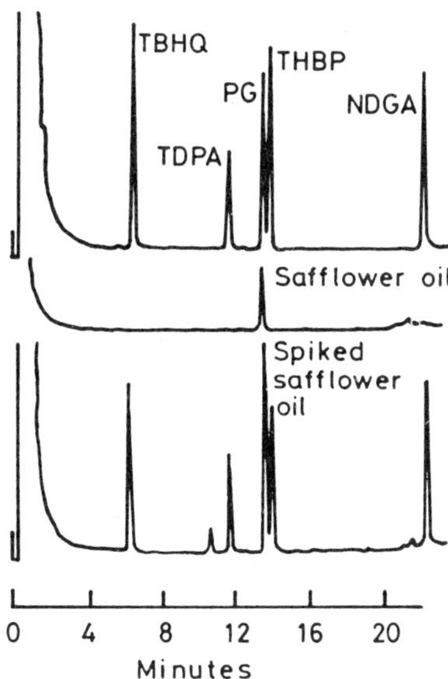

Fig. 10.6 — Gas chromatogram of antioxidants on an OV-17 column from direct injection of an oil sample. Top, antioxidant standards (0.4 mg/ml); middle, safflower oil containing PG; bottom, safflower oil fortified with 100 p.p.m. antioxidants. (Reproduced from Mancini *et al.*, 1977 with permission of *JAOAC*.)

The antioxidants eluted within 24 min. Under these conditions, propyl gallate and THBP do not quite fully resolve, but this does not present too great a problem as few samples, if any, are likely to contain both antioxidants. The recoveries obtained were not as impressive as for the previous procedure — NDGA alone gave recoveries between 90 and 100%, the others varied between 70 and 90%.

BHA and BHT have also been determined following extraction with organic solvents or liquid–liquid partition (Kitada *et al.*, 1975) prior to GC. A number of stationary phases were investigated in this study, including 5% SE-30 (at 130°C), 5% OV-17 (at 145°C), 5% XE-60 (at 140°C) and PEG 20M (at 160°C).

A more recent development of the direct injection procedure has been in the use of a capillary column (Yu *et al.*, 1984) for the determination of BHA, BHT, Ionox-100 and TBHQ in vegetable oils. In addition this procedure is capable of determining dehydroacetic acid (DHA), which is used as an antioxidant in some countries. Samples were diluted 1:10 with acetone, for injection onto the GC. Once again, a glass liner packed with silanized quartz wool was used to remove interferences from the oils. The antioxidants were separated on a 25 m×0.3 mm i.d. WCOT fused-silica column coated with a cross-linked 5% phenylmethylsilicone stationary phase (e.g. SE-52, SE-54), although other columns of low polarity such as SE-30 or OV-17 or their equivalents, have been successfully used. The column temperature was programmed from 50°C to 280°C at 15°C/min. A splitless injection technique was employed, with a linear carrier-gas velocity of 26 cm/s nitrogen. The antioxidants were all chromatographed within 16 min. Levels below 1 mg/kg for these antioxidants have been analysed using this procedure. The use of a capillary column has enabled much greater resolution of the oil components, whether naturally occurring or additives, to be obtained. Indeed, many natural oil component peaks can be very clearly resolved away from the antioxidant peaks, which results in more accurate quantification. There are indications that this procedure can be adapted for more complex matrices than oils, as the antioxidant peaks are more easily separated, thus minimizing any extraction or clean-up, although our experience indicates that a complementary technique is desirable to confirm the presence of particular compounds if very complex chromatograms are obtained.

The determination of TBHQ in products such as dried fish, dried or frozen shrimp, vegetable oils and margarines has been described (Toyoda *et al.*, 1980). In this procedure, dried or frozen samples were extracted with ethyl acetate in the presence of sodium chloride. The extract was evaporated to dryness, and the residue redissolved in hexane:ethyl acetate (99:1). Oils or fat samples were dissolved directly into the hexane:ethyl acetate solvent. TBHQ was extracted selectively from this solution using 2% aqueous sodium chloride, and then back-extracted into an organic solvent (50:50 hexane:ethyl acetate). Following drying of the extract, and reduction of the volume, this was examined by GC on a 2 m×3 mm i.d. glass column packed with 5% DEGS plus 1% orthophosphoric acid on 80–100 mesh Chromosorb W, at 196°C.

Many gas chromatographic methods for the determination of common antioxidants involve the formation of a number of derivatives prior to injection onto the gas chromatograph. Trimethylsilyl ethers have already been referred to; others include benzoyl esters, trifluoroacetates and heptafluorobutyryl derivatives.

A method for chromatographing BHA as its trifluoroacetate (Dilli and Robards, 1977) has been successfully used; this method also separates chromatographically BHT and underivatized BHA. The formation of the trifluoroacetate of BHA was achieved by heating BHA with trifluoroacetic anhydride in hexane for 1 h in a sealed ampoule. The resultant mixture was cleaned up by shaking with sodium hydroxide solution, and the organic layer taken for gas chromatographic analysis. A 1.5 m×6 mm o.d. (external diameter) glass column packed with either 5% or 10% SE-30 on 80–100 mesh Chromosorb W AW-DMCS was employed. To determine BHA, BHT and BHA trifluoroacetate (BHAT), an oven temperature of 175°C was found to be necessary, with a carrier-gas flow of 40 ml/min nitrogen, detection being achieved with an FID. Under these conditions BHAT elutes first, followed by BHA and BHT. It was noted that if the polarity of the stationary phase was increased from the non-polar SE-30, the elution order of BHA and BHT reversed; BHT elutes first on a polar phase column. However, BHAT elutes before both independent of column polarity. Using an FID, the limit of detection for these compounds was found to be about 4–5 ng. The sensitivity of the method towards BHAT can be increased by a factor of about 10 by the use of an electron-capture detector (ECD), which is more specific for the trifluoroacetate. To obtain the optimum chromatography for the use of an ECD, the column temperature was lowered to 160°C, and a carrier gas of 10% methane in argon was used. Quantification of BHAT was achieved by the use of hexachlorobenzene as an internal standard.

A more common approach has been in the formation of the heptafluorobutyrates of some antioxidants (Page and Kennedy, 1976). This method was used for the determination of BHA, TBHQ and propyl gallate, but not for the determination of BHT, in edible oils. Samples were diluted with ether:benzene (1:99) following the addition of 2,3,4,5-tetrachlorophenol as an internal standard. Derivatization was effected by adding 0.2M trimethylamine, which acts as a catalyst, into a tube, followed by 20 μl heptafluorobutyric anhydride, and then an aliquot of the sample solution. It should be noted that heptafluorobutyric anhydride is an unpleasant reagent, and needs to be used with a glove box, and manipulations carried out under dry nitrogen to prevent hydration. After derivatization, excess heptafluorobutyric anhydride was removed by shaking with pH-6 phosphate buffer, the organic layer being taken for gas chromatographic analysis. A 6′ (1.83 m)×2 mm i.d. glass column packed with 3% OV-3 on 80–100 mesh Chromosorb W HP was used at a temperature of 120°C, and eluting heptafluorobutyrates detected using an ECD employing a tritium source. Nitrogen was the carrier gas used, at a flow rate of 20 ml/min. Under these conditions TBHQ elutes first, followed by 2-BHA, its isomer 3-BHA, the internal standard and propyl gallate (see Fig. 10.7). As can be seen this method separates the two isomers of BHA. Levels of all these antioxidants down to at least 0.5 mg/kg could be detected, with recoveries of between 95 and 105%. Although the described method provides an accurate and sensitive method for these antioxidants, the lack of reactivity of BHT with heptafluorobutyric anhydride, due to steric hindrance of the *tert*-butyl groups, is something of a drawback, given the common use of this antioxidant.

The determination of other esters of gallic acid — namely octyl and dodecyl — has also been reported (Bajardi *et al.*, 1982) using a 5 m×3.2 mm i.d. glass column packed with 5% OV-101 on Chromosorb G HP 80–100 mesh. Bajardi also formed the trifluoroacetates of these gallate esters, as well as those of BHA and the *p*-hydroxybenzoate family. These trifluoroacetates were also determined on an OV-101 column (2 m×2 mm i.d., 3% on Chromosorb W HP 80–100 mesh) with an oven temperature of 240°C and a carrier-gas flow of 30 ml/min nitrogen.

The other derivatives commonly formed for antioxidant determination are the benzoyl esters (Galensa and Schäfers, 1982). This method was developed for use with foods having complex matrices, and both HPLC and GC methods have been carried out on the benzoyl esters. The antioxidants were extracted from foods using acetonitrile at 65°C in an ultrasonic bath. Following filtration, the volume of the extract was reduced and derivatization carried out with benzoyl chloride in pyridine. This was washed to remove excess reagents and extracted into a solvent of iso-octane, diethyl ether and acetonitrile (500:100:3) for chromatographic analysis. GC analysis was carried out using a WCOT capillary column, 30 m in length, coated with OV-73. A temperature programme from 120°C to 300°C was used, and BHA, BHT, all the common gallate esters and NDGA could be separated and quantified as their benzoyl esters in less than 20 min.

10.3.3 Other antioxidants

10.3.3.1 Ethoxyquin

Ethoxyquin (6-ethoxy-1,2-dihydro-2,2,4-trimethylquinoline) is used as an antioxidant in both animal feeds and to prevent degradation of apples during storage and transport. A gas chromatographic method has been described (Winell, 1976) for the determination of ethoxyquin in apples. The ethoxyquin was extracted from the

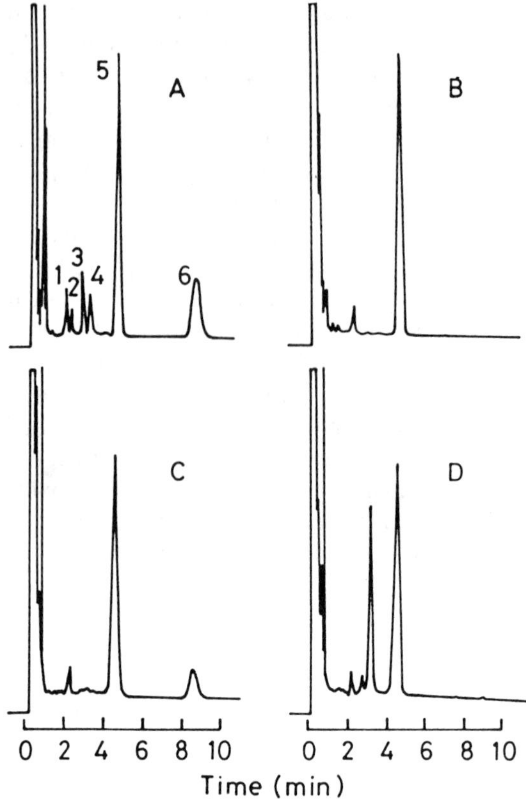

Fig. 10.7 — Gas chromatogram of the heptafluorobutyrates of BHA, TBHQ and propyl gallate using an OV-3 column and ECD. A, mixed standard: 1, TBHQ–HFB (40 pg); 2, impurity from TCP; 3, 2-BHA–HFB (230 pg); 4, 3-BHA–HFB (230 pg); 5, TCP–HFB internal standard (300 pg); 6, PG–HFB (370 pg). B, corn oil (37 μg oil injected), no detectable antioxidants; C, canbra and soya oil (22 μg oil injected) containing 5 p.p.m. PG; D, soya oil (22 μg oil injected) containing 24 p.p.m. BHA. (Reproduced from Page & Kennedy, 1976 with permission of *JAOAC*.)

homogenized apples using hexane. At this stage the internal standard, tetrahydroquinoline, was added. The extract was cleaned up by back-extraction into dilute aqueous acid, and the aqueous layer made alkaline with dilute sodium hydroxide solution. This was then again extracted into hexane. An aliquot of this extract was treated with 50 μl 5% pyridine in toluene and 50 μl heptafluorobutyric anhydride in order to form the heptafluorobutyric derivatives. Chromatography was carried out on a 1.8 m×2 mm i.d. column packed with a 1:1 mixture of 10% DC-200 and 15% QF-1, both on Chromosorb W 100–120 mesh. A column temperature of 175°C was employed, and the compounds detected with an ECD. The carrier-gas flow was 40 ml/min nitrogen. Levels of ethoxyquin down to 0.02 ng could be detected by this method. Confirmation of the presence of ethoxyquin was also carried out by GC–MS, on a column packed with 1% SE-30 on 80–100 mesh Chromosorb W, at 110°C.

10.3.3.2 Tocopherols

Tocopherols, although present as natural antioxidants, and having vitamin E activity, in many natural vegetable oils, are often added as oil-soluble antioxidants to foods such as margarines and other oil-based products. There is no method of differentiating between natural and added tocopherols, unless the L-isomers, which do not occur in nature, have been added, either specifically or as part of a racemic mixture of tocopherols. However, added tocopherols are often declared on food labels. HPLC is the method of choice nowadays for the determination of tocopherols, but GC can also be used (IUPAC, 1979). Oil or fat samples are saponified with strong methanolic sodium or potassium hydroxide in brown glass flasks, with pyrogallol added to prevent oxidation of the tocopherols. These are extracted into diethyl ether, which is in turn washed with absolute ethanol, and taken to dryness. Following clean-up on a silica gel TLC plate, to separate the tocopherols from sterols, and other unsaponifiable components, the tocopherols are derivatized with a mixture of HMDS, TMCS and pyridine to form their TMS ethers, before being injected onto a column packed with a medium polarity phase such as OV-17, OV-1701 or CP-Sil 19CB. Both packed columns and capillary columns (25 m×0.32 mm i.d.) can be used, at an oven temperature of between 230°C and 250°C (see Fig. 10.8). However, capillary GC has the advantage that it can separate β and γ tocopherols.

10.4 EMULSIFIERS AND STABILIZERS

10.4.1 Introduction

Emulsifiers and stabilizers are used in a wide range of processed foodstuffs to prevent separation of oil and aqueous fractions, ensuring stability, homogeneity and, hence, quality and consumer acceptability. There are many permitted additives of this type, which include:

alginates
lecithins
polysaccharide gums (e.g. locust bean, xanthan, etc.,)
celluloses
mono and diglycerides of fatty acids
soaps
acetic, lactic, tartaric acid and citric acid esters of mono and diglycerides
sucrose esters of fatty acids
sorbitan esters
sodium and calcium stearoyl-2-lactylate

The nature of many emulsifiers is such that they tend to be large, polyfunctional molecules (often surfactants) and are therefore non-volatile and cannot be directly determined by GC. Accordingly, many of the GC applications in the analysis of emulsifiers and stabilizers involve the breakdown of the molecule into component molecules of greater volatility, and/or derivatization.

10.4.2 Polysaccharides

Polysaccharide gums are long chains formed from various monosaccharide sugars, and are therefore analogous to starch and cellulose. They are used as thickening or

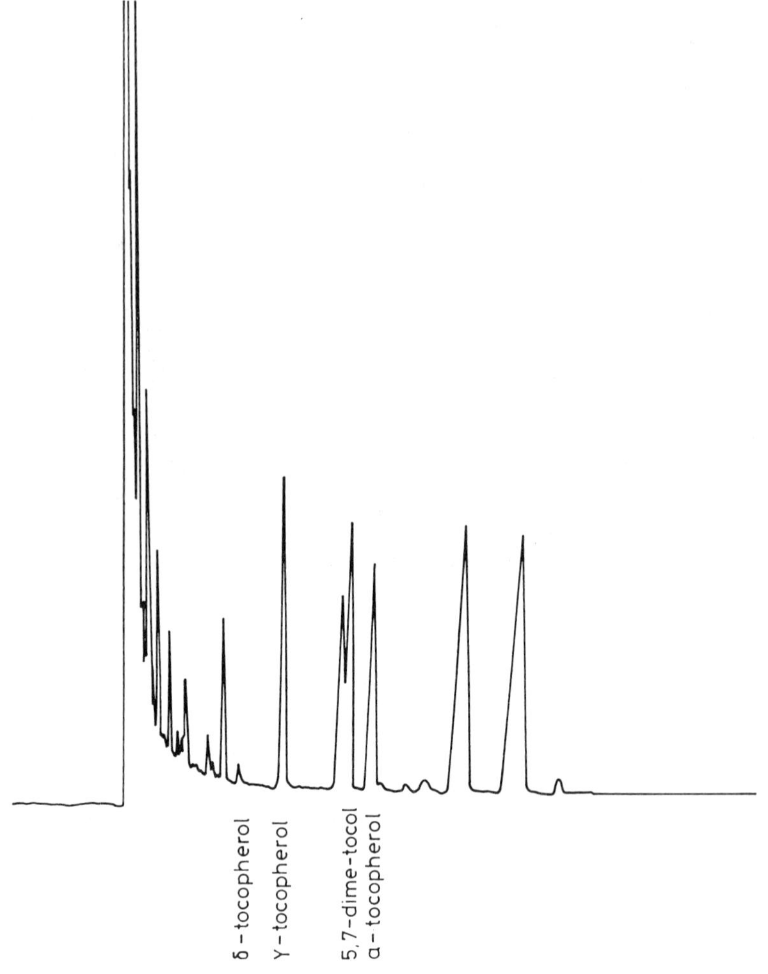

Fig. 10.8 — Gas chromatogram of α, γ and δ tocopherols on a 25-m WCOT CPSIL19 capillary column.

gelling agents in foods such as ice-cream and other processed milk-based foods. There are many of these gums — guar, carrageenan, locust bean and tragacanth are common examples — and they are natural products obtained from plant sources. As they are polymeric molecules they have very high molecular weights, and need to be reduced firstly to their component monosaccharides prior to gas chromatographic analysis. In addition, the monosaccharides are themselves non-volatile compounds, and so need to be derivatized for gas chromatographic determination.

A method for the identification, by GC, of polysaccharide gums has been described (Schmolck and Mergenthaler, 1973). The gums were hydrolysed using methanolic hydrochloric acid, which liberated methylglucosides. This methanolic solution containing the methylglucosides was taken to dryness, further dried over phosphorus pentoxide and derivatized with a mixture containing pyridine, hexamethydisilazane and trimethylchlorosilane (10:2:1) to form the trimethylsilyl ethers of the methylglucosides. The TMS ethers were chromatographed on one of two columns. These were 3.6 m×3.2 mm i.d. packed with 3% OV-17 on 80–100 mesh Chromosorb W HP AW-DMCS, or 1.8 m×3.2 mm i.d. packed with 3% SE-30 on 80–100 mesh Chromosorb G AW-DMCS. Both columns were operated at a temperature of 160°C and a flow rate of 30 ml/min helium. The gums were identified by comparison of the pattern of TMS ethers of the methylglucosides obtained with TMS ethers of methylglucosides derived from standard monosaccharides and alduronic acids, chromatographed under the same conditions. Some of the standards chromatographed gave rise to more than one peak (galacturonic acid gave five). This was found to be due to the possible number of isomers which may be formed on hydrolysis and methanolysis. Guar, carrageenan, agar, pectin, alginates, arabic, tragacanth, karaya and methylcellulose have all been differentiated by this procedure. In addition, if the identity of the particular polysaccharide gum can be established, this procedure does enable it to be quantitatively determined by using a suitable internal standard, such as β-D-phenylglucoside.

A development of this procedure (Mergenthaler and Scherz, 1976) was concerned particularly with the alduronic acid monomers found in many of the more common polysaccharide gums. As in the previously described procedure, the polysaccharide gums were hydrolysed using methanolic hydrochloric acid. The alduronic acids making up the gum matrix were converted to 1-*O*-methylalduronic acid methyl esters in the process. These were then reduced, using sodium borohydride, to give the corresponding methylglycosides, which were in turn split, forming the appropriate aldoses. These aldoses were reacted for 30 min at 90°C with hydroxylamine in pyridine, under conditions which excluded all water to form aldoximes. Further reaction with acetic anhydride for 30 min at 90°C produced aldonitrilacetates, which could be determined gas chromatographically. A 2 m×1/8″ (3 mm) column packed with 3% polyneopentylglycolsuccinate on 100–120 mesh Chromosorb W AW was used. The oven temperature was programmed from 190°C to 230°C at 2°C/min. This method has the advantage that the alduronic acids may be separated more easily as the aldonitilacetates are formed. Work carried out on the analysis of gums in the authors' own laboratory has shown that use of capillary columns can improve the resolution and decrease the analysis time in the analysis of monosaccharides, both as the TMS ethers and as aldonitrilacetates. Twenty-five-metre wall-coated open tubular (WCOT) fused-silica columns coated with either CP Sil-5CB or CP Sil-19CB have proved to be suitable for this type of work. An example is shown in Fig. 10.9.

10.4.3 Polysorbates

Polysorbates are used as emulsifying and thickening agents in food products with high oil or fat levels, such as whipped cream, or high-sugar products, such as cake icing. They are often known under commercial names such as Spans or Tweens. Chemically they are mono or tri-fatty acid esters of sorbitan (see Fig. 10.10), or polyoxyethylene derivatives of these. Commonly used polysorbates include sorbitan mono-oleate and polyoxyethylene sorbitan mono-oleate. Polysorbates are permitted for food use in the United Kingdom, but are not generally permitted within the European Community, and hence do not have 'E' numbers.

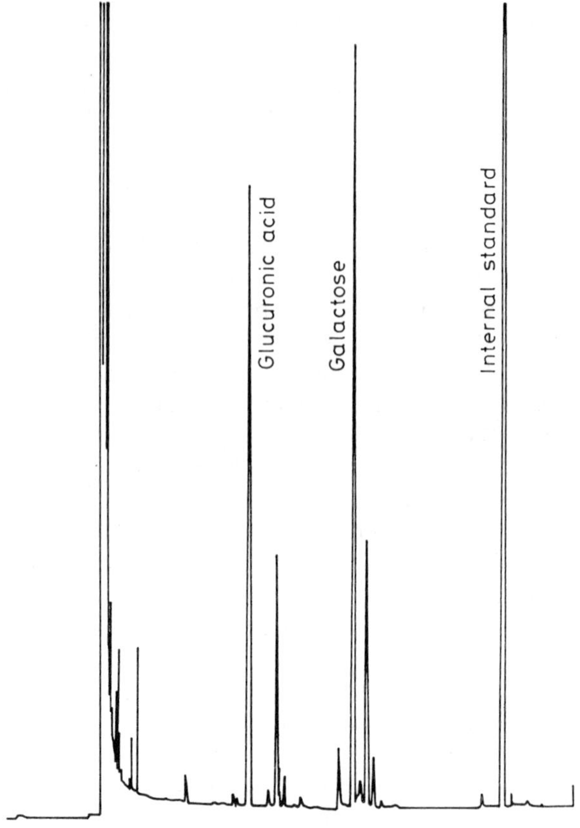

Fig. 10.9 — Gas chromatogram of the TMS ethers of sugars from the hydrolysis of a polysaccharide gum (gum arabic). Injector: 250°C; detector: 350°C; oven: 125°C, then at 4°C/min to 160°C and at 6°C/min to 290°C; carrier gas: helium 1.8 ml/min; split: 40:1; 50 m×0.2 mm WCOT column coated with CP Sil-19 CB.

Polysorbates may be determined following chemical treatment to split the molecules into their component moieties. Of these, fatty acids could be determined straightforwardly, as their methyl esters, by GC on a medium-polarity capillary column such as Silar 5 CP or Carbowax 20M. However, such an approach could lead to erroneous results unless all lipids have been previously removed. In practical terms this means that the fatty acid moieties are not used as an index in the determination of polysorbates.

An early, but simple, approach (Lundquist and Meloan, 1971) was based on the separation of the fatty acid from the sorbitan moiety of the molecule. The sample was firstly blended with water and ethanol, then extracted successively with diethyl ether and petroleum ether. The ether extracts were combined, and washed with aqueous ethanol, then reduced to a very low volume for gas chromatographic analysis. A

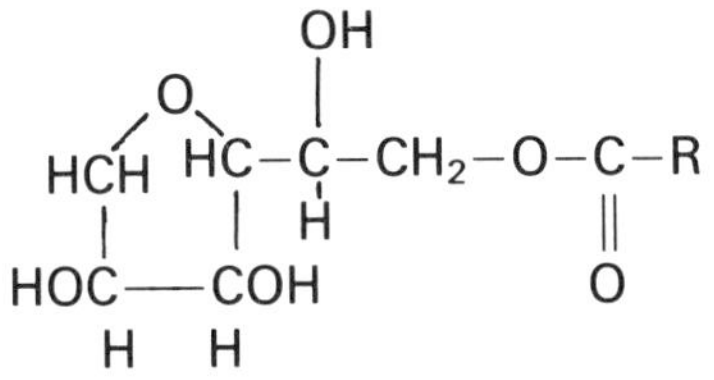

```
               OH
          O    |
        /   \  |
    HCH  HC-C-CH2-O-C-R
     |    |  H      ||
    HOC---COH       O
     H    H
```

Fig. 10.10 — Structure of Sorbitan fatty acid esters.

copper column, 6′×1/4″ i.d. (1.83 m×6 mm i.d.), packed with 15% Carbowax 20M on silanized Chromosorb T was used, at an oven temperature of 120°C, and a carrier-gas flow rate of 38 ml helium per minute. A pre-column packed with 30-mesh soda lime beads was inserted in the GC before the analytical column. This was changed after every 10–12 injections. The pre-column acted to saponify the sorbitan esters in the sample extract and retain the soaps formed as a result. The polyols liberated by this procedure passed through to the analytical column. Glycerol, resulting from the saponification of triglycerides, was separated from the polyols using this system. The polyol mixture could not be resolved into individual compounds, but was determined as ‘polyols’, and related to polysorbate concentration.

A more recent development in the determination of sorbitan fatty acid esters in confectionery products (Tsuda *et al.*, 1984) is based on the chromatographic separation of the trimethylsilyl ethers of polyols. Fat was extracted from the samples with chloroform, and saponified with ethanolic sodium hydroxide. The saponified fraction was acidified with hydrochloric acid, and the fatty acids were removed by extraction with hexane. The aqueous alcohol layer was then neutralized and dried under a stream of nitrogen at 70°C. The resulting polyols present were derivatized with a mixture of pyridine, hexamethyldisilazane and trimethylchlorosilane (10:1:1) to form the TMS ethers. These were subsequently separated on a 1.5 m×3 mm i.d. glass column packed with 2% Dexsil 300GC on 60–80 mesh Chromosorb W AW-DMCS, with the oven temperature-programmed from 120°C to 250°C at 10°C/min. The polyols isosorbide, 1,4-sorbitan and D-sorbitol could be separated under these conditions, although the isosorbide peak occurs on the trailing edge of the solvent front (see Fig. 10.11). These polyols were quantified by comparison with standard polyols, similarly derivatized and chromatographed. It was found that the TMS ethers of the common sugars fructose and glucose did not interfere chromatographically with the polyols under these conditions.

10.4.4 Mono and diglycerides of fatty acids

Mono and diglycerides of fatty acids are used as emulsifiers in many products, although they can occur naturally in high fat products due to adventitious hydrolysis of triglycerides. There are two possible routes by which they may be determined, either as the component fatty acid moieties, or directly as the mono or diglycerides.

A direct gas chromatographic approach has been described (Riva *et al.*, 1981) that uses a capillary column to determine the TMS ethers of mono- and diglycerides (MDG). Following the extraction of the fat from the sample, the MDG were

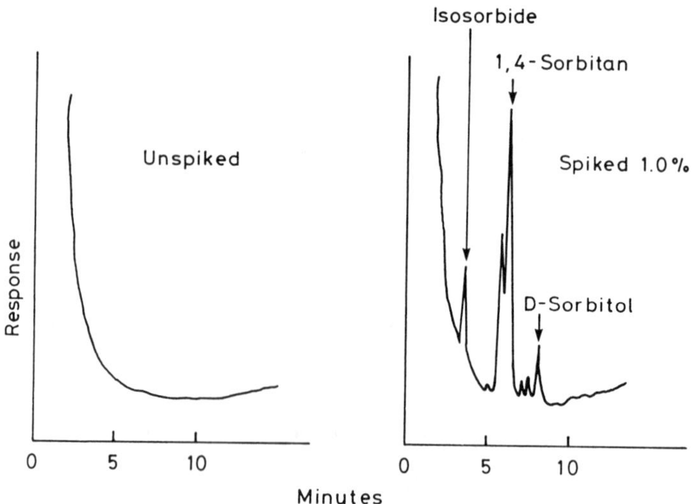

Fig. 10.11 — Gas chromatogram of the TMS derivatives of polyols on a 2% Dexsil 300GC column. (Reproduced from Tsuda *et al.*, 1984 with permission of *JAOAC*.)

separated from other lipid components by TLC on silica gel plates, with a developing solvent comprising petroleum ether, diethyl ether and formic acid (70:30:1.5). The appropriate bands were scraped off the plates, and an internal standard, betulin, added. The TMS ethers of MDG and internal standard were formed by reaction with a mixture of HMDS, TMCS and pyridine (3:1:7), and these were determined by GC on a packed or a capillary column. The packed column used was 2 m×3 mm i.d. packed with 5% OV-17, operated from 200°C to 320°C at a rate of 7°C/min. The capillary column employed was 13 m×0.3 mm i.d., persilanized, coated with SE-52 and programmed from 160°C to 350°C at a rate of 5°C/min, using a carrier-gas pressure of 11 psi (7.6×10^4 Pa) hydrogen. The α and β isomers of the monoglycerides (1- or 2-monoglycerides) were separated on a capillary column as can be seen in Fig. 10.12, but not on a packed column. The extra degree of resolution afforded by the use of a capillary column is important because, in the TLC clean-up step, one of the bands removed from the plate also contains any sterols which may be present in the sample, which do need to be chromatographically separated from the various MDG.

MDG may also be determined, following TLC clean-up, by the saponification and subsequent methylation of the appropriate bands from the TLC plate. The resultant fatty acid methyl esters are then determined on a WCOT column, 50 m in length, coated with a medium-polarity stationary phase, such as Silar 5CP, with heneicosanoic acid methyl ester as the internal standard. An oven temperature of 195°C is used for this separation, with helium as the carrier gas at 1 ml/min, with a split ratio of 80:1.

10.4.4 Other emulsifiers

Sucrose diacetate hexaisobutyrate (SAIB) is used in Canada in soft drinks as a substitute for brominated vegetable oils, which are no longer permitted for food use.

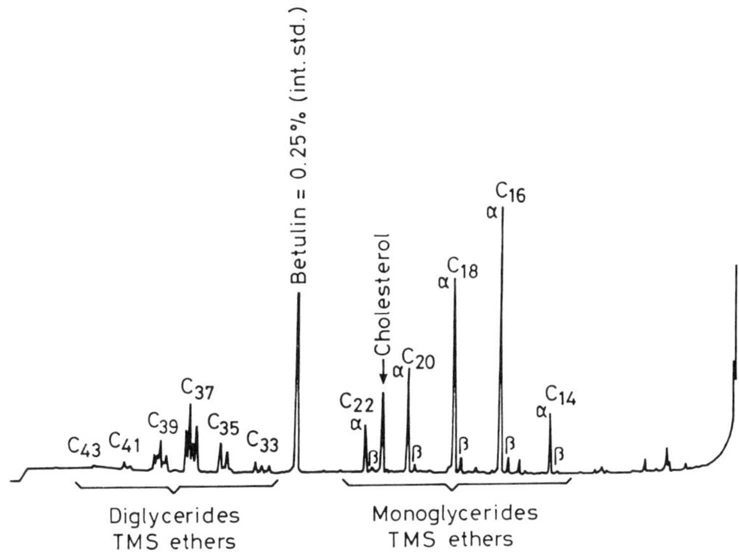

Fig. 10.12 — Gas chromatogram of the TMS ethers of mono and diglycerides on a persilanized capillary column. (Reproduced from Riva *et al.*, 1981 with permission of *La Rivista Italiana Delle Sostanze Grasse*.)

A method for the gas chromatographic determination of SAIB has been described (Conacher *et al.*, 1973) in which samples were decarbonated, and then extracted with diethyl ether after the addition of sodium chloride. These ether extracts were washed with alkali, acid and water and dried, and the solvent was evaporated. At this stage hexanoic acid was added as an internal standard. Decanol containing concentrated sulphuric acid was added, and the samples were heated to 150°C for 90 min, to form the decyl esters of acetic, isobutyric and hexanoic acids. Following neutralization of excess acid with ammonia, the decyl esters were extracted into diethyl ether for gas chromatographic analysis. A 5′×1/8″ (1.52 m×3 mm) stainless steel column packed with 3% SE-30 on 100–120 mesh Varaport 30 was used, with the oven temperature-programmed from 100°C to 290°C at 10°C/min, and a carrier-gas flow of 35 ml/min nitrogen. Peaks for decyl acetate, isobutyrate and hexanoate (internal standard) were obtained (see Fig. 10.13).

An analytical scheme for a number of emulsifiers has been proposed (Conacher and Page, 1979). This applies to a number of esters of MDG, as well as other esters, such as stearoyl lactylate. These were determined following saponification and subsequent formation of butyl esters, by reaction with butanol in the presence of an acid catalyst. The butyl esters of lactic, citric, tartaric and acetic acids were then separated on a column packed with 15% DEGS on Chromosorb W AW, with the oven temperature-programmed from 100°C to 190°C. Fatty alcohols produced from the saponification of the original esters were acetylated, and separated on a 2′ (0.61-m) column packed with a 20% SE-30 stationary phase, with diphenylmethane as the internal standard.

A scheme for the analysis of other emulsifiers was also put forward by Buxtorf *et*

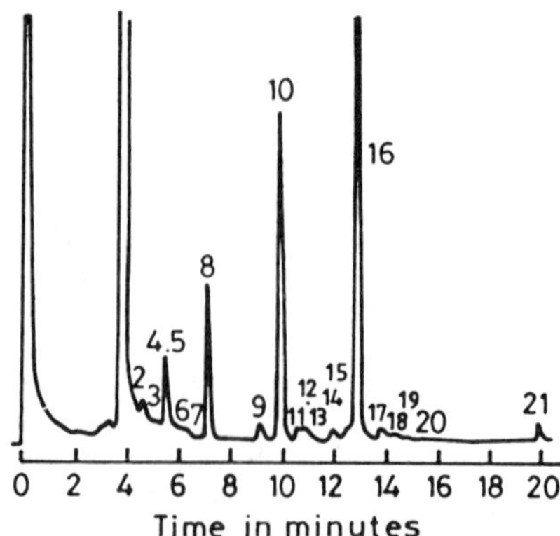

Fig. 10.13 — Gas chromatogram of the decyl esters of acetic and isobutyric acids from sucrose diacetate hexaisobutyrate. Peak 5, decyl acetate; peak 8, decyl isobutyrate; peak 10, decyl hexanoate; other peaks unidentified. (Reproduced from Conacher *et al.*, 1973 with permission of *JAOAC*.)

al., (1982), based on the determination of the products of hydrolysis of these emulsifiers. Emulsifiers were hydrolysed with ethanolic potassium hydroxide, and resulting sugars, polyols and fatty alcohols converted to their trimethylsilyl ethers for determination on a 2.1-m column packed with 5% OV-17 on Varaport 30, temperature-programmed from 80°C to 300°C. The acid moieties from the hydrolysis were determined without derivatization on a 2.1-m column packed with 5% DEGS plus 1% H_3PO_4 on 80–100-mesh Gas Chrom Q, programmed from 90°C to 180°C. These procedures have enabled a wide range of emulsifiers to be determined in a simple unified analytical scheme.

10.5 FLAVOURS

10.5.1 Introduction

A great variety of compounds and substances are added to foods and drinks to impart, enhance or modify their flavour. These compounds are often derived or extracted from natural sources, such as vanillin from vanilla beans, or are synthetic chemicals that have been previously been identified as being important in natural flavourings. Many of these are permitted for use as flavouring additives, and are listed in 'The Blue Book' (Council of Europe, 1974). Often flavours are added to food and drink as 'cocktails'; the individual concentrations of these may very often be in the sub-p.p.b. range. These cocktails are specifically designed by the flavour chemist to simulate the natural counterpart, and contain the same compounds. These are called nature-identical flavours, and differ from the corresponding natural flavour extract in fine detail and the composition of minor or trace constituents, if at all. This is an area of analytical chemistry where very little work has so far been carried out, but with the probability of forthcoming EC legislation in this area of flavours, it is to be expected that this will be remedied. In so doing, I would expect that capillary GC–MS would be the technique of choice to carry out what will be a very complex task.

Many flavours are, by their very nature, volatile or semi-volatile, and therefore amenable to gas chromatographic determination. For the more volatile flavouring compounds, headspace GC has a wide application, and for the resolution of the complex mixtures often used to flavour food, capillary columns are often essential.

In addition, another class of compounds, the determination of which is very important, are the active flavouring principles. These are present as minor constituents, usually in flavours extracted from natural sources, and have maximum limits on the concentration in which they may be present in particular foods and drinks due to their possible toxicity,

10.5.2 Artificial flavours

Many methods have been developed and published for the separation, quantification and identification by GC of naturally occurring flavour and aroma compounds in foods and drinks. These have been dealt with at length in Chapter 3. Many of these methods, both in principle and detail, can often be applied to foods which are flavoured, or which have had their flavour enhanced or modified by added artificial flavourings. A rare example of a method which has been developed to differentiate between natural and synthetic flavours (Pfannhauser *et al.*, 1982) applies to the analysis of coconut flavours using GC–MS. Flavours were extracted from coconut milk, ground coconut flesh, or other samples, with double distilled *n*-pentane, and the volume of the extract reduced under nitrogen. The extracts were subjected to gas chromatographic analysis on a capillary column (20 m×0.19 mm i.d.) coated with SE-54, the column temperature being programmed from 100°C to 200°C at 4°C/min (see Fig. 10.14). The separated compounds were identified by MS. The major

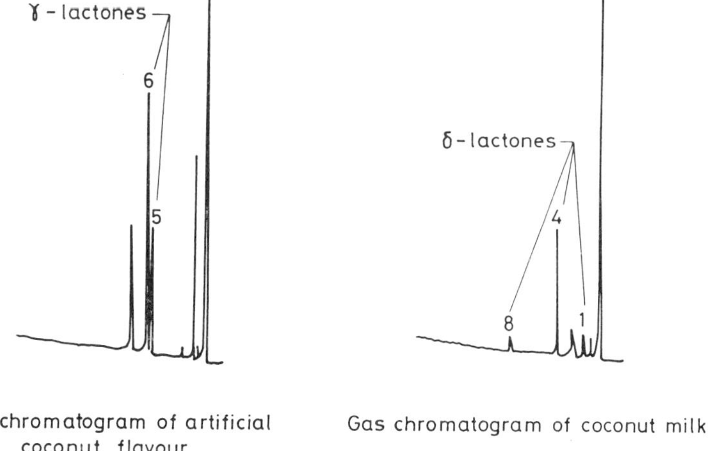

Fig. 10.14 — Gas chromatograms showing the differences between synthetic (left) and natural (right) coconut flavours, using a 20-m WCOT SE-54 capillary column. (Reproduced from Pfannhauser *et al.*, 1982 with permission of Springer-Verlag.)

components from the extract of natural coconut milk were identified as being the δ-lactones octalactone, decalactone and undecalactone. Samples which had been artificially flavoured, or had their natural flavour enhanced, were found to contain γ-lactones, such as γ-nonalactone. In the case of a product with an enhanced flavour, these appear on the chromatogram in addition to the naturally-occurring δ-lactones. In addition, for products which contained *only* synthetic coconut flavour, glycerol triacetate (triacetin) could also be identified. This compound is used in the flavour industry as a diluent or solvent for flavouring compounds, but imparts no flavour or aroma to the product. Therefore, it can be used as an index for the addition of synthetic, or even nature-identical flavourings, as it would be present in comparatively high concentrations. Glycerol diacetate (diacetin) and propane-1,2-diol (propylene glycol) are also used as flavour diluents.

Many published techniques which use GC for the determination of added flavours concentrate on specific compounds and/or products. An early example of this considered the flavours used in cola beverages (Young *et al.*, 1970). The cola samples were extracted using the Likens and Nickerson apparatus, with pentane as the solvent. The extract was dried and reduced in volume for gas chromatographic analysis. A stainless steel column, 20 m×1/4″ (6 mm) i.d., was used, packed with 10% Triton X-305 on 60–80 mesh Diatoport S, in an oven temperature-programmed from 160°C to 210°C at 6°C/min and a carrier-gas flow rate of 30 ml/min nitrogen. Under these conditions, several important compounds could be separated, including benzaldehyde, amyl butyrate, isovaleric aldehyde, ethyl butyrate and ethyl heptanoate. By measurement of the relative concentrations of these components, it was possible to distinguish easily between the two leading brands of cola drink on retail sale.

Ammonium glycyrrhizinate, prepared from the hot water extract of liquorice root, is used as a flavouring in sweet desserts and soft drinks because of its high sweetening property. A method has been described for its determination, as the silyl ether (Larry *et al.*, 1970). The ammonium glycyrrhizinate was hydrolysed with acid in a water–dioxane medium, to split the aglycone glycyrrhetic acid from the disaccharide moiety. The chloroform extract was dried, cholesterol added as an internal standard, and silylated with a mixture of pyridine, trimethylchlorosilane and *N,O*-bis(trimethylsilyl)acetamide (10:2:3). The silyl ethers were chromatographed on a 4′ (1.22 m)×4 mm i.d. glass column packed with 1.5% OV-1 on 60–80 mesh Gas Chrom Q. The column temperature was programmed from 200°C to 260°C at 6°C/min, with nitrogen as the carrier-gas at a flow rate of 75 ml/min.

Maltol (2-hydroxy,3-methyl 4-pyrone) is frequently used as a flavouring or flavour enhancer, and smells like candyfloss. Ethyl maltol (3-ethyl,2-hydroxy 4-pyrone) is also used, and is a more powerful flavour enhancer. These two compounds have been determined gas chromatographically in apple juice (Gunner *et al.*, 1967). The samples were extracted with ethyl acetate, 2,6-dimethoxyphenol added as the internal standard, and the solution taken to dryness on a rotary evaporator. The residue was silylated using a mixture of trimethylchlorosilane, hexamethyldisilazane and pyridine for GC analysis. A 6′×1/8″ o.d. (1.83 m×3 mm o.d.) stainless steel column was used, packed with 10% UCW-98 on 80–100 mesh Diatoport S, and operated at a temperature of 120°C. Work has recently been carried out in the author's laboratory using a 25-m WCOT capillary column coated with CPSil-5CB at

130°C to determine the TMS ethers of maltol and ethyl maltol in samples of edible oils (see Fig. 10.15). The oils were reacted with the silylating medium, and injected

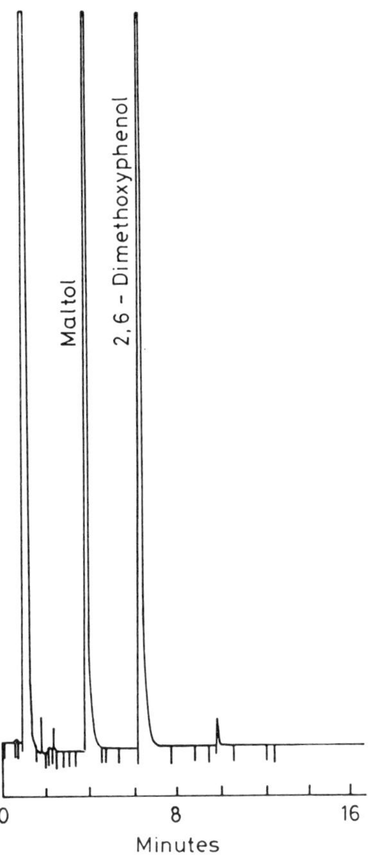

Fig. 10.15 — Gas chromatogram of maltol TMS ether on a 25-m WCOT CP Sil5-CB column.

directly onto the GC, with a glass liner packed with silanized glass wool in the injector to absorb the underivatized oil components.

Vanillin and ethyl vanillin are added to many foods, particularly confectionary and dessert products, to impart vanilla flavour, as an alternative to natural vanilla essence. Although many methods for the determination of vanillins involve the use of HPLC, GC has been successfully used in the determination of whether vanilla flavours are of natural or synthetic origin. Headspace GC has been used, with the samples equilibrated at 80°C prior to injection onto a 25-m WCOT column coated with SE-30, and temperature-programmed from 60°C to 195°C, with a carrier-gas flow of 1 ml/min helium.

Propylene glycol is used as a solvent for artificial flavourings, and has been determined by GC (Kajiwara *et al.*, 1981) in a variety of foods. Samples were extracted by refluxing with methanol, filtered, centrifuged and concentrated down to low volume. An aliquot of the resulting solution was injected onto a 1.5 m×3 mm column packed with 20% PEG-20M on Celite 545, AW-DMCS 60–80 mesh, with an oven temperature of 160°C and a carrier-gas flow of 50 ml/min nitrogen. Diethylene glycol–monoethyl ether (DEG-MEE) is used for a similar purpose, and has been determined in ice-cream by GC (Ogawa *et al.*, 1986). Samples were refluxed with sodium hydroxide, and extracted with dichloromethane following addition of sodium chloride. An improved procedure involved washing with hexane after the initial addition of sodium hydroxide. The dichloromethane layer was dried over anhydrous sodium sulphate and reduced in volume. If high levels of propylene glycol are present, this solution should be further washed with water. The DEG–MEE was determined on a 3 m×3 mm glass column packed with 25% PEG-20M on 60–80 mesh Chromosorb W AW-DMCS, at an oven temperature of 137°C and a carrier-gas flow rate of 40 ml/min nitrogen. Under these conditions, DEG–MEE was separated completely from any propylene glycol present.

10.5.3 Active principles

Active principles are alkaloid, terpenoid or aromatic components found in flavours, many of which are used to flavour alcoholic drinks. Many methods using GC have been developed for some of these: coumarin, safrole, thujone, pulegone and β-asarone.

10.5.3.1 β-Asarone

β-Asarone (*cis*-2,4,5-trimethoxy-1-propenylbenzene) is found in wines flavoured with calamus (wild sweet flag). A method based on GC following distillation has been described (Dyer *et al.*, 1976) in which samples were distilled after the addition of sodium chloride. Ethyl palmitate was added as an internal standard, and an aliquot was chromatographed on a 6′ (1.83 m)×2 mm i.d. column packed with 10% SP-1000 on 80–100 mesh Chromosorb W HP. The oven temperature employed was 180°C, with a carrier-gas flow of 40 ml/min helium. This procedure can determine β-asarone concentrations down to 0.5 mg/l, but the introduction of a step to reduce the volume of the distillate could improve this.

An alternative procedure (Larry, 1973) involves the steam distillation of the sample followed by solvent extraction with diethyl ether: hexane (1:1). The concentrated extract was determined by GC on a column packed with 5% Reoplex-400 on 60–80 mesh Chromosorb G, at 180°C and a carrier-gas flow of 90 ml/min nitrogen.

10.5.3.2 Coumarin

Coumarin (1,2-benzopyrone) is also found in wine that has been flavoured with woodruff. Some confectionary products may also contain coumarin. Dyer *et al.*, (1975) developed a GC procedure for the determination of coumarin in May wine. Samples were extracted with chloroform, and following centrifugation, the chloroform extract was subjected to gas chromatographic analysis on a 6′ (1.83 m)×2 mm i.d. glass column packed with 10% SP-1000 on 100–120 mesh Chromosorb W AW at 170°C. This procedure enabled levels of coumarin below the FDA (Food and Drug Administration) limit of 5 mg/kg to be rapidly detected. A development of this procedure forms the basis of the IOFI recommended method (Dyer and Martin, 1976). In this method samples are neutralized with sodium hydrogen carbonate and saturated with anhydrous sodium sulphate before extraction with diethyl ether. Solid samples such as ice-cream, chocolate confectionery, etc., are steam-distilled using a modified Schmidt apparatus and 250 ml distillate is collected and extracted with diethyl ether. The ether extract is dried over anhydrous sodium sulphate, and evaporated to dryness prior to addition of benzyl ether as the internal standard. This solution is diluted to 2 ml with methanol and determined gas chromatographically on a 2 m×4 mm i.d. glass column packed with 10% Carbowax 20M on 60–80 mesh Chromosorb W, at an oven temperature of 180°C and a carrier-gas flow of 30 ml helium per minute. Work carried out by the author (1982) has shown that better results can be obtained if samples, of any type, are steam-distilled prior to extraction, and the internal standard added before steam distilling, to improve the accuracy of the method. It was also found that the oven temperature of the GC could be increased to 210°C to speed up the chromatography. (See Fig. 10.16.)

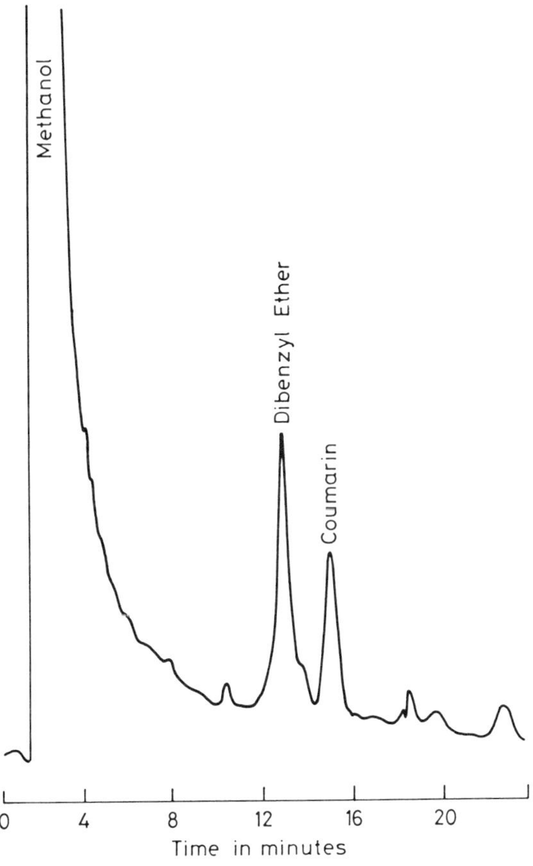

Fig. 10.16 — Gas chromatogram of extract of sweets spiked with coumarin at 210°C on a 10% Carbowax 20M column.

10.5.3.3 Pulegone

Pulegone derives from pennyroyal, peppermint, spearmint and other botanical sources and is found in mint-flavoured or peppermint beverages, both alcoholic and non-alcoholic, and in mint-flavoured confectionary products, such as extra-strong mints. The IOFI-recommended method for the determination of pulegone is that of the Lausanne Cantonal laboratory in Switzerland (1971). Fat-containing samples are steam-distilled using the modified Schmidt apparatus to give 150 ml distillate, which is extracted with diethyl ether following the addition of sodium chloride. Other sample types are dissolved in or diluted with water, and similarly extracted with diethyl ether after addition of sodium chloride. The ether extracts are dried and ethyl decanaote added as an internal standard, prior to injection onto a glass column, 2.5 m×4 mm i.d., packed with Carbowax 20M on 80–100 mesh Chromosorb G, at an oven temperature of 125°C. The presence of pulegone needs to be confirmed by GC–MS. There are some drawbacks to this method, however, in that the internal standard and pulegone elute very close to each other, and can often co-elute. Additionally, under these chromatographic conditions pulegone and menthol, which is a considerably more common component of mint flavourings, also co-elute, which explains the need for GC–MS confirmation. Boley & Cohen (1983) developed a modified chromatographic procedure, using a capillary column. A 25 m×0.2 mm i.d. WCOT fused-silica column coated with SE-30 was used, with the oven temperature-programmed from 90°C to 135°C at 5°C/min, after a 5-min hold. Helium was used as carrier gas, at 2 ml per minute, and a split injection technique used, with a 20:1 split ratio. As can be seen from Fig. 10.17 pulegone and menthol are well resolved under these conditions, thereby reducing the need for GC–MS confirmation.

10.5.3.4 Safrole and isosafrole

Safrole and its isomer, iso-safrole, are found in wines and vermouths that have been flavoured with wormwood, nutmeg or sassafras. An early gas chromatographic method for the determination of these compounds, as well as others, such as methyl salicylate, was described by Larry (1971). In this method samples were subjected to steam distillation (solid samples being dissolved initially in an appropriate volume of water) following addition of an internal standard, *m*-tolyl acetate. The distillate was extracted with chloroform, dried by filtration through anhydrous sodium sulphate and evaporated to a low volume for GC analysis. Separation was achieved on a 2.5 m×4 mm i.d. glass column packed with 4% Carbowax 20M on 80–100 mesh Chromosorb G at a temperature of 125°C, and a flow rate of 40 ml/min helium. Recoveries obtained using this procedure varied from 77% to 98% for safrole, but were less satisfactory for isosafrole. The proposed IOFI method is based upon this procedure, but utilizes dichloromethane or pentane as the extracting medium, although pentane is preferred as dichloromethane often produces emulsions during extraction. Recoveries obtained using this method varied from 95% to 100% for both safrole and iso-safrole. A chromatogram showing this separation is shown in Fig. 10.18.

10.5.3.5 Thujone

Thujone is present as an active principle in many plants, of which sage is one of the more common, and is found in sage-containing foods, alcoholic drinks, and bitters.

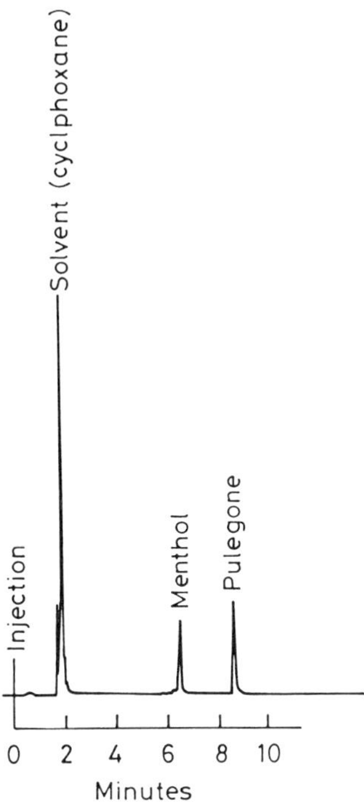

Fig. 10.17 — Gas chromatogram showing separation of pulegone and menthol on a 25-m WCOT OV-351 column.

The proposed IOFI procedure for the determination of α and β thujone involves steam distillation into pentane, the pentane layer being dried over anhydrous magnesium sulphate and evaporated to dryness. The residue was dissolved in cyclohexane for GC analysis on a 2.5 m×4 mm i.d. glass column packed with 4% OV-210 on 80–100 mesh Chromosorb G at an oven temperature of 90°C and a carrier-gas flow of 35 ml/min helium. Quantification was achieved by use of a standard addition technique. The α and β forms of thujone could not be resolved.

The use of capillary-column GLC has been favoured for the determination of thujone. An early use of this type of column (Liddle and de Smedt, 1976) involves the use of pentane to extract the thujone from the sample, and the reduction of the extract volume on a Kuderna–Danish apparatus. The concentrated pentane extract was examined chromatographically using a 30-m SCOT column coated with FFAP or modified Carbowax 20M. The oven temperature was programmed from 80°C to 210°C at 4°C/min, in order to separate thujones effectively from co-extractants (other flavours and volatiles) in the sample. MS was used to confirm the identity of α and β thujones. Safrole, β-asarone and coumarin could also be determined using this procedure. More recently Boley and Cohen (1983) chromatographed thujone successfully on a 25-m WCOT fused-silica column coated with OV-351. A temperature programmed from 60°C to 135°C was employed, with helium as carrier gas at

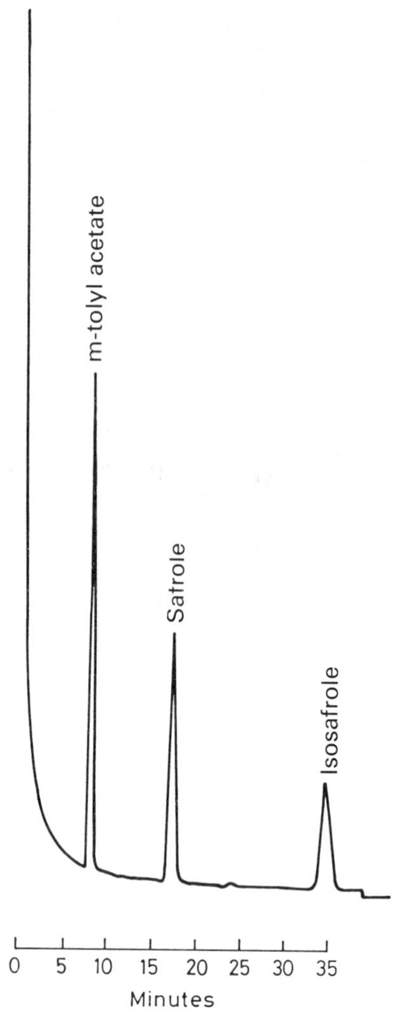

Fig. 10.18 — Gas chromatogram of safole and isosafrole. Column: 2.5 m×4 mm i.d. 4% Carbowax 20M on chromosorb G, 80–100 mesh; flow rate: 40 ml/min helium; injector temperature: 170°C; detector temperature: 170°C (FID); oven temperature: 125°C; attenuation: 1×16; chart speed: 2 mm/min; injection volume: 5 μl.

1 ml/min. Again, both α and β thujones were separated, and resolved from co-extractants from the sample matrix.

10.6 ADDITIVES IN WINE

10.6.1 Introduction

Many of the classes of additives found in other foods, and non-alcoholic drinks, are also used in wines — preservatives, for example.

However, some additives considerably less desirable than these can also be found in wine. This is a problem which has come into sharper public focus in recent years with scandals regarding the addition of diethylene glycol and methanol to wine, in particular lower-quality wines.

10.6.2 Diethylene glycol

The development of methodology, using GC and other techniques, has only really come about, to any serious degree, since the recent scandal involving the adulteration of wine, particularly from Austria, with diethylene glycol (DEG). This additive, harmful if ingested in sufficiently large quantities, imparts a degree of sweetness to the wine, which is often expected, but not naturally present. The resulting concern in the wake of this scandal led, by necessity, to the rapid development of sensitive and accurate methods, often using GC, for the determination of DEG in wines.

Some of these methods were developed by government regulatory authorities, in both the UK and other European countries. There are two general approaches to this particular analytical problem: direct determination of DEG on a polar phase column, or derivatization followed by separation on an apolar phase column.

The former approach was developed initially in Austria (1985). In this procedure the wine was diluted 1:1 with absolute ethanol prior to injection onto a 25 m×0.3 mm i.d. fused-silica capillary column coated with Carbowax 20M. A split-injection technique was employed (20:1 split ratio), with helium as carrier gas at 2 ml/min, and an isothermal oven temperature of 150°C. Quantification was achieved by use of butane-1,4-diol as an internal standard, and levels of DEG down to 5 mg/l could be determined. Confirmation of the presence of DEG was carried out by MS of the GC effluent over the *m/z* range 35 to 110, as possible chromatographic interferences, which could result in false positives, have been detected using this procedure.

Methodology developed in UK regulatory laboratories has followed the derivatization route. Using this procedure, the wine (5 μl was derivatized by reaction with bis (TMS)trifluoroacetamide (BSTFA) to form the TMS ethers of any compounds containing hydroxyl groups. A range of derivatization conditions have been successfully employed, ranging from heating at 60°C for 15 min, to 90°C for 45 min. The derivatized sample was diluted with acetonitrile (100 μl) and injected onto the GC. A capillary column, coated with CPSil-5CB, or equivalent was used. A 25-m column has been used with some success, but a 50-m column does give slightly improved resolution, which is important as a very complex chromatogram is almost always obtained. A split injection was used, with either hydrogen as the carrier gas at 1.8 ml/min and a split ratio of 20:1, or helium at 30 cm/s and a split ratio of 40:1. The oven temperature was set at 100°C and the TMS ethers detected either by FID, or using an MS (e.g. an ion-trap detector), carrying out SIM at *m/z* values of 191 and 235. The limit of detection obtained using this procedure was 10 mg/l, but this could only be obtained using MS detection — it was significantly higher using an FID. It has been recommended that, in order to check whether there may be possible chromatographic interferences from wine under these conditions, a 'blank' wine, known to be free of DEG, should be run initially. Fig. 10.19 shows a chromatogram of DEG in wine using this procedure.

A more recent development from Japan (Hori *et al.*, 1986) employed an

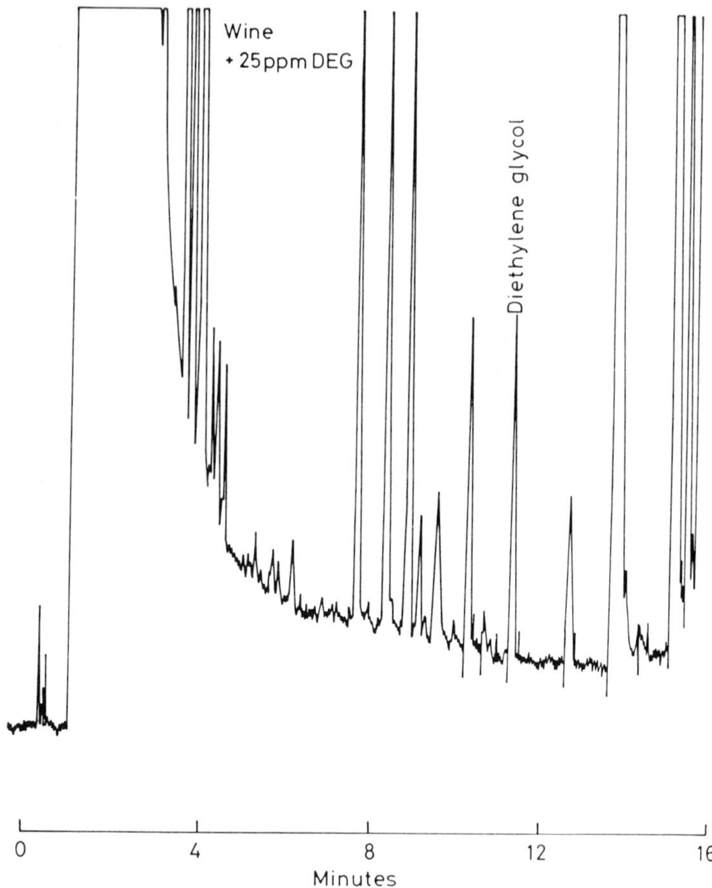

Fig. 10.19 — Gas chromatogram of diethylene glycol in wine as the TMS ether on a 50-m WCOT CPSil-5 column using an ion-trap detector.

additional clean-up step. The wine was cleaned up by passing through an alumina chromatography column, the DEG being retained and eluted subsequently with methanol. A packed column was used, 1.5 m×3 mm, packed with 20% PEG 20M on Chromosorb W AW-DMCS, 80–100 mesh, at 205°C and a flow rate of 40 ml/min helium. A capillary column using a similar stationary phase (e.g. CW20M) could also be used. The use of the clean-up step removes interferences and results in more certain identification of DEG and a reduced analysis time. DEG levels of 10 mg/l could be detected by this procedure.

10.6.3 Methanol

Methanol has recently been added as an adulterant to some wines. Most of the existing gas chromatographic methods for the determination of methanol (which occurs naturally in trace levels as a fermentation by-product) pre-date the recent scandal. An early procedure (Lee *et al.*, 1975) involved the direct injection of the wine sample onto the gas chromatograph. A stainless steel column, 2 m×2 mm i.d. was used, packed with Porapak QS, operated at 115°C with a carrier-gas flow of 40 ml/min nitrogen. The injection volume was 3 μl, and levels of methanol down to 5 mg/l could be detected. Quantification was achieved by reference to standard aqueous solutions of methanol, similarly chromatographed. A more recent approach (Rottsahl & Jessen, 1987) utilizes headspace GC. The wine samples were placed in headspace vials, and equilibrated at 80°C for 20 min, prior to injection, which was carried out automatically. It was found to be necessary to add alkaline silver nitrate to the vial before equilibration in order to remove any acetaldehyde, which would otherwise interfere chromatographically with the methanol. A capillary column, fused silica coated with OV-101, film thickness 0.4 μm, was used, at a flow rate of 0.76 ml/min helium and a split ratio of 60:1. The oven temperature was programmed from 50°C to 200°C at 40°C/min. *tert*-Butanol was added to the sample as an internal standard, and levels of methanol down to 10 mg/l could be detected. A typical chromatogram is shown in Fig. 10.20.

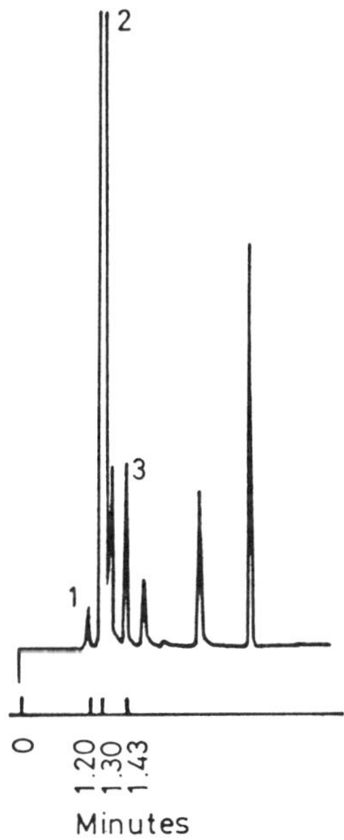

Fig. 10.20 — Gas chromatogram of methanol in wine on a 40-m WCOT carbowax 20M column. (Reproduced from Rottsahl & Jessen, 1987 with permission of *Deutsche Lebensmitteln Rundschau*.)

10.6.4 Other additives in wine

Some other additives found specifically in wines have been determined by GC. Diethyl pyrocarbonate has been added to wines to prevent fermentation, and is hydrolysed *in situ* to form diethyl carbonate (Wunderlich, 1972). The diethyl carbonate was extracted into carbon disulphide for GC determination on a 6′×1/8″ (1.83 m×3 mm) stainless steel column packed with 15% trimethylolpropantripelargonate on 60–100 mesh Celite 545. A column temperature of 80°C was employed with a carrier-gas flow of 35 ml/min nitrogen. Levels of diethyl carbonate down to 1 mg/l in wine could be determined.

Urethane can also be formed from the breakdown of diethyl pyrocarbonate at neutral or alkaline pH conditions (Joe *et al.*, 1977). The wine samples were centrifuged, hot sodium borate added to bring the pH to 7.5, and then sodium chloride added and dissolved. This was extracted with dichloromethane, the dichloromethane layer dried over anhydrous sodium sulphate and passed through a chromatography column packed with Celite or alumina. The cleaned-up extract was subjected to gas chromatographic analysis on a 12′ (3.66 m)×4 mm glass column packed with 10% ethyleneglycol adipate on 80–100 mesh Chromosorb W AW, at 150°C and a helium carrier-gas flow of 40 ml/min. Detection was achieved using either an FID or an alkali flame ionisation detector (KCl type). With the latter detector a 9′ (2.74-m) column was used, and the oven temperature-programmed from 100°C to 200°C at 5°C/min, using argon as the carrier gas at 70 ml/min.

10.7 REFERENCES

Bajardi, M. L., Giannola, L. I. & Giammona, G. (1982) Determinazione di conservati fenolici e loro derivati mediante GC-MS, *Riv. Soc. Ital. Sci. Aliment.*, **11**, 105–110.

Barnett, D. & Davis, E. G. (1983) A GC method for the determination of sulphur dioxide in food headspaces., *J. Chrom. Sci.*, **21**, 205–208.

Bertrand, A. & Sarre, C. (1978) Dosage de l'acide benzoique dans les sodas et autres produits alimentaires liquides par chromatographie en phase gazeuse, *Ann. Fals. Exp. Chim.*, **71**, 35–39

Boley, N. P. & Cohen, I. C. (1983) Evaluation of methods for the determination of active flavouring principles, (Internal Report).

Buxtorf, U. P., Brïschweiler, H. & Martin, E. (1982) Identifizierung von emulgatoren mittels gaschromatographie, *Z Lebensm. Unters. Forsch.*, **175**, 25–28.

Ciudad, C. (1975) Determinación de conservadores en alimentos medoiante cromatografia gas líquida. *Agricultura Tecnica (Chile)*, **35**, 209–215.

Conacher, H. B. S., Chadha, R. K. & Iyengar, J. R. (1973) Determination of sucrose diacetate hexaisobutyrate in soft drinks by gas–liquid chromatographic analysis of isobutyric and acetic acid components as decyl esters. *JAOAC*, **56**, 1264–1266.

Conacher, H. B. S. and Page, B. D. (1979). Derivative formation in the chromatographic analysis of food additives. *J. Chromatogr. Sci.*, **17**, 188–195.

Council of Europe (1974) *Natural Flavouring Substances, their Sources, and added artificial flavouring substances,* Maisonneuve, Sainte-Ruffine, France.

Daenens, P. & Laruelle, L. (1973) Column chromatographic clean-up and gas–liquid chromatographic determination of hydroxybenzoic esters in food. *JAOAC*, **56**, 1515–1517.

Davis, P. L. & Munroe, K. A. (1977) Determination of biphenyl by gas and liquid chromatography *J. Agric. Food. Chem.*, **25**, 426–428.

Dilli, S. & Robards, K. (1977) Comparative gas chromatographic behaviour and detection limits of 2,6-di-tert butyl-4-methylphenol, 3-*tert*-butyl-4-hydroxyanisole (BHA) and the trifluoroacetate of BHA. *J. Chromatogr.*, **133**, 363–366.

Dyer, R. H., Martin, G. E. & Butts, B. B. (1975) Gas–liquid chromatographic determination of coumarin (1,2-benzopyrone) in May wine. *JAOAC*, **58**, 140–143.

Dyer, R. H., Martin, G. E. & Buscemi, P. C. (1976) Gas–liquid chromatographic determination of beta-asarone in wines and flavours. *JAOAC*, **59**, 675–677.

Dyer, R. H. & Martin, G. E. (1976) Collaborative study of the gas–liquid chromatographic determination of coumarin (1,2-benzopyrone) in May wine. *JAOAC*, **59**, 780–782.

Dyer, R. H. (1977) Gas–liquid chromatographic determination of beta-asarone in wine: collaborative study. *JAOAC*, **60**, 1041–1043. *Emulsifiers and Stabilisers in Food Regulations* (1980), HMSO.

Fogden, E., Fryer, M. & Urry, S. (1974) The detection and determination of some food preservatives by gas chromatography. *J. Assoc. Publ. Analysts*, **12**, 93–102.

Galensa, R. & Schäfers, F. I. (1982) Hochleistungflüssigkeits- und kapillargaschromatographische bestimmung von phenolischen antioxidanten in lebensmitteln mit komplexer matrix. *Deutsche Lebensmitteln-Rundschau*, **78**, 258–260.

Gosselé, J. A. W. (1971) Gas chromatographic determination of preservatives in foods. *J. Chromatogr.*, **63**, 429–432.

Graveland, A. (1972) Gas chromatographic determination of propionic, sorbic and benzoic acids in rye bread and margarine. *JAOAC*, **55**, 1024–1026.

Gunner, S. W., Hand, B. & Sahasrabudhe, A. (1968) Determination of maltol and ethyl maltol in apple juice by gas–liquid chromatography. *JAOAC*, **51**, 959–962.

Hild, J. & Gertz, C. (1980) Möglichkeiten der analytischen erfassung von konservierungsstoffen in lebensmitteln. *Z. Lebensm. Unters. Forsch.*, **170**, 100–114.

Hori, Y., Chonan, T. & Nishizawama, M. (1986) Rapid determination of diethylene glycol in wine by gas chromatography using alumina column clean-up. *J. Food Hygienic Society of Japan*, **27**, 187–189.

International Union of Pure and Applied Chemistry (1987) *Standard Methods for the Analysis of Oils, Fats and Derivatives, 7th Edition*, Pergamon Press, Oxford, p 123.

Isshiki, K., Tsumura, S. & Watanabe, T. (1980a) Determination of diphenyl, *o*-phenylphenol and thiabendazole in citrus fruits. *Nippon Nôgeikagaku Kaishi*, **54**, 1045 1050.

Isshiki, K., Tsumura, S. & Watanabe, T. (1980b) Determination of preservatives butylhydroxyanisole and dibutylhydroxytoluene. *Agric. Biol. Chem.*, **44**, 1601–1607.

Iwaida, M., Ito, Y., Toyoda, M. & Suzuki, H. (1977) Detection and determination of *p*-hydroxybenzoates in cold-pressed wheat germ oil. *Bulletin of National Institute of Hygienic Sciences Eisei Shikenjo Hokuku*, **95**, 54–56.

Jedrych, Z. Karlowski, K. (1979) Determination of BHA and BHT in certain food products by GC. *Roczniki Panstwowego Zakladu Higieny*, **30**, 39–42.

Joe, F. L., Kline, D. A., Miletta, E. M., Roach, J. A. G., Roseboro, E. L. & Fazio, T. (1977) Determination of urethane in wines by gas–liquid chromatography and its confirmation by mass spectrometry. *JAOAC*, **60**, 509–516.

Kajiwara, N., Kawai, H. & Hosogai, Y. (1981) Determination of propylene glycol in commercial foods by gas chromatography. *Nippon Shokuhin Kogyo Gakkaishi,* **28**, 471–475.

Kline, D. A., Joe, F. L. & Fazio, T. (1978) A rapid gas–liquid chromatographic method for the multi-determination of antioxidants in fats, oils and dried food products. *JAOAC,* **61**, 514–519.

Larry, D. (1971) Gas chromatographic determination of safrole and related compounds in nonalcoholic beverages: collaborative study. *JAOAC,* **54**, 903–905.

Larry, D. (1973) Gas chromatographic determination of beta-asarone, a component of oil of calamus, in flavours and beverages. *JAOAC,* **56**, 1281–1283.

Larry, D., Fuller, M. J. & Harrill, P. G. (1970) Quantitation of ammonium glycyrrhizate by gas–liquid chromatography of the 4 silyl ether ester derivative of the aglycone. *JAOAC,* **53**, 698–700.

Larsson, B. K., (1983) Gas–liquid chromatographic determination of benzoic acid and sorbic acid in foods: NMKL Collaborative study. *JAOAC,* **66**, 775–780.

Larsson, B. K. & Fuchs, G. (1974) Quantitative determination of benzoic acid, sorbic acid and esters of 4-hydroxybenzoic acid in food products by gas–liquid chromatography. *Swedish J. Agric. Res.,* **4**, 109–110.

Lee, C. Y., Acree, T. E. & Butts, R. M. (1975) Determination of methyl alcohol in wine by gas chromatography. *Anal. Chem.,* **47**, 747–748.

Liddle, P. A. P. & de Smedt, P. (1976) Détection et dosage de quatre composés limites par diverse législations dans les boissons alcooliques. *Ann. Fals. Exp. Chim.,* **69**, 857–864.

Lord, E., Bunton, N. G. & Crosby, N. T. (1978) The determination of biphenyl and 2-hydroxybiphenyl in citrus fruits. *J. Assoc. Public Analysts,* **16**, 25–32.

Lundquist, G. & Meloan, C. E. (1971) Determination of polysorbates in food products by reaction gas chromatography. *Anal. Chem.,* **43**, 1122–1123.

Mergenthaler, E. & Scherz, H. (1976) Beträge zur Analytik von als lebensmittelzusatzstoffe verwendten polysacchariden. *Z. Lebensm. Unters. Forsch.,* **162**, 159–162.

Neale, M. E. & Ridlington, J. (1978) A rapid GLC method for the determination of benzoic acid in soft drinks and similar products. *J. Assoc. Public Analysts,* **16**, 135–139.

Nose, N., Koboyashi, S., Tanaka, A., Hirose, A. & Watanabe, A. (1977) Determination of thiabendazole by electron capture gas–liquid chromatography after reaction with pentafluorobenzoyl chloride. *J. Chromatogr.,* **130**, 410–413.

Ogawa, T., Fuji, R., Tanaka, K., Morita, I., Ono, K., Harada, T. & Sakaguchi, H. (1986) Gas chromatographic determination of diethylene glycol monoethyl ether in ice-cream. *J. Food Hygienic Society of Japan,* **27**, 656–661.

Page, B. D. & Kennedy, B. C. P. (1976) Rapid determination of butylated hydroxyanisole, *tert*-butylhydroquinone and propyl gallate in edible oils by electron capture gas chromatography. *JAOAC,* **59**, 1208–1212.

Pfannhauser, W., Eberhardt, R. & Woidich, H. (1982) Aroma analysis in food chemistry by GC/MS techniques: determination of natural and artificial coconut flavour. *Mikrochimica Acta (Wien),* **I**, 159–167.

Player, R. B. & Wood, R. (1980) Methods of analysis — Collaborative studies, Part III: determination of biphenyl and 2-hydroxybiphenyl in citrus fruits. *J. Assoc. Public Analysts,* **18**, 109–117.

Preservatives in Food Regulations (1989), HMSO.

Riva, M., Daghetta, A. & Galli, M. (1981) Analisi gascromatografica ad alta risoluzione dei mono e dei digliceridi con colona capillare persilanizzata associata al sistema di introdzione diretta in colonna. *La Rivista Italiana della Sostanza Grasse*, **58**, 432–444.

Ro, H. S. (1972) Simultaneous gas chromatographic determination of sorbic acid, dehydroacetic acid, benzoic acid, butyl *p*-hydroxybenzoate. *Korean J. Food Sci. Technol.*, **4**, 24–28.

Ro, C. B. (1979) Gas–liquid chromatographic determination of preservatives *Report of NIH, Korea*, **16**, 425–428.

Rottsahl, H. & Jessen, T. (1987) Gaschromatographische bestimmung von methanol in wein mit der headspace-technik. *Deutsche Lebensm-Rundschau*, **83**, 42–47.

Schmolck, W. & Mergenthaler, E. (1973) On the analysis of polysaccharides used as food additives: gas chromatographic identification after methanolysis and trimethylsilylation. *Z. Lebensm. Unters. Forschung*, **152**, 263–273.

Stafford, A. E. & Black, D. R. (1978) Analysis of sorbic acid in dried prunes by gas chromatography *J. Agric. Food Chem.*, **26**, 1142–1144.

Takalashi, D. M. (1970) GLC determination of BHA and BHT in breakfast cereals. *JAOAC*, **53**, 39–43.

Tanaka, A. & Fujimoto, Y. (1976) Gas chromatographic determination of thiabendazole in fruits as its methyl derivative *J. Chromatogr.*, **117**, 149–160.

Tanaka, A., Nose, N., Suzuki, T., Hirose, A. & Watanabe, A. (1978) Determination of biphenyl and 2-phenylphenol in citrus fruits. *Analyst*, **103**, 851–855.

Toyoda, M., Ogawa, S., Suzuki, T., Ito, Y. & Iwaida, M. (1980) Gas–liquid chromatographic determination of tertiary-butylhydroquinone (TBHQ) in dried fish, frozen shrimp, vegetable oils, butter and margarine. *JAOAC*, **63**, 1135–1137.

Tsuda, T., Nakanishi, H., Kobayashi, S. & Morita, T. (1984) Improved gas chromatographic determination of sorbitan fatty acid esters in confectionery products. *JAOAC*, **67**, 1149–1151.

Westöö, G. & Andersson, A. (1975) Determination of biphenyl and 2-phenylphenol in citrus fruits by gas–liquid chromatography. *Analyst*, **100**, 173–177.

Wu, W. S. & Saschenbrecker, P. W. (1977) Nitration of benzene as method for determining nitrites and nitrates in meat and meat products. *JAOAC*, **60**, 1137–1141.

Wunderlich, H. (1972) Collaborative study of a gas liquid chromatographic method for the determination of diethyl carbonate in wine. *JAOAC*, **55**, 557–558.

Young, L. L., Bergmann, R. E. & Powers, J. J. (1970) Correlation between gas-chromatographic patterns and flavour evaluation of chemical mixtures and of cola beverages. *Journal of Food Science*, **35**, 219–223.

Yu, L. Z., Inoko, M. & Matsuno, T. (1984) A gas chromatographic method for the rapid determination of food additives in vegetable oils. *J. Agric. Food Chem.*, **32**, 681–683.

Winell, B. (1976) Quantitative determination of ethoxyquin in apples by gas chromatography. *Analyst*, **101**, 883–886.

Abbreviations and symbols

α	Separation factor
ABS	Acrylonitrile butadiene styrene
AED	Atomic emission detector
AFID	Alkali flame ionization detector
Ala	Alanine
amu	Atomic mass units
AOAC	Association of official analytical chemists
Arg	Arginine
Asn	Asparagine
Asp	Aspartic acid
ATBC	Acetyltributyl citrate
AW–DMCS	Acid washed–dimethylchlorosilane treated
ß	Phase ratio
BHA	Butylated hydroxyanisole
BHAT	Butylated hydroxytoluene trifluoroacetate
BHT	Butylated hydroxytoluene
BSA	Bis(trimethylsilyl)acetamide
BSTFA	Bis(trimethylsilyl)trifluoroacetamide
BTFA	Bis(trifluoroacetamide)
C_g	Mass transfer term for gas phase
C_l	Mass transfer term for liquid phase
CAPA	Computer-aided pesticide analysis
CI	Chemical ionization
CP	Chlorinated pesticide
CV	Coefficient of variation
Cys	Cysteine
2,4-D	2,4-Dioxin
D_g	Diffusion coefficient in gas phase

D_l	Diffusion coefficient in liquid phase
DA	Dopamine
DDD	Dichlorodiphenyldichloroethane
DDE	Dichlorodiphenylethane
DDT	Dichlorodiphenyltrichloroethane
DEG	Diethyleneglycol
DEG–MEE	Diethyleneglycol monoethyl ether
DEGS	Diethyleneglycol succinate
DEHA	Di-(2-ethylhexyl)adipate
DES	Diethylstilboestrol
DHA	Dehydroacetic acid
DHPG	3,4-Dihydroxyphenylethyleneglycol
DMF–DMA	Dimethylformamide–dimethylacetal
DON	4-Deoxynivalenol
DOPAC	3,4-Dihydroxyphenylacetic acid
DP	Degree of polymerization
DPA	α,ε-diaminopemelic acid
E	Electric field
ECCI	Electron-capture chemical ionization
ECD	Electron-capture detector
ECL	Equivalent chain length
ED	Electrochemical detector
EDB	Ethylene dibromide
EI	Electron impact
ELCD	Electrolytic conductivity detector
ELISA	Enzyme-linked immunosorbent assay
EPA	U.S. Environmental Protection Agency
ESA	Electrostatic analyser
ESBO	Epoxidized soyabean oil
FAME	Fatty acid methyl esters
FID	Flame ionization detector
FPD	Flame photometric detector
FTIR	Fourier transform infrared
GC	Gas chromatography
GLC	Gas–liquid chromatography
Gln	Glutamine
Glu	Glutamic acid
Gly	Glycine
h	Height equivalent to a theoretical plate
H	Height equivalent to an effective theroretical plate
ΔH	Enthalpy of solvation
HCH	Hexachlorohexane
HFB	Heptafluorobutyryl
His	Histidine
HMDS	Hexamethyldisilazane
HPLC	High-performance liquid chromatography
ΔI	McReynolds' constants

i.d.	Internal diameter
I_k	Retention index
Ile	Isoleucine
IR	Infrared
IS	Internal standard
ITD	Ion trap detector
k	Solute capacity factor
K_D	Distribution constant
L	Column length
LAL	Lysinoalanine
Leu	Leucine
Lys	Lysine
m	nominal mass of an ion
MAPA	Multiplan-aided pesticide analysis
MBTFA	*N*-Methylbis(trifluoroacetamide)
MDG	Mono- and diglycerides
MDGC	Multidimensional gas chromatography
MDS	Multidimensional scaling
MEG	Monoethylene glycol
Met	Methionine
MHPG	3-Methyl-4-hydroxyphenylglycol
MS	Mass spectrometry
MSD	Mass selective detection
MSHFBA	*N*-Methyl,*N*-trimethylsilylheptafluorobutyramide
MSTFA	*N*-Methyl,*N*-trimethylsilyltrifluoroacetamide
m/z	Mass : charge ratio
n	Number of theoretical plates
N	Number of effective theoretical plates
NCI	Negative chemical ionization
NDBA	*N*-nitrosodibutylamine
NDEA	*N*-nitrosodiethylamine
NDGA	Nordihydroguaiaretic acid
NDMA	*N*-nitrosodimethylamine
NDPA	*N*-nitrosodipropylamine
NPIP	*N*-nitrosopiperidine
NPYR	*N*-nitrosopyrrolidine
NT	O-2-Naphthyl dimethylthiophosphinate
OP	Organophosphorus pesticide
PBM	Probability-based matching
PCB	Polychlorinated biphenyl
PCDD	Polychlorinated dibenzodioxin
PCI	Positive chemical ionization
PCP	Pentachlorophenol
PEG	Polyethylene glycol
PET	Polyethylene terephthalate
PFP	Pentafluoropropionyl
Phe	Phenylalanine

PID	Photoionization detector
PLOT	Porous-layer open tubular
Pro	Proline
PT	O-Phenyl dimethylthiophosphinate
PTV	Programmed temperature vaporizer
PVC	Polyvinyl chloride
PVDC	Polyvinylidene chloride
r	Internal column radius
r_s	Radius of curved ESA plates
R_S	Resolution
RCF	Regenerated cellulose film
Rf	Radiofrequency
RI	Refractive index
RRT	Relative retention time
RSD	Relative standard deviation
s	seconds
σ	Standard deviation
SAIB	Sucrose diacetate hexaisobutyrate
SCOT	Surface-coated open tubular
SEC	Size-exclusion chromatography
SECAT	Serial-coupled columns at different temperatures
Ser	Serine
SIM	Selected-ion monitoring
sn	Stereospecific numbering
S/N	Signal : noise ratio
STIRS	Self-training interpretive and retrieval system
T	Temperature
t_m	Elution time for non-retained molecule
t_R	Retention time
t'_R	Adjusted retention time
T_Z	Separation number or Trennzahl
TBHQ	Tert-butylhydroquinone
TCD	Thermal conductivity detector
TD	Thermionic detector
TFA	Trifluoroacetic acid; Trifluoroacetyl
THBP	2,4,5-Trihydroxybutyrophenone
Thr	Threonine
TIC	Total ion current chromatogram
TMCS	Trimethylchlorosilane
TMS	Trimethylsilyl
TMSIM	Trimethylsilylimidazole
TMSN	Tetramethylsuccinonitrile
Trp	Tryptophan
TSİM	Trimethylsilylimidazole
Tyr	Tyrosine
$\bar{u}$	Mean linear gas velocity
UV	Ultraviolet

V_3	Volume of column occupied by gas phase
V_L	Volume of column occupied by liquid phase
Val	Valine
$w_{0.5}$	Width of peak at half height
w_b	Width of peak at base
WCOT	Wall-coated open tubular
z	Number of carbon atoms of a hydrocarbon; charge

Index